The U.S. Forest Service in the Pacific Northwest

A History

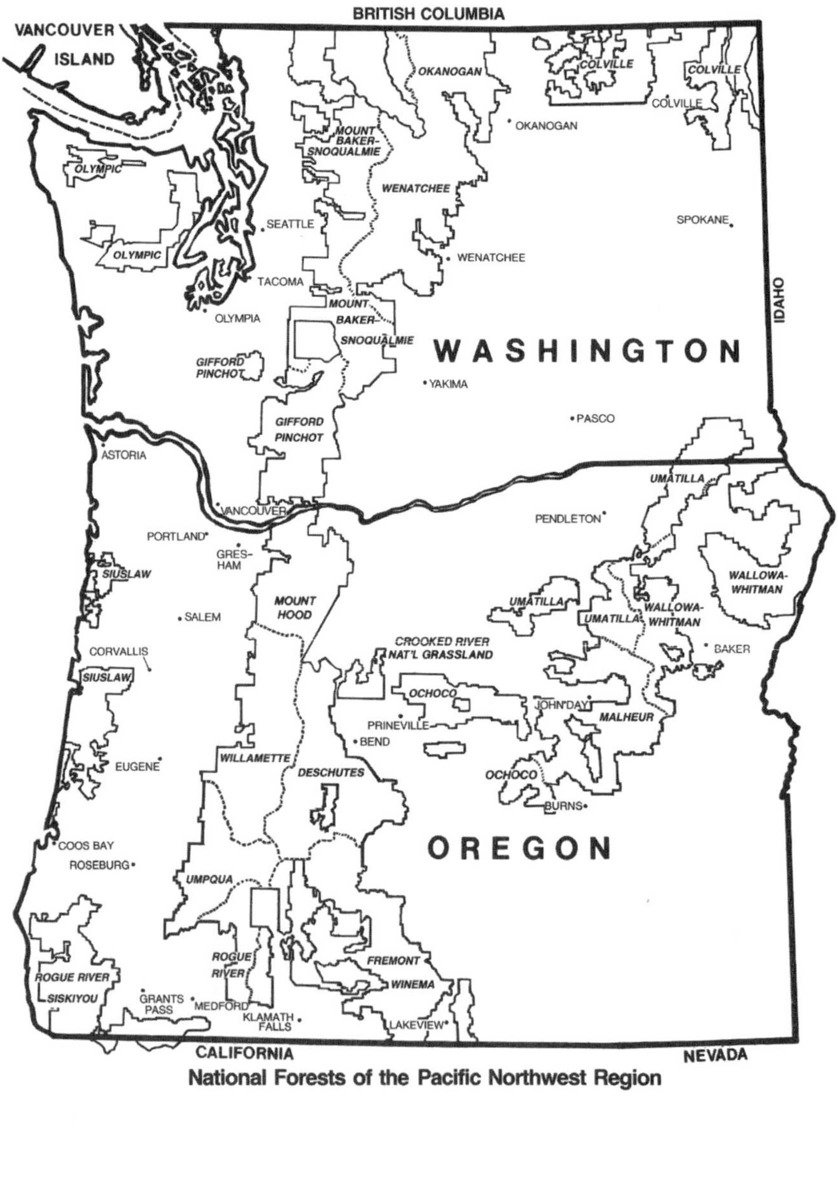

National Forests of the Pacific Northwest Region

The U.S. Forest Service in the Pacific Northwest

A History

Gerald W. Williams

Foreword by **Mike Dombeck,**
Chief Emeritus, USDA Forest Service

Oregon State University Press • Corvallis

Cover photograph: CCC enrollee Thales Bay at work on a fire tower, Camp Ingram, Fremont National Forest, Oregon. USFS photo #281497. Back cover photograph: Fire tower construction, Fremont National Forest. USFS photo #281493. Both images courtesy Gerald W. Williams Collection, Oregon State University Archives.

The paper in this book meets the guidelines for permanence and durability of the Committee on Production Guidelines for Book Longevity of the Council on Library Resources and the minimum requirements of the American National Standard for Permanence of Paper for Printed Library Materials Z39.48-1984.

Library of Congress Cataloging-in-Publication Data

Williams, Gerald W.
The U.S. Forest Service in the Pacific Northwest : a history / Gerald W. Williams.
p. cm.
Includes bibliographical references and index.
ISBN 978-0-87071-572-3 (alk. paper)
1. United States. Forest Service. Pacific Northwest Region--History. 2. Forest rangers--Northwest, Pacific--History. I. United States. Forest Service. Pacific Northwest Region. II. Title.
SD565.W558 2009
333.7509759--dc22
2009011337

Oregon State University Press
121 The Valley Library
Corvallis OR 97331-4501
541-737-3166 • fax 541-737-3170
http://oregonstate.edu/dept/press

Table of Contents

Foreword

The Pacific Northwest region of the United States encompasses nearly every type of ecosystem in the country with the exception of arctic tundra and tropical forests. You will read in the pages that follow that over the past one hundred years this region has experienced nearly every conceivable socio-political and environmental controversy. The employees of the United States Forest Service have been the stewards of millions of acres of the region's public land and central players in much of its history.

As the Civil War came to an end, few expected the United States would soon be positioned to become an international leader among nations. A frontier country of immigrants, rich in natural resources, America already was a land of opportunity. But this young nation lacked the cultural and scientific achievements of its Eastern Hemisphere counterparts. Europe had majestic cathedrals and museums filled with great works of art. Asia had great dynasties that embodied great traditions and cultures of the past. What America did have was a powerful national spirit and seemingly endless land, much of which was in the public domain, west of the Mississippi—the great western frontier that historian Fredrick Jackson Turner referred to as the "greatest free gift ever bestowed on mankind."

Many view the story of conservation and the national forests as beginning with the presidency of Theodore Roosevelt. Yet many earlier events, too many to mention here, set the stage for this great conservationist and president. The Lewis and Clark Expedition of 1804-1806, for example, demonstrated the vastness of the West and highlighted the grandeur and beauty of the landscape, as well as the abundance and variety of plants and animals.

By the mid-1800s many influential Americans became concerned that the country's lands were being plundered. For example, in a 1850 Report of the Commissioner of Patents, Thomas Ewing said: "The waste of valuable timber in the U.S. . . . will hardly be appreciated until our population reaches 50 million. Then the folly and shortsightedness of this age will meet with a degree of censure and reproach not pleasant to contemplate." Ewing went on to warn that "the vast multitudes of bisons [sic] slain yearly, the ceaseless war carried out against them, if continued, threatens their extermination, and must hereafter cause deep regret. . . . It should never be said that the noblest of American indigenous ruminants have become extinct." Concerns regarding the long-term harm of the deforestation of the eastern U.S. appeared in government reports from 1850 through 1880 and the devastation of the Appalachian and

Lake States' forests prompted interest in and appreciation for the scenic and wilderness qualities of the landscape. Henry David Thoreau's address to the Concord, Massachusetts, Lyceum declared that "in wildness is the preservation of the world." In his 1864 book *The Maine Woods*, Thoreau called for the establishment of "national preserves" of virgin forests, "not for idle sport or food, but for inspiration and our own true recreation." Near the end of the Civil War, George Perkins Marsh published the seminal *Man and Nature; or, Physical Geography as Modified by Human Action* (revised in 1874 as *The Earth as Modified by Human Action*)—the first systematic analysis of the human impact on the environment.

Carl Schurz, interior secretary from 1877 to 1881, was the first cabinet member with an active interest in the conservation of natural resources and recommended the establishment of forest reserves to protect the land. In 1885, six bills were introduced in Congress to create public forest reserves. None of them passed. Similar bills failed to pass in 1888, 1889, and 1890. Finally, on March 3, 1891, the Forest Reserve "Creative" Act passed, empowering the president to create "forest reserves" by withdrawing forestlands from the public domain.

As the nineteenth century came to a close, four trends signaled the need for change in the management of the public domain: growing western communities were increasingly demanding dependable water supplies; there was increasing public distrust of large timber companies; sustainable timber supplies were needed to replace cutover forest of the Great Lakes region; and the accelerating general demand for timber products. But the reserves were, at this time, little more than lines drawn on a map without managers, regulations, or a budget. Interest in the application of science-based forest management continued to grow.

The stage had been set for Theodore Roosevelt to thrust the nation into its first conservation movement and aggressively expand federal forest reserves. In 1905, with Gifford Pinchot at the helm, the U.S. Forest Service was established to manage the national forests "for the greatest good for the greatest number for the long run."

This book is not a comprehensive history, but instead it chronicles the story of the agency and its employees striving to bring order and scientific management to the majestic national forests of the Pacific Northwest over more than a century. It is a story about people and land.

Dr. Mike Dombeck, Chief Emeritus, USDA Forest Service

Acknowledgements

This book would not have been possible without the strong support, and money, from the Region 6 top staff, especially Regional Forester Linda Goodman and Deputy Regional Forester Liz Agpaoa. The Forest Service, through a cooperating agreement with the retiree group OldSmokeys, provided the funding for this project. The OldSmokeys contributed time in a dollar amount equal to the Regional funds, spending countless hours meeting and discussing the book, as well as finding photographs.

I would like to thank many people who were important in the production of this book, particularly the OldSmokeys review committee of five Forest Service retirees, Rolf Anderson, Phil Hirl, Wendall Jones, Dave Scott, and Ron Walters, who helped me in many ways. They were especially good at pointing out the obvious errors, filling in details, and locating other people who could add needed material; and they tried to keep me honest in my interpretations of events.

I would be remiss if I didn't include information about the OldSmokeys. In 1945, the first club of Forest Service retirees began operations in the Pacific Northwest. It was called the Thirty-Year Club and was open to Forest Service employees and retirees who had served this many years. Many of the original Thirty-Year Club members had worked for the Forest Service since the early 1900s, and they felt it important to record their experiences. In June of 1947, the club began publishing an annual newsletter, called Timber Lines. The Pacific Northwest Forest Service Association (PNWFSA), the official name of the club after it gained nonprofit status in 2001, decided these and other stories describing the Forest Service in the Pacific Northwest from 1905 to 2005 should be shared with a wide audience. The book We Had an Objective in Mind was published to celebrate the one-hundredth anniversary of the Forest Service and to add to the festivities of the Forest Service National Reunion in Portland in 2005.

In addition, the U.S. Forest Service Wives Club, renamed the Portland Forest Service Women's Club in 1974, was originally founded in 1940 at the suggestion of Regional Forester Lyle Watts. Rita Glazebrook noted that the purpose of the club was to welcome women who had come to the Portland area with their Forest Service families. As times changed the club became inactive by the late 1990s. However, its members have left a legacy of supporting and often directly contributing to the everyday management of the agency in the Pacific Northwest.

My family tolerated the typing of the book, expenses related to the production of the same, several trips to various parts of the Pacific Northwest Region to take photographs and talk with Forest Service employees, and trips to the Forest History Society in Durham, North Carolina. My wife Joyce Casey, a Forest Service employee in the Regional Office, helped me understand current events. Our kids at home, Justin and Tristan, as well as those at college, Colin and Maggie, had no active part in the production of the book, but they are part of the wonderful family that was supportive of the book effort. Then, of course, I can't forget our golden retriever dogs Barlow and Copper, who died while I was typing the book, our new black lab Lolo, and cat Sauvie who were my constant companions while I was working on the book.

The U.S. Forest Service in the Pacific Northwest

A History

Introduction

Nine U.S. Forest Service regions cover all of the national forests and national grasslands in the United States. The national forest system contains 193 million acres of land.

The U.S. Forest Service in the Pacific Northwest (Region 6), as well as the other western regions, is celebrating a century of service to the nation in 2008-2009. Region 6 over the last one hundred years has often served as a lightning rod for changing uses of the national forests across the country. Prior to the 1940s, Region 6 could be characterized as managing the forests through custodial management. This involved establishing a management system, building the infrastructure for management of the national forests, and building a professional work force. Region 6 took the lead early in the 1940s as the location of intensive management through timber harvesting. Timber harvesting rapidly increased during the next three decades with employees who were imbued with a high sense of purpose and superior ability to get "the cut out." But change was coming. New laws in the 1970s, coupled with wilderness debates, public interest groups, congressional actions, and court challenges, meant that the old ways of doing business on the national forests would have to change. It was not easy.

The national forests in Washington and Oregon are important to the life and lifestyles of the people who reside in the Pacific Northwest. The national forests in Region 6 sit atop the heavily forested mountain ranges and provide scenic settings, recreation, water, fishing, timber, grazing, and many other opportunities. They define the Pacific Northwest.

The population of the two states is approximately 9.9 million (6.4 million in Washington and 3.6 million in Oregon) and growing. The lifestyles of these Northwesterners are often defined by what they do out-of-doors: hiking, swimming, salmon fishing, boating, hunting, golfing, trail biking, sight-seeing, camping, picnicking, wind surfing, river rafting, and many other outdoor recreational activities. The abundant rain west of the Cascades and the dry climate east of the mountains create two different places in which to live and enjoy the out-of-doors. Even the fast-growing metropolitan areas in the western parts of the two states are framed by the snow-covered mountains to the east and the Pacific Ocean to the west. People living in the central and eastern regions of the two states are also framed by the Cascades and other

- Region 1, the Northern Region, manages the national forests and grasslands in Montana, North Dakota, northern Idaho, and northwest South Dakota.
- Region 2, the Rocky Mountain Region, oversees the national forests and grasslands in Colorado, Kansas, Nebraska, South Dakota, and Wyoming.
- Region 3, the Southwestern Region, is responsible for national forests and grasslands in Arizona and New Mexico.
- Region 4, the Intermountain Region, manages the national forests and grasslands in southern Idaho, Nevada, Utah, and western Wyoming.
- Region 5, the Pacific Southwest Region, administers the national forests and one grassland in California, Hawaii, Guam, and the Trust Territories of the Pacific Islands.
- Region 6, the Pacific Northwest Region, manages the national forests and one grassland in Washington and Oregon.
- There is no Region 7 anymore; it was separated into Region 8 and Region 9.
- Region 8, the Southern Region, oversees of the national forests and grasslands in the southern states.
- Region 9, the Eastern Region, is responsible for management of the national forests of the states east of the Mississippi River and generally north of the Mason-Dixon line.
- Finally, Region 10, the Alaska Region, handles the national forests in just the state of Alaska.

prominent mountain ranges; the drier conditions here help define a different, more rural way of life.

This book covers many topics, starting with physical descriptions of the land and waters in the Pacific Northwest and background historical information in Chapter 1. The next chapter recounts the significant role that the late nineteenth-century conservationists played in establishing forest reserves, now called national forests, and national parks in the Pacific Northwest, which is not well known. Massive land frauds also marked this era. Judge John Waldo, William Steel, and Gifford Pinchot stand out as important figures in support of the forest reserves. Chapter 3 discusses the establishment of federal control over the forest reserves/national forests through the General Land Office and the fledgling Forest Service. The following chapter describes custodial management such as fire fighting and building ranger stations became the preferred method of management. This "white hat" era took the Forest Service through the Stock Market Crash of 1929.

Chapter 5 deals with the federal and Region 6 responses to meeting national needs during the Great Depression, with the many "alphabet" agencies, and World War II, as well as battles over the establishment of Olympic National Park. This significant era was followed in Chapter 6 by increasing demand to open the national forests to extensive and intensive efforts to boost timber production from the middle 1940s until the end of the 1950s. Chapter 7,

Oregon	
Forest Service	15,180,164 acres (47.5 percent)
Bureau of Land Management	15,712,468 acres (49.1 percent).
federal government total	31,982,600 acres (51.9 percent)
Washington	
Forest Service	8,902,434 acres (68.7 percent)
Bureau of Land Management	370,872 acres (2.8 percent)
federal government total	12,952,900 acres (30.4 percent)

Atlas of the Pacific Northwest (2003)

which begins in the 1960s and lasts through the middle of the 1970s, describes how timber production was increasing while other resources were coming into ascendancy. Wilderness and the roadless area review and evaluation (RARE) process began affecting how the national forests in the Pacific Northwest would be managed in the future, and these are covered in Chapter 8. Beginning in 1976 and ending in 1989, a significant change in Forest Service management occurred that was in part caused by an ever-increasing number of environmental and forest management laws. We see how the environmental and timber special-interest groups, as well as the federal courts and Congress, began to play a more significant role in the management of the national forests in Region 6. Finally, in the 1990s and into the new century the Forest Service in Region 6 was struggling to survive in the midst of massive fires, changing direction instigated by Congress and the courts, internal morale problems through huge employee reductions, and a changing workforce, and these changes are covered.

But make no mistake when reading this book. This is not a "critical history" (to use Frederich Nietzsche's phrase) of the Forest Service in the Pacific Northwest, although at times the material delves into past mistakes and over-zealousness which were not considered that way at the time. Nor is the book a "monumental history" filled with the wonderful and glorious accomplishments of a flawless Forest Service over the past one hundred years. Lastly, this book should not be considered to be an "antiquarian history" where the historical detail overwhelms the reader. This book combines all three approaches as well as a little humor, and I hope that it will give you, the reader, an understanding of the complexity and changes in management of the national forests in the Pacific Northwest and the men and women who have the assignment to protect and manage the national forests for all people.

Gerald (Jerry) Williams
June 2009

Chapter 1

Environment and Early History

The Pacific Northwest

The Pacific Northwest, broadly defined, extends from the Pacific Ocean to the continental divide and includes all of Washington, Oregon, Idaho, and western Montana. The Washington and Oregon portion of the region is diverse, dominated by numerous mountain ranges, the most important of which is the Cascade Range. It is situated along the Pacific rim of fire that is characterized by historic and modern volcanism as the various land and ocean geologic plates collide. The Cascades and other mountain ranges are, for the most part, where the region's national forests and parks are located. Immediately inland from the Cascade Range is the broad Columbia Plateau, which extends from almost the Canadian border on the north to the California border on the south and is more than 150 miles wide in places.

All of the rivers east of the Cascades in Washington are tributary to the Columbia River, except for those in a small portion of the Okanogan-Wenatchee National Forest (NF), which drains northward into the Fraser River system in British Columbia. Streams that flow from the Mt. Baker-Snoqualmie and Olympic NFs drain either directly into the Pacific Ocean or into Puget Sound. Rivers and creeks from the Gifford Pinchot NF flow into the Columbia River system.

In Oregon, all of the rivers and creeks on the national forests drain into the Columbia River system with the following exceptions: The rivers and streams from the Rogue River-Siskiyou NF flow directly into the Pacific Ocean, as do some in the Siuslaw and Umpqua NFs, while others flow into the Willamette River and hence to the Columbia River system. Some of the streams flowing from the Ochoco, Malheur, and Fremont-Winema NFs drain into the Malheur Basin or Goose Lake, which have no outlets to the ocean. Other streams from the Fremont-Winema NF also flow into the Klamath River, which continues to the Pacific Ocean through northwestern California.

Because many areas have plentiful rainfall, especially west of the Cascades, the Pacific Northwest provides clean and abundant water. A number of cities have municipal watershed agreements with the Forest Service to protect their

National parks: Oregon—Crater Lake; Washington—Olympic, Mount Rainier, and North Cascades.
National historical parks: Nez Perce—eastern Washington and Oregon; Lewis and Clark—near the mouth of the Columbia River in both states.
National monuments: Washington—Mt. St. Helens (a national volcanic monument) and the Hanford Reach; Oregon—John Day Fossil Beds, Oregon Caves, and Newberry Crater (a national volcanic monument). Both states share the Columbia River Gorge National Scenic Area, while Oregon and Idaho share the Hells Canyon National Recreation Area (NRA), and Oregon alone has the Oregon Dunes NRA.
National forests: Washington—Colville, Gifford Pinchot, Kaniksu (a portion of this forest is in Idaho), Mt. Baker, Okanogan, Olympic, Snoqualmie, Umatilla (part in Oregon), and Wenatchee. Oregon—Deschutes, Fremont, Klamath (mostly in California), Malheur, Mt. Hood, Ochoco, Rogue River, Siskiyou, Siuslaw, Umatilla (part in Washington), Umpqua, Wallowa, Whitman, Willamette, and Winema. Note that several of these forests, like the Rogue River-Siskiyou, are currently managed together. Also, the Kaniksu NF (in Washington) and the Klamath NF (in Oregon) are managed by other Forest Service Regions. The Rogue River and Siskiyou NFs have lands located in California while the Wallowa-Whitman NF has land in Idaho.

city watersheds. These include Seattle (Cedar River) and Portland (Bull Run), as well as Cottage Grove (Layng Creek) and Ashland (Ashland Creek).

The Pacific Northwest climate is different on the east- and westside of the Cascades. The western area has a mild coastal climate, while the eastside is colder and drier. The climate also shows a great deal of year-to-year variability. These variations are caused by two large-scale weather patterns that appear over the Pacific and vary on a yearly basis: La Niña, which is characterized by unusually cold ocean temperatures in the eastern equatorial Pacific, and El Niño, which is characterized by unusually warm ocean temperatures in the same ocean areas. Another ocean pattern that can affect the region's climate is the Pacific decadal oscillation, characterized by wet and dry cycles that last 20 to 30 years.

Warm tropical storms, often referred to as a"Pineapple Express" or "Chinook Winds," sometimes occur in the winter months. They are often accompanied by warm rains that melt the mountain snowpacks, increasing greatly the runoff in rain-swollen rivers, and have in the past led to severe flooding. Recent examples of this are the Pacific Northwest flood in December-January 1964 and the Willamette Valley Flood of February 1996 that had massive consequences for the national forests in western Washington and Oregon (see Chapters 7 and 9). The latter flood almost drowned downtown Portland, but the river wall was extended upward several feet, which prevented the fast-moving flood waters from entering the city. The Regional Office of the Forest Service closed for several days until the flood threat was over.

In most of the Northwest precipitation (usually rain on the westside and often snow on the east) falls in the months of October through May. The usually dry summers, on both sides of the Cascades, make water a much-needed resource for many ecosystems and human needs. The availability of water is already stressed by multiple growing demands, including irrigation, drinking water, hydroelectric-power generation, and fish populations.

The coniferous forests in both states are quite sensitive to climate variation. Warm, dry summers stress the trees, as well as creating conditions favorable to forest pests and fire. Although there were large fires before 1900 (such as the Yaquina Fire of 1853 that burned some 480,000 acres), probably the first well-documented fire in the Pacific Northwest was the 1902 Yacolt Burn in southwest Washington. Before this fire was put out, it burned some 240,000 acres.

The extent, species mix, and productivity of Northwest forests are likely to be different in the future due to global climate change. Thus far, the effects have been relatively minor, but drastic changes are forecast. Projected heavier winter rain- and snowfall are likely to increase the frequency of winter flooding and landslides. Drier summers will likely increase water shortages. Climate change is very likely to stress presently healthy fish stocks and to hamper efforts to restore already depleted native salmon stocks. There will be an increasing risk of the total numbers, size, and severity of wildfires. Also, there is likely to be an increase in destructive insect populations affecting the forests and grasslands in the Pacific Northwest and elsewhere.

Forest Service scientists have been working on basic climate-change research for more than 20 years. Coordination on climate-change issues is occurring at all levels. The US Geological Survey (USGS) has a Western Mountain Initiative that is collaborating with the Forest Service, National Park Service, and university partners to further understand and predict the physical responses of the forests to climate change (www.usgs.gov/global_change/). The Forest Service's Region 6 also has a climate-change strategy group comprising R-6 and Pacific Northwest Research Station staff. They are spearheading efforts to share information, coordinate with other agencies and universities, and incorporate climate-change issues into environmental documents, as well as to reduce the Region's ecological footprint on the land and in communities (see www.fs.fed.us/ccrc/).

Forest Species

Forests in Oregon cover about 30 million acres or 49 percent of the land base. The federal government, including Indian tribal reservations, holds 59 percent of the forest lands, private industry another 37 percent, the state 3 percent, and other public ownership 1 percent. In Washington, more than half the state is covered in forests. The federal government, including Indian tribal reservations, holds about 57 percent of the forests, while Washington State owns about 13 percent, while another 36 percent are privately owned. Thirty species of conifers are found in Pacific Northwest forests. Near the Pacific coast, where rainfall is higher and fogs cool the summer heat, a mixture of western hemlock, Douglas-fir, Sitka spruce, and western red cedar predominate. In the valley areas west of the Cascades, the forest is predominantly Douglas-fir and western hemlock. In southwest Oregon and northwest California are small patches of redwoods and Port Orford cedar, which is highly valued for Japanese homes, but is threatened by a root rot disease (*Phytophthora lateralis*), an introduced pathogen. Briefly in the 1980s, there was a flurry of interest and activity around the diminutive Pacific yew tree that grows in the westside evergreen forests (see Chapter 9).

At higher elevations on the westside of the Cascades, several species of true fir, mountain hemlock, and lodgepole pine often join the mix of Douglas-fir and western hemlock. The forests of central and eastern Oregon and Washington are mostly ponderosa pine and lodgepole pine. As elevation and rainfall increase, ponderosa pine gives way to a mixed-conifer forest of dry-site Douglas-fir, lodgepole pine, western larch, Englemann spruce, and a mixture of true firs. Juniper forests, where the trees tend to be widely spaced, occupy many sites on the Columbia Plateau. In many areas, junipers have been burned, plowed, and chained to provide additional grazing area for cattle operations.

In 1826, explorer David Douglas noted that oaks covered the low hills of western Oregon. They were a product of frequent fires set by Indians to improve browse for deer and elk. Over the past 100 years most of the open oak forests were removed to create farms, ranches, cities, and for use as firewood. Despite this, the white or garry oak is still a feature in portions of the interior valleys from California to British Columbia, but could be in danger of extinction because of "sudden oak death" disease, which has been recently found in Oregon.

Salmon

There are seven species of Pacific salmon, but only five (chinook, coho, sockeye, chum, and pink) are found in the North American rivers, where there are also two species of sea-run trout, steelhead/rainbow and cutthroat. A major difference between salmon and sea-run trout is that salmon nearly always die shortly after spawning, but many sea-run trout do not, returning to the ocean to go through another cycle of migration and spawning.

Because of salmon hatcheries, no *species* of Pacific salmon is near extinction. Nonetheless, many *wild stocks* of salmon in the Pacific Northwest have been extirpated or are experiencing population declines. These declines have been especially noticed in California, Oregon, and Washington, with smaller reductions found in British Columbia, but fewer reductions have been noted in Alaska.

The question of whether wild salmon will continue to exist in the western United States is not a new one. The decline started a century and a half ago in the aftermath of the 1849 California gold rush. By the 1850s, excessive harvest and the impacts of mining activities were decimating salmon in the Sacramento and San Joaquin rivers, as well as streams surrounding California's Central Valley. By the late 1880s, the U.S. Commission of Fish and Fisheries (National Marine Fisheries Service today) became concerned about the declining Pacific salmon stocks. Fish Commission biologist Livingston Stone was disillusioned about the future prospects for the still-abundant Pacific salmon. Fearing they would "go the way of the buffalo," he proposed a national salmon park for Alaska. On the basis of his recommendation, Alaska's Afognak Island was set aside as a "Forest and Fish Cultural Reserve" in late 1892, one of the first forest reserves and the only one set aside for salmon (Rakestraw 1981). Two years later, the Fish Commission proclaimed to Congress that the Columbia River runs were much reduced and still declining. The most serious decline of Columbia River salmon took place in the nineteenth century—not the twentieth—though that is not to imply that the last century was a favorable one for salmon.

High rates of salmon harvesting were not the only major cause of salmon decline. Dams built on many rivers and streams in the Pacific Northwest began to interfere with salmon reaching their spawning areas. But by 1933, the year the first main-stem dam on the Columbia was finished, the total Columbia salmon run had already been reduced to a fifth or less. Flood control, and associated dam, levee, and channel construction, has been a societal priority for over a century, often at the expense of the salmon and steelhead. Numbers

The author with a wild steelhead caught on the North Umpqua River. Author's collection.

of Pacific harbor seals and California sea lions have increased greatly in recent years because killing of these animals has been prohibited by their protection under the Marine Mammal Protection Act of 1972. These mammals are especially efficient at eating salmon congregated at river mouths, dams, and fish ladders. In early 2009, the rule against killing the salmon-eating sea lions was relaxed and by the spring several of the most egregious sea lions were killed. This followed individual, and illegal, killing by salmon anglers. Several species of birds near the mouth of the Columbia River have also been especially hard on small migrating salmon in the last decade.

Historically, hatcheries have been used as tools to stock streams to maintain declining salmon runs and harvest levels. Many believe there is an over-reliance on the use of hatcheries, contending that the more than 100 hatcheries releasing salmon into the Pacific Ocean actually worsen conditions for wild salmon (Lackey 1997 and 2003).

State fish and game departments in the two states have tended to focus most of their management efforts on fishable species—salmon, trout, sturgeon, bass, and a handful of other fishes. State rules governing the catch of fish have been around since the early 1900s. Catchable fish, usually rainbow trout, have been planted in lakes, ponds, and streams throughout Washington and Oregon, usually in lower-elevation water bodies, but beginning in the 1920s, high lakes in the Cascades and other mountain ranges within the national forests have been regularly stocked with fingerlings and larger fish. Planting fish in these areas has required transporting fish in specially built trucks, then in water cans on horseback, or in recent times dropping the fingerlings from airplanes and helicopters.

Another development has been the introduction of non-native fishes including walleye, striped bass, American shad, brown and brook trout, small- and largemouth bass, bluegill, northern pike, crappie, yellow perch, channel catfish, and carp, as well as the expansion of native species such as squawfish. Some lakes, like Diamond Lake on the Umpqua NF, have been beset by the accidental or intentional release of tui chub, used as a bait fish. In many cases, the application of a killing agent like Rotenone has been effective at reducing or eliminating the exotic species. In 2006, for the second time in 52 years, Diamond Lake was treated successfully with Rotenone to remove the tui chub (see Chapter 9).

American Indians in the Pacific Northwest

American Indian tribes occupied all the land in the Pacific Northwest, including the area now in national forests, national parks, and managed by the Bureau of Land Management, for thousands of years. For the most part, Indians lived peacefully on the land, and several thousands were living in the two states at the time of European contact in the late 1700s and early 1800s (Boyd 1999).

From the mid-1830s, a series of human disease epidemics (caused by Old World diseases, principally smallpox, measles, whooping cough, mumps, cholera, and gonorrhea) led to a significant decline in Indian populations and in some cases utter destruction of villages and entire bands (Boyd 1999). With the advent of settlers, wars, treaties, and removal to reservations, the Indians essentially lost most of their world (including their land) in a short 100-year period. European religions were often used to separate the Indians from their traditions and even languages through various missionary practices. Indian ways, by the early twentieth century, were essentially gone. Remaining tribes survived and in recent years asserted their rights established by treaties from the middle 1800s (Ruby and Brown 1981).

INDIAN WARS

Missionaries Dr. Marcus Whitman and his wife Narcissa Whitman, along with 13 others at their mission at Waiilatpu near Walla Walla were murdered in late November 1847. During a measles outbreak among the Indians, Dr. Whitman was unable to halt the epidemic and the quickly spreading disease killed half the tribe. The Indians held Whitman and the missionaries responsible for the deaths. The incident began the Cayuse War (1847-1855) against the tribes in eastern Washington and Oregon, retribution by early settlers in the Oregon country, and the start of civil government in Oregon. This was the first war with the Indians in the Pacific Northwest.

Indian-white conflicts erupted in 1855 with confrontations with miners traveling to the Colville mines. In 1856, members of the Yakama, Klickitat, and Cascades Chinookans swept through the Columbia Gorge. As a result, the U.S. Army retreated down the Columbia to regroup and mount a counter offensive. After numerous attacks on Indian villages by the miners, Indian bands along the southern coast of Oregon went to war in early 1856. Fighting continued in small battles that became known as the Rogue Indian War. Hostilities ceased in the summer of 1856 with the surrender of all major bands of Indians. The conflicts continued in 1858 when the Spokane Indians turned back Army troops. More soldiers were brought into the region, killing Indian horses and destroying caches of winter food. During the Civil War a series of engagements pitted the Northern Paiute of central and eastern Oregon against U.S. Army regulars. The continuing conflicts wore down the Indian resistance and in 1873 led to confinement of many Northern Paiute on the Malheur Reservation. Ranchers and cattlemen were successful in disbanding the reservation in 1882 and removing the Northern Paiutes to other reservations.

The Modoc War that erupted in 1872 was the result of long-smoldering hostilities between travelers on the Applegate Trail through southern Oregon and Indians residing in the marsh and lake region along the Oregon-California border. The war pitted Captain Jack and his small band of Modocs against the U.S. Army from Ft. Klamath. Ultimately forced to surrender, the Modoc leaders were hung and the surviving tribal members removed to Oklahoma. Since then some Modoc families have moved back to the Klamath area, just west of the Fremont-Winema NF. In 1877 the U.S. Army attempted to force Chief Joseph's band of Nez Perce in the Wallowa Valley to move to the Nez Perce Reservation along the Clearwater River in Idaho. He and his war chiefs then led the band on a circuitous route trying to escape the Army. Surrounded within a few miles of the Canadian border, the refugees were compelled to surrender. The Nez Perce were moved to Kansas and finally to the Indian Territory (Oklahoma). Many survivors were moved to the Colville Reservation, but Chief Joseph was never allowed to return to his ancestral Wallowa Valley and mountains of the Wallowa-Whitman NF. Members of the tribe live on the Colville and Umatilla Reservations, while only a few reside in their ancestral home in the Wallowas.

Also during 1877-78, members of the Bannock Tribe in southern Idaho marched against the white settlers. A war party of about 200 Indians rode across southern Idaho, stealing horses, killing settlers, and burning ranches. The war party rode northward to the Umatilla Reservation, but they were not warmly received. The Army then forced the war party into the Wallowa Mountains in the current Wallowa-Whitman NF. The Bannocks surrendered in mid-September 1878, after which they were returned to their reservation in southern Idaho.

Indians fishing at Celilo Falls, Columbia River, Oregon, c. 1930. Author's collection.

INDIAN TREATIES

As the valleys and Puget Sound areas were filling with settlers, there was a need to solve the Indian land ownership question. This dilemma was resolved by forcing the tribes to cede their land to the federal government that could then be granted or sold to settlers. For the most part, treaties in Washington were negotiated and signed by Isaac Stevens, while in Oregon Joel Palmer was the responsible official (Ruby and Brown 1981). Most of the treaties were never ratified by the U.S. Senate. However, various laws and regulations require that the Forest Service treat the tribes as equals (a "government-to-government" relationship) in land management decisions, regardless of a signed treaty or not. More than 100 years after signing, the Cow Creek treaty became the focal point of tribal recognition and cooperation with the Umpqua NF.

The Columbia Basin treaty, as well as all treaties in Washington State, provided for numerous reserved rights where the federal government acknowledged that the tribes possessed the right to fish in all "usual and accustomed grounds and stations," of "gathering roots and berries," of hunting, of erecting temporary buildings for curing fish, and of pasturing their horses and cattle "upon open and unclaimed land." These treaty rights would become in the 1980s a cornerstone of salmon management in the Pacific Northwest (see Chapter 7).

INDIAN RESERVATIONS

Originally, the term reservation meant places to which American Indian tribes were assigned to get them out of the way of settlers, miners, and railroads. Indian reservations were usually situated on lands that were considered less desirable. The federal government retained the right to renegotiate treaties, and it almost never renegotiated them to benefit the Indians.

Most of the treaties designated the tribal reservations. For example, the Siletz people were provided with the Siletz or Oregon Coast Reservation in 1855, but the reservation was later disbanded and much of it sold to settlers. Part of this reservation eventually became part of the present-day Siuslaw NF. The Middle Oregon Treaty created the Warm Springs Reservation, although controversy over the exact boundaries of the reservation adjacent to the Mt. Hood NF (the McQuinn Strip) would be debated for more than a century (see Chapter 6). Tribes north of the Yakima Reservation in central Washington were moved to the Colville and Spokane reservations. The treaties with the Klamath, Modoc, Yahuksin Band of Northern Paiute, and the Walpapi Band also set up the 1.1-million-acre Klamath Reservation, which would remain intact for almost a century (see Chapter 6).

At the end of the nineteenth century laws were passed that broke many reservations into individual homesteads, encouraging occupants to sell their property to white settlers. Land within the reservation not claimed by Indians was then put up for sale. As a result Indian ownership of reservations shrank dramatically (Ruby and Brown 1981). One example came from western Oregon where the Oregon Coast or Siletz Reservation of 1855 was progressively opened to settlement by a presidential order in 1865, then by acts of Congress in 1875 and 1894. The original reservation no longer exists. The Klamath Reservation was eliminated through the Klamath Termination Act of 1954 which stipulated that tribal members would have to sell their interests in the reservation. In 1958 the federal government purchased about 15,000 acres of the Klamath Marsh to establish a migratory bird refuge under the jurisdiction of the U.S. Fish & Wildlife Service. In 1961 and again in 1973, portions of the former Klamath Reservation became part of the Winema NF. With these purchases, the federal government became the owner of approximately 70 percent of former reservation lands.

Oregon was unique in the mid-1950s when former Governor Douglas McKay, then secretary of the Department of the Interior, decided to "mainstream" tribes and get rid of the Indian reservations under the auspices of the Indian Reorganization Act of 1934. Following the intent of the act, the Bureau of Indian Affairs in 1954 began to terminate many tribes and sell the reservations. The terminations were intended to show that Indians could be integrated into larger American society and in doing so would not need their reservations. Of the 109 tribes and bands terminated by the U.S. government in the 1950s, 62 tribes were native to Oregon.

In 1975, the federal government recognized the failure of its termination policy and passed the Indian Self Determination and Education Assistance Act. Shortly after passage several tribes mounted efforts to restore their sovereign nation status. The Confederated Tribes of Siletz became the second tribe in the nation to achieve restoration in 1977, followed by the Cow Creek Band of Umpqua Indians in 1982, the Confederated Tribes of Grand Ronde in 1983, the Confederated Tribes of Coos, Lower Umpqua, and Siuslaw in 1984, the Klamath Tribe in 1986, and the Coquille Tribe in 1989.

In a land deal brokered in December 2008 by the Trust for Public Lands and the Cascade Timberlands, a new Klamath Reservation gained around 90,000 acres. The tribe and the federal government agree to purchase the land from the timber company. This area, known as the Mazama Tract, is located in the northwest corner of the old reservation adjacent to the Fremont-Winema NF.

The balance of the old reservation is in Indian and non-Indian ownership (see Chapter 6). Another land transfer in 2009 has recently enabled the Wallowa band of the Nez Perce Tribe to acquire 15,325 acres as the Precious Lands Wildlife Area in northern Wallowa County in Oregon. This area, which is not reservation land, adjoins the northern border of the Wallowa-Whitman NF and the Hells Canyon National Recreation Area, was developed under the Pacific Northwest Electric Power Planning and Conservation Act of 1980. Funding came from the Bonneville Power Administration to mitigate salmon habitat that was lost to the dams blocking the Snake River. (*Statesman Journal* March 10, 2009)

The land areas of the reservations in Oregon and Washington vary tremendously. The largest Oregon reservations are the Warm Springs with 644,000 acres and the Umatilla with 172,882 acres in tribal lands. The smallest reservations are the Coos, Lower Umpqua, and Siuslaw, which have less than 200,000 acres (Bradbury 2007-8). In Washington State, the Colville Reservation has the largest land area with 1,300,000 acres, followed by the Yakima Reservation with 837,735 acres. The smallest reservations are the Muckleshoot Reservation with 169 acres and the Ozette Reservation with only 64 acres (Washington State 2007).

INDIAN USE OF THE FORESTS

Many of these reservations adjoin national forests in the Region 6; the Warm Springs Reservation abuts the Mt. Hood, Willamette, and Deschutes NFs and the Quinault Reservation borders the Olympic NF. The tribes still have a great affection for (and in some cases reserved treaty rights) on lands that were once theirs, but now are part of the national forests. Even after the Indians were confined to the reservations they continued to use the national forest lands. For example, the Willamette NF was used by the Warm Springs people during the late 1800s and early 1900s for grazing horses, picking berries, killing deer, and catching salmon.

Along the west side of the Cascades, Indians relied upon western red cedar trees to build their large, solid, split-plank winter quarters called longhouses. Cedar trees were felled by fire and were also used to make canoes and long strips of cedar bark were the preferred material for making berry containers, temporary shelter, and dresses and hats. The Makah tribe used masks of carved and painted western red cedar in ceremonies and storytelling. Oak trees in westside valleys were an important food source; acorns were a primary food in many areas for both Indians and deer. Various species of seed grasses (like tarweed) were gathered in the fall after fires had been set.

East of the Cascades, ponderosa pines, lodgepole pines, and junipers were used extensively for shelters, travois, canoes, and food-serving utensils. Pitch from the ponderosa pine was sometimes used as a sweet gum for chewing. Open, wet prairie areas on both sides of the Cascades were important sources for the starchy bulbs of the blue camas, an important food.

INDIAN USE OF FIRE

By the time European explorers, fur traders, and settlers arrived, millions of acres of so-called wilderness in many parts of North America were already manipulated and maintained for human use. The early observers did not often recognize the signs that the native people were managing the landscape through the use of fire. Explorers and fur trappers often observed huge burned-over or cleared areas with dead trees. Many settlers noted evidence of burned or scorched trees and open prairies or savannas with tall grass in every river basin. The abundance of rich prairie land ready for the plow was one of the primary reasons settlers headed west to the Oregon Territory (the current states of Oregon, Washington, and parts of Idaho).

There are numerous accounts of the Kalapuya tribe burning the bottom lands of the Willamette Valley. "Lack of horses and adequate weapons limited the hunting activities of the Willamette Valley Indians and caused them to invent ingenious methods of killing game. ... By burning the prairies the Indians forced the deer to graze on convenient hunting grounds, and they by this method also made it easy to collect wild honey, grasshoppers and crickets." (Clark 1927: I-52) The Kalapuya people also used fire to ensure the continual renewal of their lands, plants, and animals (Williams 2000a, Williams 2000b, Williams 2000c). An account by an early settler, described the widespread use of fire by the Indians to keep the brush down and the streams open, and to collect seeds in the Umpqua Valley to the south: "Before the white settlers came to the valley, the Indians would set fire and burn off one side of the valley in the fall of each year. This kept the brush from crowding in and the streams open." (Cornutt 1971: 36)

Documentary evidence about Indian use of fire to change or improve the conditions in the forested mountains is fragmentary but important. An early Oregon pioneer related that the Molalla tribe in Oregon often set fires in the Cascade Range, noting that "fire was their agency [method] in improving game range and berry picking." (Minto 1908: 153) He also remarked that small prairies could be found on all the western slopes of the Cascades from the valley floor to almost the crest (4,000 to 6,000 feet). These open prairie areas would, in the 1890s and early 1900s, become a source of controversy over sheep grazing in the forest reserves (see Chapter 3). Another Oregon author describing himself as "Pioneer of 1847" related that Indians burned portions of the mountain forests in the northern Cascades:

> Indians and early settlers had complete charge and Summer fires were a rare thing, for the reason there was nothing to burn ... The Indian method was to burn the old burns about every three years or as soon as there was growth enough to make a good fire. They would burn early in the Summer before the logs and old stumps were dry enough to burn. This light growth would make a light fire and would hardly smoke the old logs, and just as soon as the fire reached the green timber it stopped, as everything in the timber was too wet to burn (Pioneer of 1847 1911:10).

Meanwhile in the Blue Mountains of northeastern Oregon, especially in the Grande Ronde and Powder River country, fires set by Indians were a common occurrence as late as the mid-1800s. "The Cayuse, Nez Perce, Paiute, Umatilla, and Shoshone tribes had heavily used the Blue Mountains for centuries and had altered the landscape accordingly. Native Americans had traveled, traded, hunted, fished, gathered roots and berries, maintained herds of horses [sometimes numbering in the thousands], burned the hills to improve hunting and grazing, and fought wars in the Blues for centuries before whites showed up." (Langston 1995: 44)

The early demise of many of the Indian inhabitants, the removal of the remaining Indian people to distant reservations, and their replacement by white farmers by the 1850s meant that an entire way of managing the land was never to be restored at the same scale.

> [T]he modification of the American continent by fire … was the result of repeated, controlled, surface burns on a cycle of one to three years, broken by occasional holocausts from escape fires and periodic conflagrations during times of drought, … So extensive were the cumulative effects of these modifications that it may be said that the general consequence of the Indian occupation of the New World was to replace forested land with grassland or savannah, or, where the forest persisted, to open it up and free it from underbrush ... almost wherever the European went, forests followed. The Great American Forest may be more a product of settlement than a victim of it (Pyne 1982: 79-80).

Without periodic fire, ecosystems have changed in many parts of the West. Native trees are dying from the effects of insects and disease, and highly flammable woody vegetation has built up in forest understories. Whether still growing or dead, these materials are fueling uncharacteristically severe fires in the West, often with catastrophic results.

In recent decades, several Indian tribes have begun working closely with the national forests in the Region to successfully use prescribed fires to rehabilitate grass and huckleberry areas. The Willamette NF over the last decade has been working closely with the Confederated Tribes of Siletz and the Confederated Tribes of Grand Ronde to rehabilitate at least one small prairie area on the forest with prescribed fire.

> The Forest Service acquired the site in 1994 from private landowners. The project was initiated in 1996 as an interagency and intergovernmental endeavor involving the Siletz and Grand Ronde tribes, Bureau of Land Management (Eugene District), private nursery contractors, along with the Forest Service, Sweet Home Ranger District. Personnel from the Sweet Home and Detroit Ranger Districts conducted the first prescribed burn at the site in September of 1998. The prairie has been burned five times since then, at two year intervals. The Siletz and Grand Ronde tribes provided fire engines and a fire crew to assist Forest Service personnel with the burns (Healthy Forests and Rangelands Web site, http://www.forestsandrangelands.gov/).

INDIAN INVOLVEMENT

The Mt. Baker-Snoqualmie NF since the late 1990s has had a memorandum of agreement (MOA) to work with the Muckleshoot on a number of issues concerning the tribe's historic rights and traditional uses on the forest. The MOA provides free use permits for tribal firewood, cedar logs, and other items, as well as access for gathering plant materials; addresses concerns over water quality by summer home users; and includes design of an elk forage project; and in 2007 the forest with the State of Washington conducted a huckleberry symposium with the intent to restore and enhance huckleberry patches on the forest.

Several reservations in the Pacific Northwest are fortunate to have large forested areas from which they can sustainably manage timber harvesting: the Colville Reservation in Washington has well over 950,000 forested acres; the Quinault Reservation, also in Washington, has over 200,000 acres; in Oregon, the Warm Springs Reservation has some 330,000 acres of forest, and Yakama

Reservation has almost a 150,000 acres. Several of the tribes have built sawmills to manufacture wood products on the reservation. Such enterprises have made these tribes some of the largest employers in their respective counties, hiring employees who contribute to the local communities through taxes and purchasing power.

A number of tribes have taken advantage of opportunities under the National Indian Gaming Regulatory Act of 1988 to operate gaming centers and casinos on reservation or trust land. Besides providing employment opportunities for tribal members and citizens of surrounding communities, revenues from these tribal enterprises go towards health clinics, education, scholarships, housing, and other services. The tribes have made a tremendous comeback from near extinction caused by disease and war.

European Discovery, Fur Trade, and Settlement

The search for the fabled Northwest Passage in the mid and late 1700s led maritime explorers to sail the coast of the Pacific Northwest. Although the passage was never found, important discoveries were made regarding the land and resources, especially beavers, that eventually drew maritime and overland explorers, as well as missionaries and settlers, to the far reaches of western North America. Expansion of the United States and opportunities for other countries to colonize the Pacific Northwest was the source of many confrontations until the issue of ownership of the area was settled by international treaty and purchase.

MARITIME HISTORY

In the 1774-1792 era, Spanish, French, English, and United States ships sailed along the coast of what is now Oregon and Washington. Few ventured beyond looking at the land through telescopes or briefly walking along the shoreline. The Nootka Convention allowed the British and Americans to battle it out for possession of the Pacific Northwest. In the summer of 1791, English Captain George Vancouver arrived off the coast of Oregon and Washington.

At the same time, Captain Robert Gray, a New England merchant ship captain who was the first American to circumnavigate the globe was also exploring along the coast. In the summer of 1791, Gray encountered a strong river current off the northern Oregon coast, spending nine days trying to enter the river before abandoning the effort. After wintering on Vancouver Island, Gray again attempted to enter the mouth of the river, but bad weather forced him to give up. Finally on the evening of May 11, 1792, Gray's men found a safe channel, and the ship and crew sailed into the estuary of the

Columbia River. He raised the American flag and planted some coins under a large pine tree. Also on this day Gray officially named the river Columbia after his ship *Columbia Rediviva*. Gray traded goods with the natives, met Captain Vancouver again (informing him of the Columbia River entrance), then sailed to China to sell pelts before returning to Boston in July 1793.

Captain Vancouver sent Lt. William Broughton to investigate in October 1792. Broughton sailed several miles up the Columbia and continued up river in a small boat to the head of tide at the western end of the Columbia River Gorge. Broughton went ashore where he formally claimed for England the river, its huge watershed, and the adjacent Pacific Coast line. This claim would set in motion a controversy about the "ownership" of the Pacific Northwest. He also made the first European sighting of a tall, snow capped mountain that he named Mt. Hood (McArthur and McArthur 2007).

The Treaty of Ghent on December 24, 1814, ended the War of 1812. The Treaty of Joint Occupation October 20, 1818, stated, in part, that the Pacific Northwest was to be jointly claimed by the United States and Great Britain for a 10-year period. Yet the Oregon boundary dispute was not settled until the 1846 Oregon Treaty, partitioning the region along the 49th parallel with the British possessions (Canada) to the north and the U.S. possessions (Oregon country) to the south. This established U.S. control over the Oregon country. But the occupation of the land by Indian tribes remained a problem for settlement.

The Pacific Coast mariners also left a legacy in the form of naming landscape features. As mentioned above, Captain Gray named the Columbia River, as well as Grays Harbor. Captain Vancouver named Mt. Baker, Mt. Rainier, Mount St. Helens, and Mt. Hood, as well as Hood Canal, Puget Sound, and a number of other locations. Spanish Captain Juan Perez named Mt. Olympus. English explorer Captain James Cook named Cape Flattery and Cape Perpetua, while Spanish explorer Bruno de Hezeta named Cape Disappointment and Heceta Head is named after him (McArthur and McArthur 2007).

OVERLAND HISTORY

The first overland exploration expedition to the Pacific Coast from the United States was sent by President Thomas Jefferson, not long after the Louisiana Purchase from France in 1803. Captains Meriwether Lewis and William Clark left St. Louis, Missouri, on May 14, 1804, with 31 men including York, Captain Clark's slave. After over-wintering in what is now North Dakota, the party added Sacajawea (also spelled Sacagawea and Sakakwea), wife of Touissant Charbonneau, the expedition interpreter. After crossing the Rockies they canoed down the Columbia River arriving at the mouth of the Columbia on November 7, 1805. The party selected the location for Ft. Clatsop (now a part of the Lewis & Clark National Historical Park) on Youngs Bay, just south of present-day Astoria. By December 12 they finished construction of three log cabins in which they were to spend a very wet, miserable winter. The Lewis and Clark party began their return journey eastward to the United States on March 23, 1806. The party passed but then returned to explore the mouth of the Willamette River on April 2 and 3. They arrived at St. Louis on September 23, 1806. News of their journey to the Pacific Ocean and return soon energized hundreds, then thousands of fur traders, Indian traders, and emigrants to the new country. The party named Mt. Jefferson, Wenatchee, Walla Walla, as well as a variety of other geographic features, but many were later renamed (McArthur and McArthur 2007).

Captain Nathan Winship plus about 30 others from Boston aboard the vessel *Albatross* entered the Columbia River on May 26, 1810. Their settlement lasted a mere eight days because of flooding. Floods of the Columbia River would continue to cause problems for settlers and cities for more than 150 years until flood control dams were built across the Columbia River and its tributaries. Hall Jackson Kelley was a fanatical promoter for American settlement of the Oregon country. His party arrived in California from the United States in the summer of 1834 which then headed northward to Oregon. Kelley became quite ill, but was nursed back to health by the Hudson's Bay Company fur brigade who then assisted the Kelly party northward to Ft. Vancouver where they stayed until the spring. Interestingly, Kelley in 1839 wanted to name the peaks of the Cascade Range as the "Presidents' Range" with each major mountain named for a former president. He thought that Mount St. Helens should be called Mt. Washington, Mt. Hood to be renamed as Mt. Adams in honor of John Adams, and Mt. McLoughlin to be called Mt. J.Q. Adams. The Pacific Railroad Survey mapmaker in 1853 placed the Mt. Adams name on an unnamed peak that was northeast of Mt. Hood. The renaming of the range never happened, but the Mt. Adams name stuck (Meany 1923).

Nathaniel Wyeth in the spring of 1834 organized an overland party to venture to the Columbia River. Along the way they established Ft. Hall in Idaho. The party arrived at Ft. Vancouver in mid-September. The Wyeth party was supplied by the brig *May Dacre*. The ship held goods and equipment for the Wyeth expedition, Methodist missionaries led by Jason Lee, and two noted naturalists, Professor Thomas Nuttall of Harvard University and John Kirk Townsend. While on the trip, Nuttall collected over 110 species of western plants. Wyeth set up a salmon and fur trade business at what he called Ft.

William on the southwest side of Sauvie Island. The Columbia River business failed after floods ruined the fort site and 14 of his men drowned or died at the hands of Indians. After returning to the States, Wyeth sold Ft. William and Ft. Hall to the Hudson's Bay Company. Most of his men remained, moving into the Willamette Valley where they began some of the first successful farms in the Oregon country.

The "Peoria Party" of emigrants under the leadership of Thomas J. Farnham traveled westward from Independence, Missouri, in the late spring of 1839 over the new Oregon Trail. Part of the Oregon Trail crosses the Wallowa-Whitman NF and Umatilla NF in northeastern Oregon. Another overland settler group, called the Bidwell-Bartleson Party, arrived in the Willamette Valley two years later. This was the first attempt to bring wagons over the full length of the Oregon Trail. The first big settler party arrived in the Willamette Valley in the fall of 1842, led by Dr. Elijah White from the Willamette Mission. The "Great Migration" of American settlers arrived in the Willamette Valley during the fall of 1843, bringing between 875 and 1,000 emigrants. The party was able to partially excavate a wagon road through the Blue Mountains northeast of La Grande in the current Umatilla NF and Wallowa-Whitman NF, as well as build a crude trail along the south bank of the Columbia River westward from The Dalles to avoid the treacherous rapids in the Columbia.

Samuel Barlow and at least 15 families made a trail around Mt. Hood in the fall of 1845. The route called the Barlow Road crossed the southern slopes of Mt. Hood. This route soon became a toll road and was opened for wagon travel the following year. Previously, the emigrants would have to pay for boats and barges that would float them down the dangerous Columbia River, through the rapids and swift flowing waters. The river cost the lives of many would-be

Covered wagon on Santiam Wagon Road, and Mt. Washington, Santiam National Forest, Oregon, c. 1905. Author's collection.

settlers on the last leg of the long, tiring journey across the Great Plains to fabled Oregon. Emigrants to the Pacific Northwest named many geographic features, including Applegate River, Barlow, Battle Ground, Bend, Concrete, Crater Lake, Detroit, Diamond Lake, Eugene, Mt. Adams, Mt. Hebo, Mt. Thielsen, Olympia, Portland, Seattle, Shelton, and Three Sisters (McArthur and McArthur 2007).

FUR TRADE ERA

The North West Fur Company and Hudson's Bay Company, British fur companies, penetrated the Oregon Country from the present-day Canada. At the same time, John Jacob Astor founded the Pacific Fur Company that established a fur-trading post at Astoria, Oregon in 1811, beginning an era of competition between American and British fur companies trading in the region. During the War of 1812, the arrival of a British warship on the Columbia River in 1813 prompted the Astorians to salvage what they could by selling the Pacific Fur Company to their British rival – the North West Fur Company. The British then took control of the Pacific Northwest's fur trade. Astoria was renamed Ft. George.

The North West Fur Company merged with the Hudson's Bay Company (HBC) in 1821. John McLoughlin was appointed chief factor of the region in 1824. He moved the HBC headquarters from Ft. George to Ft. Vancouver in 1825. The HBC sent fur trappers throughout the Pacific Northwest and by the 1840s they had almost totally decimated native beaver populations and other fur-bearing species from the rivers and streams. How many of the watershed environments were changed forever because of this is unknown.

The fur trappers left a legacy of early settlements in Washington and Oregon such as Astoria, Vancouver, Nisqually, and French Prairie. Several fur trappers left the fur trade, deciding to stay and farm in the Pacific Northwest. They also bestowed place names throughout the valleys and mountains such as Blue Mountains, Cape Horn, Chelan, Colville (originally spelled Colvile), Cowlitz, Grand Coulee, John Day River, Kettle Falls, McKenzie River, Rogue River, Snake River, Spokane, Willamette River, and Yakima. Many of these place names were derived from earlier Indian words, but others such as McKenzie River was named after Donald McKenzie of the Pacific Fur Company.

David Douglas, a famous Scottish botanist, arrived in the Pacific Northwest in 1825 under the auspices of the Royal Horticultural Society of London. In April 1826, an HBC party with Douglas arrived at Ft. Colvile. The Douglas party then began exploring the Blue Mountains, traveling to the Clearwater River, then crossing overland to the Spokane House and descending the

Columbia some 800 miles to Ft. Vancouver. Douglas returned to Ft. Vancouver three years later and carried out further work along the main stem of the Columbia in 1833. Douglas discovered and described more than 50 species of trees including the ponderosa pine (*Pinus ponderosa*), sugar pine (*Pinus lambertiana*) and Douglas-fir (*Pseudotsuga menziesii*). He also named more than 100 species of flowers, grasses, shrubs, and ferns. His legacy in the Pacific Northwest has been cemented forever because the Douglas-fir tree is named for him. A story about his collection of the ponderosa pine came from his journals:

> He is credited with indentifying or introducing to England several conifers, including the Noble Silver fir, Grand Silver fir, and Lodepole pine. Douglas made his other great discovery while…in northeast Washington. On a visit from Fort Colvile to Spokane House on 8-9 May 1826, he collected specimens of a tree with bark cracked like an alligator's hide. "I saw a new pinus," he casually noted in his journal, apparently unaware of the significance of his new discovery. When specimens of the tree were analyzed by botanists in England, they were found to be of a previously unidentified species, that of Pinus ponderosa, or ponderosa pine (Holstine 1987: 8).

Douglas is also credited with the first description of the sugar pine. On October 26, 1826, Douglas was in the vicinity of present-day Roseburg, Oregon. He had seen the giant cone, but not the tree from which it came. An Indian took him to a site where the tree could be found.

> At miday I reached my long-wished *Pinus* (called by the Umpqua tribe *Natale*), and lost no time in examining and endeavoring to collect specimens and seeds….the dimensions of the largest one I could find was that was blown down by the wind: Three feet from the ground, 57 feet 9 inches in circumference; 134 feet from the ground, 17 feet 5 inches; extreme length, 215 feet….Being unable to climb or hew [cut] down any, I took my gun and was busy clipping them from the branches with ball when eight Indians came at the report of my gun [he left as soon as possible (*Forest Worker* 1931: 17).

MISSIONARY HISTORY

In the early spring of 1833, four Flathead Indians visited General William Clark in St. Louis inquiring about the white-man's God. As churches in the eastern United States heard the news about the Indians wanting religious instruction, several denominations sent missionaries to convert the Indians. Jason Lee, a Methodist minister, in 1834 selected the site for the Willamette Mission located just north of present Salem that became the first permanent American structure in Oregon. Catholic missionary Father Francois Norbert Blanchet arrived at Ft. Vancouver in late November 1838, taking up residence in the Willamette Valley the following year. Father Modeste Demers went north into present Washington, where he established the Mission St. Francis Xavier in the Cowlitz Valley. An eruption of Mount St. Helens recorded by Father John Baptist Bolduc on December 5, 1842, read:

> towards three o'clock in the afternoon, one of its sides opened and there was an eruption of smoke such that our old voyagers have never seen anything equal to it. These eruptions of smoke took place for several days at intervals not far apart, after which eruptions of flame began. They take place almost continually, but with an intensity that varies greatly from time to time.... After the volcano manifested itself, almost all the fish that it [Cowlitz River] used to feed died—which is attributed to the quantity of cinders with which the waters are affected.

Competing Land Claims and Eventual Statehood

In 1843, settlers in the Willamette Valley established a provisional government at Champoeg. On August 14, 1848, the Oregon country was formally organized as the Oregon Territory. The territory extended from the California border on the south, Pacific Ocean on the west, Rocky Mountains on the east, and the Canadian border on the north. Joseph Lane was appointed by President Polk to be governor of Oregon Territory in 1848. Interestingly, the job of governor of the Oregon Territory was offered to Abraham Lincoln, but he turned down the appointment. The territorial capital was originally located at Oregon City just south of present-day Portland but was moved to Salem in 1851. The State of Oregon contains 97,131 square miles of land and water which makes it the ninth largest state in the Union.

The Washington Territory was formed on February 8, 1853, from the portion of the Oregon Territory north of the lower Columbia River and south of the 49th parallel and north of the 46th parallel. The original boundaries of the territory included all of the present-day State of Washington, as well as northern Idaho and Montana west of the continental divide. The Washington Territory name was chosen to honor George Washington. On March 2, 1853, Isaac Stevens was named by President Franklin Pierce as governor of the new territory.

Upon the admission of the State of Oregon to the union on February 14, 1859, the eastern portions of the Oregon Territory, including southern Idaho and portions of Wyoming west of the continental divide (then Nebraska Territory) were annexed to the Washington Territory. In 1863, the area of Washington Territory east of the Snake River and the 117th meridian was reorganized as part of the newly created Idaho Territory. Washington Territory with the current boundaries of the State of Washington was admitted to the United States on November 11, 1889, as the 42nd state. The State of Washington contains 71,265 square miles of land and water, making it the 18th largest state in the Union.

Each new state was granted a large amount of land for schools—sections 16 and 36 in each unsurveyed township, each section a nominal square mile. The idea was that the state would sell the sections, known for many years as school lands, and keep the money to fund schools and agricultural colleges in the state. By 1881, school lands in the Northwest amounted to some 5.8 million acres (2.488 million acres in Washington and 3.330 million acres in Oregon). Another 92,000 acres in the two states were added for land grant colleges. Trading national forest land for those school sections within the national forests became a priority for the Forest Service in the early twentieth century. This decades-long effort was referred to as "blocking up" the land, thus reducing the number of checkerboard ownerships to a minimum.

With the issue of Indian land ownership settled through treaties and war, as well as statehood for both territories, the nature of land tenure and management changed. While the valleys were taken up by land claims, the ownership of vast millions of acres of public domain land was still unresolved. The federal government began to take an active role in disposing of these lands. Surveying the land became a high priority for the federal government.

The General Land Office (now the Bureau of Land Management) administered programs created by Congress for the disposition of the public domain (see Chapter 2). Several homestead laws proved very important to the future in the Pacific Northwest. The first of these was the Donation Land Claim Act of 1850 (DLC) intended to promote homestead settlement in the Oregon Territory. By the time the law expired on December 1, 1855, over 8,000 patents had been issued, covering nearly 2.54 million acres. Most of the DLCs were taken up in the lush, grass-filled valleys of Washington and Oregon; later emigrants would have to be satisfied with foothill land, often covered with evergreen trees, to make a living.

The Homestead Act of 1862 was another law that proved to be highly popular. Another land law, the Desert Land Act of 1877, permitted filing if the claimant could bring the land under irrigation within three years; most applications were in the plateau area of central and eastern Washington and Oregon. In addition, several more land acts passed Congress. The amounts of each were relatively small, but sometimes significant. The Timber and Stone Act of 1878 allowed persons to claim land at $2.50 for its timber or stone that was otherwise unfit for cultivation. By 1881, almost 5,000 acres had been claimed in Washington and almost 3,000 acres in Oregon. This act would have a huge impact on the work load of early forest rangers in Region 6 as there was great potential for land fraud.

The Forest Homestead Act of 1906 permitted settlers who could find lands within forest reserves with demonstrated agricultural potential to settle upon these lands under the Homestead Act. As a result, the forests and the Region had to make extensive surveys and maps to verify the claims (see Chapter 3). These acts created a massive backlog of claims on the national forests in Region 6 during the 1900s and 1910s. Much time and effort was expended in trying to resolve them. Rangers made personal inspections of each claim, then had to interview the residents (if any), conduct at least a preliminary survey, take photographs of any improvements, and file illustrated reports. Most of the claims were rejected because of widespread fraud.

With the creation of the first national parks, national forests, and wildlife refuges (see Chapter 2) in the late nineteenth and early twentieth centuries, there was marked a shift in federal land management. By withdrawing these lands from settlement, Congress signaled a shift from using the public domain land to promote settlement to deciding that some lands should be held in public ownership because of their resource values. This shift was reinforced in 1946 when the General Land Office (GLO) was combined with the U.S. Grazing Service to establish the Bureau of Land Management (BLM). Today, the BLM manages more than 15 million acres of federal land in Oregon and another 400,000 acres in Washington.

US Geological Survey

In the 1870s the U.S. Geological Survey (USGS) began work in the Pacific Northwest. For the next 30 years the agency initiated studies of existing mining districts to assess coal, gold, silver, and other valuable mineral deposits. The USGS began mapping and publishing topographic maps in conjunction with the U.S. Geodetic Survey. These USGS maps, revised every 20 to 30 years for more than a century, provided vital information on terrain, rivers, roads, and settlements. The USGS compiled the first comprehensive assessment of forest reserves in the West; the major ones appeared in 1899, 1900, and 1901.

Measuring Douglas-firs. Washington Forest Reserve, 1899. USFS photograph.

They estimated standing volumes of hardwoods and softwoods, mapped areas ravaged by forest fires, photographed the terrain, and published technical reports on forest conditions and grazing impacts. Other maps provided by the USGS were important for early forest rangers as they served as a base for more accurate maps of the new national forests. By 1909-1910, many national forests in the Pacific Northwest produced oversized map volumes called forest atlases.

CORPS OF ENGINEERS

The U.S. Army Corps of Engineers was established in 1866 with regional headquarters in San Francisco. The Corps' most controversial projects, in retrospect, were the numerous flood-control, hydroelectric, and irrigation dams on the Columbia River and its tributaries. Starting in 1926 with the 308 Report to Congress, the Corps has been a strong advocate of dam building. The long-term effects of these dams on native salmon runs are only recently being understood. In recent years there have been active proposals to breach several of the dams along the Snake River to increase salmon runs, but the dams remain in place. (Willingham 1983)

Roads and Highways

Many roads were proposed over the years in both states, but few were actually built as roads were not a high priority for federal or state governments. Many of the roads that did exist were simply converted Indian and trapper trails supplemented by emigrant wagon routes. The trails were not improved—mud was the calling card every winter for half a century. Most Northwesterners were not connected with cities and towns. Travel was difficult at best and in the winter months impossible. It was only the beginnings of the automobile in the early 1900s that brought about a dramatic improvement in the roads. Major roads and highways were built through the western valleys that connected Seattle with Portland and Eugene. East and west were slowly connected through mountain passes. The federal government through the Bureau of Public Roads set standards, built or improved highways and bridges, and greatly improved the movement of automobiles, people, and freight across both states.

The Columbia River Highway was a technical and civic achievement of its time, successfully combining traditional and innovative engineering designs. Portions of the Columbia River Highway, now designated a national scenic highway, are located on the northern edge of the Mt. Hood NF.

The Cascade River Road in northern Washington was conceived as a major transportation corridor for public travel and commerce between western and central parts of the state. The road crosses state land, enters the Mt. Baker-

Mitchell Point Tunnel along the Columbia River Highway, Oregon, c. 1915. Author's collection.

"Old Scout," the first automobile to cross the Cascades at Prineville, Oregon, 1905. Author's collection.

Snoqualmie NF, extends eastward through the North Cascades National Park, and over the Okanogan and Wenatchee NF.

In 1893, the first official state road in Washington designated by the state legislature crossed the Cascade Range roughly where State Route 20 now crosses on the Mt. Baker-Snoqualmie NF. Today it is called the North Cascades Highway. The first Snoqualmie Pass road was completed in 1867. A private company in Ellensburg decided to make a better quality road, and received permission to charge a toll for using it. When the railroad came the road fell into disrepair again. But the road was improved to be able to handle increasing automobile traffic and the first automobile crossed Snoqualmie Pass in 1905. Today the route is designated as I-90.

Meanwhile in Oregon three significant roads were opened: the Santiam, McKenzie, and Willamette Pass highways. The Santiam Pass route crosses the crest of the Cascades south of Mt. Jefferson and north of the Three Sisters. The Santiam Highway has its roots in the old Willamette Valley and Cascade Mountain Military Wagon Road (see below). This important highway crosses the Willamette and Deschutes NFs.

The McKenzie Pass Highway (State Route 242) has a long and varied history. The first recorded trip over the Cascades in this region was in 1853 by a group of emigrants on horseback who were lost in the mountainous area near the Three Sisters. The first emigrant wagon train crossed in October 1874 on the McKenzie Pass road, which was by then a toll road. The road over the summit later languished until reconstruction began in 1920 on the old Cascade NF. The highway is closed during the winter months. When the South Santiam Highway (U.S. 20) was completed in 1964, the old McKenzie Pass route crossing the Willamette and Deschutes NFs, was designated as a scenic highway.

The Willamette Pass Highway (State Route 58) is the third major highway crossing the central Cascades in Oregon. The highway, which was completed in 1940, follows part of the old Oregon Central Military Wagon Road (see below), but deviates at Oakridge to head southeasterly over the Willamette Pass. Oregon Route 58 is highly important as a freight corridor between the Willamette Valley, Klamath Falls, and California.

Railroads and Military Wagon Road Land Grants

The railroad era of the mid and late 1800s has particular relevance today. During and after the Civil War, the federal government sought to expand the population across the United States by investing in railroads. Land grants were used as enticements for private companies to build railroads in areas where at the time there was little need for passenger travel or freight needs. The discovery of gold in California helped spur the need for a railroad across the continent. In 1862, President Abraham Lincoln signed the Pacific Railroad Act, which named and directed two companies, the Union Pacific and the Central Pacific, to construct a transcontinental railroad that was completed in 1869. Railroads to the Pacific Northwest started later, but were nonetheless important for the growth and development of the region. There were many smaller railroad lines built in the two states in the 1880s and '90s. Several extended north and south in the two states, mostly through the western valleys and the Columbia Plateau. Other shorter rail lines were built to access markets or timber operations, like the Southern Pacific line to Booth Kelly Lumber Company lands in the Mohawk Valley north of Springfield. None of these lines had federal land grants.

Two railroad lines received huge land grants—the Northern Pacific and the Oregon and California. In addition, Oregon had the distinction of being the only state to have military wagon road land grants—Willamette Valley and Cascade Mountain (Santiam), Oregon Central (Willamette), The Dalles

Fort Vancouver from U. S Railroad Survey, Washington. Author's collection.

(John Day), and the Coos Bay-Roseburg. The idea was that the grant land received by the investors was subject to several conditions—land was to be sold to settlers not investors, land was to be sold at usually $2.50 per acre, and land sold could not exceed 160 acres per person. Monies received from the land sales were to offset the costs of construction of the railroad. In every case, the companies violated the conditions of the congressional land grants, but only the Oregon and California Railroad in Oregon and Coos Bay Wagon Road lands were returned to the government. Many of the land grants were sold to timber and land companies, which have left a checkerboard ownership pattern found in many of the national forests and BLM lands in Washington and Oregon.

Oregon and California Railroad—Congress granted specific lands to the States of Oregon and California in 1866 for the construction of a railroad from Portland, Oregon, southward to California. The land grant gave both states the authority to designate a company to construct the railroad and receive the land grant as a subsidy to offset construction costs. The land grant included all odd-numbered sections of public domain land, non-mineral in character, within 20 miles of each side of the proposed railroad line. The total acreage of the Oregon portion of the land grant was 4.22 million acres. In 1869, Congress amended the granting act to permit the Oregon and California Railroad Company (O&C) to meet a new deadline. By this time, so much land had already gone into private ownership that the grant contained only 3.728 million acres.

Construction of the rail line was begun by the O&C Railroad, but completed by the Southern Pacific Railroad Company, which acquired the O&C Railroad in 1887. Both companies violated all three conditions of the land disposition rules. The Supreme Court issued an injunction in 1915 forbidding the company to make further land sales, but also stating that the actual disposition of the unsold O&C land was a legislative rather than a judicial function (Ellis 1948).

Congress passed the O&C Revestment Act on June 9, 1916, returning 2.4 million acres of unsold O&C and Coos Bay Wagon Road grant lands in Oregon, but not California, to federal ownership. The O&C Act of 1937 placed management of the O&C lands under the General Land Office, although hundreds of thousands of acres were and still are within national forest boundaries. The 1937 Act directed that timber thereon be managed for permanent forest production using the principle of sustained yield. The O&C Act also required that 50 percent of the revenue generated from management be returned to the counties that contain O&C lands. Since the Coos Bay Wagon Road (CBWR) was not specifically mentioned in the 1937 act for revenue sharing, another act two years later specified that the CBWR and O&C lands were to be managed together and revenues shared with the same formula.

The Southern Pacific Railroad soon brought about various changes. Previous to the mid-1920s, the railroad ran south from Portland to Eugene, then to Medford and south to Redding over Siskiyou Summit. The railroad built a new main line, called the Natron Cutoff at the time, from Eugene, southeast over Willamette Pass, then south to Klamath Falls and Redding. On August 7, 1926, the new rail line was officially completed. The railroad line crosses the present-day Willamette, Deschutes, and Fremont-Winema NFs. The old route crosses near the current Willamette, Umpqua, and Rogue River-Siskiyou NFs. The O&C land grant encompasses forest land in the three national forests as well as the primary source of lands managed by the BLM in western Oregon. The railroad named the community of Chemult as well as scores of railroad crossings, creeks, and other nearby geographic features. (McArthur and McArthur 2007)

Northern Pacific Railroad—President Abraham Lincoln signed an act creating the Northern Pacific Railroad in 1864 and authorizing construction of the first of the northern transcontinental railroad and telegraph lines between Lake Superior and Puget Sound. The act provided for a right of way and land grant of 38.6 million acres in a checkerboard pattern, 20 miles on each side of the railroad main line. The railroad land grant extended along the northern tier states from Lake Superior to Puget Sound. As with the other land grant rail and wagon roads, the Northern Pacific was required to sell the land to homesteaders. The company stopped selling the land to individuals and instead sold huge blocks to private timber and land companies (Jensen and Graffan 1995).

Tacoma in Washington Territory was named as Northern Pacific's western terminus. The main line was completed in August 1883 running through Richland, Pasco, and Kennewick (Tri-Cities) to Vancouver then north to Tacoma. Almost as soon as the main line was completed, the Northern Pacific began construction of the Cascade Branch to reduce the travel time. This new line extended from the Tri-Cities through Yakima, Ellensburg, and via a tunnel under Stampede Pass to western Washington. By the end of May 1888, the tunnel was open to traffic and the Northern Pacific's main line was once again complete. This railroad line crosses portions of the Mt. Baker-Snoqualmie NF and Okanogan-Wenatchee NF. The old land grant included portions of the Gifford Pinchot NF and Mt. Hood NF.

Today, much of the land grant is controlled by Plum Creek, Weyerhaeuser, Potlatch, and Boise Cascade timber companies. Plum Creek is a direct spin-off company from the Northern Pacific/Great Northern Railroad. President Calvin Coolidge requested a congressional investigation of the land grant in 1924. Four years later, Congress directed the Department of Justice to take action against the Northern Pacific for violating terms of the land grant. There was a partial settlement in 1941, but the major timber holdings remaining from the land grant continue to be in private ownership.

Oregon Central Military Wagon Road—In 1864, Congress passed an act to grant the State of Oregon 806,000 acres of land for the building of the wagon road. The Oregon Central Military Wagon Road Company surveyed a route that crossed the Cascades at Willamette Pass to the headwaters of the Deschutes River. Rather than proceeding directly east toward Ft. Boise the surveyors turned south into the Klamath Basin, then east and northeast. The route through the Klamath Indian Reservation and the Klamath Marsh areas gave the company an opportunity to claim valuable grazing lands. By August 1875, the road was completed, as far as the government was concerned. The crude road barely made it from Eugene to Klamath Marsh. The old wagon road was used sparingly until the all-weather Willamette Highway 58 opened in 1940. The old land grant crosses the Willamette, Deschutes, and Fremont-Winema NFs.

Willamette Valley and Cascade Mountain Military Wagon Road—The Willamette Valley and Cascade Mountain Military Wagon Road, or the Santiam Wagon Road as it was called, originated in the Willamette Valley, crossing the Cascade Range at Santiam Pass, and extending to eastern Oregon and Idaho. A land grant of 860,000 acres made to the State of Oregon which then deeded portions of the grant to the company upon completion of continuous sections of the wagon road. By 1868 the road was pronounced open to Ontario, the eastern terminus. Later testimony before several congressional committees stated that the toll road was never constructed past the Deschutes River, a distance of about 80 miles (Clark 1987).

The Santiam Wagon Road on the Cascade NF was nearly the last leg of a transcontinental automobile race from New York City to Portland, Oregon. Two seven-horsepower Oldsmobiles raced from May 8 to June 21, 1905, to the opening of the Lewis and Clark Exposition. This was the first recorded passage of any automobiles over the Santiam Wagon Road and the first autos over the Cascades, and the Rockies from east to west. The winners, driving "Old Scout," completed the cross-country journey in the then amazing time of 44 days, three hours, and 25 minutes. Both cars were returned to Fish Lake in 1995 to drive on portions of the old Santiam Wagon Road on the Willamette NF. The land grant crosses the Willamette, Deschutes, and Ochoco NFs.

Coos Bay Military Wagon Road—Congress by an act of March 3, 1869, granted to the State of Oregon lands to aid in the construction of a military wagon road from Coos Bay to Roseburg. In February 1908, the federal government brought suit against the Coos Bay Wagon Road Company, or rather its successor the Southern Pacific Railroad Company, for failing to live up to provisions of the land grant. The suit wanted the lands embraced within this grant to be returned to the government. Congress revested about 93,000 acres of Coos Bay Wagon Road lands in 1919 (see the O&C Railroad above for further discussion).

The Dalles Military Wagon Road—Another military wagon road was to connect The Dalles to Ft. Boise. The land grant to the State of Oregon passed Congress on February 25, 1867. The wagon road often traveled over the old emigrant road, county-built roads, and at times faced fences built by settlers. Little of the road or bridges were actually constructed, yet thousands of acres of land were conveyed to the company. The Dalles Military Road Company stopped maintaining the road by 1869 and cancelled land sales. On March 2, 1889, the Oregon legislature asked the U.S. Attorney General to look into the matter and bring a suit of forfeiture to all the lands granted on the grounds that the terms of the grant had not been complied with. This act also called for the cancellation of all patents issued by the state. The old land grant crosses parts of the Malheur and Wallowa-Whitman NFs. In the 1980s, some of the old grant lands, then owned by the Hines Lumber Company, were exchanged with the Malheur NF in another effort to block up ownership.

Chapter 2

Early Conservation Efforts 1880s–1905

The first forest reserves in the Pacific Northwest were established in 1892-93, although actual management did not begin until the summer of 1898. Control and management of the forest reserves passed from the Department of the Interior to the Department of Agriculture (Bureau of Forestry) on February 1, 1905. Later that summer, the Bureau of Forestry was renamed as the Forest Service. The federal forest areas of the Pacific Northwest have historically been either the focus of controversies or part of the solution to national problems. The major problems and controversies that faced the fledgling Forest Service included establishment of the forest reserves, land frauds, revestment of the railroad land grants, and early on-the-ground management. An undercurrent throughout the last 100 years has been a tension between the Departments of Agriculture and Interior. The way in which these events and opportunities were handled sheds light on the reason for high federal ownership in the two states and continued issues in federal forest management today.

Creating Forest Reserves

At the national level, efforts were underway in the 1880s to persuade the president and Congress to allow for the creation of forest reservations from public domain land, that is, land that was not under private, Indian reservation, or other federal/state ownership. In 1889 a small group from the American Forestry Association met with President Benjamin Harrison and presented a petition that advocated adoption of a government forest policy, but the President took no action. The following year, the American Forestry Association sent a petition to Congress to make forest reservations and to provide a commission to administer them, but once again, nothing happened (Steen 1991).

Prior to 1891, there were several partially successful efforts by individuals and groups to preserve the remaining public domain forestland in both Oregon and Washington for future generations (Rakestraw 1955). The origin of the Cascade Range Forest Reserve (FR) dates from the summer of 1885, after William G. Steel, from Portland, visited the wondrous and stunning Crater Lake near the southern end of the 300-mile Cascade Range. He believed the U.S. Department of the Interior's General Land Office (GLO, which was established in 1812) was inappropriately selling the remaining timber-covered public domain lands to homesteaders, timber companies, and miners, as well as providing free land for state schools and huge land grants for the construction of railroads and military wagon roads (Williams and Mark 1995). On his return from the lake, Steel visited Roseburg to confer with Congressman Binger Hermann, who encouraged him to seek federal protection for the area. On his way to Portland, Steel stopped in Salem to meet with Judge Waldo, who also suggested that the best course of action was to seek federal protection for the entire Cascade Range. Steel began with a petition to make Crater Lake into a national park.

This Crater Lake petition movement was successful. On February 1, 1886, President Grover Cleveland by executive order suspended homesteading in 10 townships around Crater Lake and northward to encompass the Diamond Lake area. This was the first withdrawal of public land in the state for scenic or forestry purposes. However, it was a tenuous triumph because without congressional designation as a national park, the land could be restored to entry by a future president. It wasn't until 1902 that Crater Lake National Park was established by Congress.

Former Oregon Supreme Court judge John B. Waldo led an effort aimed at obtaining a much larger forest reserve along the crest of the Oregon Cascade Range. In 1888, shortly after completing the first documented journey along the crest of the Cascades from Mt. Jefferson to Mt. Shasta, Waldo was elected to the Oregon Legislature (Williams 1992). He introduced House Joint Memorial No. 8 when the legislature convened in January 1889, a memorial to Congress urging the establishment of a forest reserve or preserve along the crest of the Cascade Range. It was to be 12 miles on either side of the divide from the Columbia River on the north almost to the California border on the south. A joint state and federal commission was to manage the reserve.

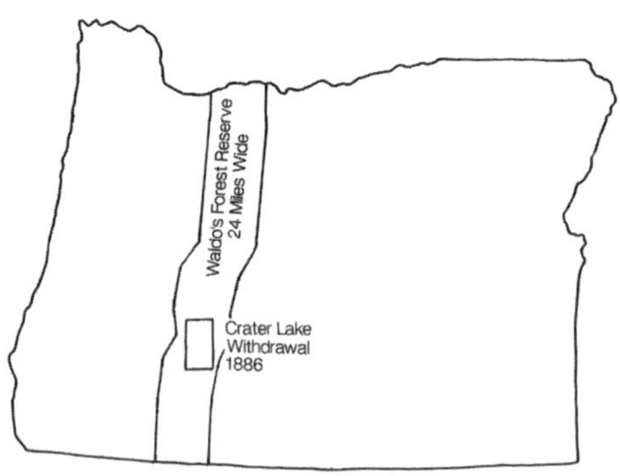

Judge John B. Waldo's proposed forest reserve for Oregon, 1889. Author's collection.

After passing the House, with amendments, it was defeated in the Oregon Senate by sheep interests in north-central Oregon, who felt such a reserve would eliminate sheep grazing from the alpine meadows during the critical summer months.

The pivotal year in the fight for the protection of federal forestlands was 1891. The western states held the only remaining large blocks of public domain land still under federal ownership. Many people became concerned that the land and forests were being transferred to the private sector in a reckless manner. The earlier Donation Land Claim Act of 1850 and Timber-Culture Act of 1873 were felt to be especially generous. In addition, many settlers and "land grabbers" were making a mockery of the legislation by using illegal schemes to take more land than they were allowed and by not following the intent of the laws. In the spring of 1891, while Congress was debating the repeal or modification of these liberal homesteading laws, an amendment or rider was attached to the legislation which would allow the president to establish forest reserves. This important amendment (section 24) to the Act of March 3, 1891 (26 Stat. 1095), afterward referred to as the Creative Act or Forest Reserve Act, was a long sentence that read: "Sec. 24. The President of the United States may, from time to time, set apart and reserve, in any State or Territory having public land bearing forests, in any part of the public lands wholly or in part covered with timber or undergrowth, whether of commercial value or not, as public reservations, and the President shall, by public proclamation, declare the establishment of such reservations and the limits thereof."

The first attempt to use this law in the Pacific Northwest was by the City of Portland. In the early 1890s, the city filed a claim for the water rights in the Bull Run watershed, west of Mt. Hood, as it was and remains the main source for the city water supply. Henry Failing, chair of the Portland Water Commission, wrote in January 1892 to Thomas H. Carter, commissioner of the General Land Office (GLO), asking that a forest reserve be created around the municipal watershed. After his request was endorsed by state congressional representatives, Carter wrote to the Secretary of the Interior recommending the withdrawal under the provisions of the Forest Reserve Act. The Bull Run Timberland Reserve, covering 142,080 acres, was established by presidential proclamation on June 17, 1892. This was the first forest reserve in the Pacific Northwest (Rakestraw 1955).

The next forest reserve to be created was the Pacific FR in the State of Washington. Much like the earlier Crater Lake withdrawal, it was designed to protect the scenic and recreational values around Mt. Rainier. Only one objection was noted: the Commercial Club of Tacoma did not want the word

Drinking fountain with water from the Bull Run watershed, Portland, Oregon, 1916. USFS photograph.

Rainier appearing in the name, as a controversy was raging regarding the name of the mountain. The City of Tacoma took the position that Mt. Tacoma was the proper name for this feature while those in Seattle thought Mt. Rainier should be retained. The GLO took a safe position and named it the Pacific FR when the 967,680-acre reserve was established on February 20, 1893 (Rakestraw 1955).

The third and fourth forest reserves in the Pacific Northwest were established in the fall of 1893. The Ashland FR was created in southern Oregon to protect the city watershed, much like the earlier Bull Run Reserve. The Ashland City Council and Board of Trade petitioned the president through Senator Joseph N. Dolph to withdraw about two townships for their water supply. As no protests arose, President Cleveland established 18,560 acres as the Ashland FR on September 28, 1893 (Rakestraw 1955).

Another reserve created on the same day was the Cascade Range FR, which covered basically the area that Judge Waldo had proposed four years previously. With the Forest Reserve Act as the basis, William G. Steel, with encouragement from Judge Waldo, undertook to renew the effort to create a reserve along the backbone of the Cascades. In 1891 and 1892 the prevailing mood in the state was for protection of the high Cascades. Steel was able to gather many signatures on his petition, but the process was halted in the Department of the Interior as there was a strong suspicion that the effort was being undermined by men from the state who had been very successful in obtaining land and timber under fraudulent circumstances. After more than a year of investigation, President Cleveland proclaimed the Cascade Range FR on September 28, 1893. Whereas the other reserves were small in size, the Cascade Range FR encompassed 4,492,800 acres and was 235 miles in length—the largest forest reserve in the nation (Rakestraw 1955; Steel 1932; Williams and Mark 1995).

Petitions to dismantle the reserve, which had not been an issue for the other reserves, began to come into the GLO office in Washington, DC, during 1895-96. Protests came from sheep owners in north-central Oregon as well as a few homesteaders in or near the reserve and miners in the Bohemia Mining District. John Minto, a prominent Salem-area resident, state legislator, and sheep farmer, began writing numerous lengthy newspaper articles against the reserve. Minto advocated that the Cascade Range FR allow grazing homesteads on the reserve, where no one would live, but where ranchers could graze their sheep without problem or regulation (Williams and Mark 1995).

Management of the new forest reserves was practically nonexistent. But there had been an executive order in 1894 to bar sheep from forest reserves. For a while, it was barely enforced. But in 1896 two cases against sheep owners found their way into the U.S. Circuit Court in Portland (Langille 1956).

In the winter of 1895-96, the entire Oregon congressional delegation was ready to eliminate or severely reduce the Cascade Range FR. The outcry for elimination was averted only by William Steel traveling to Washington, DC, and spending several months convincing the president and bureaucrats to keep the reserve intact. Meanwhile, Waldo and others from the Mazamas, a local mountaineering group, kept up a steady letter, petition, and telegram campaign to convince the president, Congress, and the GLO that the reserve should be kept intact. By the late spring of 1896, there was no doubt that their combined efforts were successful. Had the Cascade Range FR been dismantled, the fledgling national forest system in Oregon might have never developed into its present form (Williams and Mark 1995).

The large Cascade Range FR was eventually split into the Oregon (now Mt. Hood), Cascade (now Willamette), Umpqua, and Crater (now Rogue River-Siskiyou) NFs on July 1, 1908. Another division followed three years later when the Santiam NF was carved from portions of the Oregon and Cascade NFs and the new Deschutes and Paulina NFs from the land area east of the crest of the Cascade Range. The Cascade NF name disappeared as a distinct entity in 1933 when it was combined with the Santiam NF to form the Willamette NF.

From 1893 to 1897 no additional forest reserves were established in the Pacific Northwest or elsewhere in the country. President Cleveland was awaiting action by Congress to pass legislation that would authorize the investigation and management of these federal forests.

Meanwhile efforts were underway in Congress to change the procedures for establishing federal forests. The National Forest Commission, funded by Congress through the National Academy of Sciences, was sent in 1896 to investigate the forest reserves in the West and to report on their findings to Congress (Williams and Mark 1995). William G. Steel was invited to guide the commission through the mountains of Oregon, and took them to Crater Lake.

At the same time as the commission was visiting the lake, Henry Graves, who in 1910 would become the second chief of the Forest Service, was paid by Commission member Gifford Pinchot to study portions of the Cascade Range FR. Although not officially part of the commission, Graves was there "to see parts of the Reserve we had missed, study the rate of growth of the Douglas-fir, that most wonderful tree, and make himself familiar with the effect of sheep grazing on the forest ... it was original firsthand information" (Pinchot 1947: 102). After the National Forest Commission left Crater Lake, they traveled to the Sierras in California, then on to southern California and Arizona New Mexico, Colorado, and home. John Muir accompanied the commission as an unpaid member. The members disagreed much of the time but they did agree on the need for a Mt. Rainier National Park and a Grand Canyon National Park. However, their establishment would have to wait for several years (Williams and Mark 1995).

The commission also proposed that 13 new forest reserves be created or expanded, totaling 21,378,840 acres. On February 22, 1897, President Cleveland issued the proclamations to establish these reserves, which were afterward known as the "Washington's Birthday Reserves." Several new forest reserves were created in Washington State including the Washington FR (3,594,240 acres), Olympic FR (2,188,800 acres), Mt. Rainier FR, which changed the name of and enlarged the old Pacific FR (1,267,000 acres), and Priest River FR (645,120 acres mostly in Idaho). Establishment of the Mt. Rainier National Park came on March 2, 1899, when it was created by Act of Congress from part of the Mt. Rainier FR.

News of the Washington's Birthday Reserves caused a furor in the West. An immediate outcry to have the reserves canceled came from the affected states, ranchers, stockowners, and miners. Arguments against them were almost identical to the earlier protests about the Cascade Range FR; some wanted all the reserves eliminated. As before, much of the opposition was led by western stockowners, who "did not command broad-based support outside their region. Rather, they spoke for a narrow, but highly organized user group whose activities inspired suspicion on the part of conservationists ... [However,] the political power of the grazing interests in the West compelled attention" (Rowley 1985: 24). In Washington, the Seattle Chamber of Commerce and miners near Monte Cristo were among the most vocal against the Washington FR. The staff of the American Forestry Association's publication summarized the numerous objections:

> (a) Unnatural irritation at the idea that Eastern influences are presuming to assert themselves in regard to the Western states. (b) Natural irritation at the manner in which the reservations were made, without consultation with Western Representatives. (c) Reasonable objection to the inclusion of agricultural lands within the bounds of the reservation. (d) Unreasonable objection to the whole forest reservation idea as impeding licentious use of the public domain by everybody (*The Forester* 1898: 96).

Many were favorable to the forest reserves, but they were barely heard over the opponents' letters to editors, Congress, and the president. A bill that would have eliminated the Washington's Birthday Reserves was pocket vetoed by outgoing President Cleveland in early March (the old inauguration time) of 1897, leaving the final disposition to the new president, McKinley. The bill was reintroduced after cooler heads prevailed but with several important changes. One was the suspension of the new forest reserves for nine months to allow bona fide settlers to homestead portions of the new reserves. The suspension quelled the western demands for total elimination. The revised bill that was signed into law on June 4, 1897, has since been referred to as the "Organic Act of 1897."

The 1897 act had other provisions that would affect the forest reserves for many years. One of the clauses, the so-called "Pettigrew Amendment," provided that any new reserves would have to meet three criteria— forest improvement and protection, securing favorable conditions for water flows, and furnishing a continuous supply of timber—setting the basis for federal forest management for the next 63 years (Williams 1997).

In addition, the act allowed, for the first time, for organization and management of the reserves by forest rangers. Gifford Pinchot was perhaps the first forest reserve employee in the GLO when he was hired in July 1897 as a "special forestry agent" to study the new FRs and find a means to manage them.

That summer, the GLO also appointed superintendents for each state where there were forest reserves. By the following year, supervisors were also appointed for each forest reserve and forest rangers to patrol them (see Chapter 3).

Gifford Pinchot, probably because of his experience in seeing GLO rangers at work (see next chapter), rather bitterly noted that "unfortunately the Interior Department, with its tradition of political toadeating and executive incompetence, was incapable of employing the powers the act gave it" (Pinchot 1947: 116). This frustration would set the stage for the transfer (February 1, 1905) of the forest reserves to the Department of Agriculture under Pinchot's own inspired leadership. Another provision in the act was the "Forest Lieu" clause, which allowed homesteaders within the proclaimed forest reserve boundary to exchange their claims for public domain land outside the boundary. This provision would lead to massive land frauds in the Pacific Northwest and elsewhere.

Coville Grazing Investigations in the Summer of 1897

In 1897, the Secretary of the Interior shut down grazing in all forest reserves, which quickly raised the anger of sheep owners in the West. At the time, the Cascades of Washington and Oregon supported well over 600,000 sheep that grazed the alpine meadows every summer. During the summers of 1897 and 1898, as arguments against the new forest reserves were creating headlines in the region, USDA botanist Frederick V. Coville quietly studied the grazing situation in the Cascade Range FR to scientifically measure the damage caused by sheep. His report of the following year suggested that sheep grazing was a danger to the forest only if it were unregulated. Coville also "contradicted the earlier [National Forest Commission] report's main assertions regarding the commercial value of the sheep-raising industry to the region, fires caused by herders, and the extent to which erosion problems occurred because of overgrazing" (Rowley 1985: 33).

Soon after Coville's report was issued, the Cascade Range FR and the forest reserves in Washington were allowed regulated grazing, often at reduced numbers, although the order came too late in the season for any grazing to occur in the summer of 1898. Reserves in other states would have to wait until new grazing studies were initiated and reported, but they too were allowed to have regulated grazing after the turn of the twentieth century.

An interesting and sometimes violent controversy developed in the central Oregon area during the late 1890s, pitting the sheep grazers against the cattle owners—the range wars. In the spring of each year, the first group to the mountain meadows could and often did secure the best grass and camp

Sheep grazing in alpine meadow, Oregon National Forest, 1916. USFS photograph.

sites. Often the first on these areas were the sheep. The Crook County Sheep Shooting Association was formed in July 1896 to remedy the situation. The association's written agreement or compact said, in part:

> If in the accomplishment of a mission—the killing of sheep—it becomes necessary for the groups to shoot a herder or camp tender, the victim is to be buried where he falls. Should any member of the sheep-killing group fall in a range battle, he is to be brought home for burial and nothing is to be said about the manner in which he met his death. In the event any member of the association is brought to trial for sheep killing, other members must be willing to go on the witness stand and swear to lies to obtain an acquittal.

Tens of thousands of sheep and a few sheep herders were killed. The association reported in the *Morning Oregonian* on December 29, 1904, that their members had shot some 8,000 to 10,000 sheep during the summer. The sheep growers associations were enraged, but there was little they could do. By 1906, the controversy simmered at a low boil as the Forest Service rangers from the new Blue Mountains FR (now the Ochoco and Malheur NFs) entered the fray, establishing range allotments and generally keeping a lid on the problem.

Land Frauds in the Pacific Northwest

Public land frauds linked to the establishment of new forest reserves were common after the passage of the Organic Act of 1897. As new lands were evaluated for possible inclusion into the forest reserve system, numerous "land grabbers" paid people by the hundreds to file fraudulent land claims and then quickly "sell" their claims to the men who had paid them to file. These speculators would hold on to the claims, and, when a forest reserve was proclaimed, they would hurry to the GLO and request that the government trade the claimed land inside the forest reserve boundary for land of equal value outside the boundary. As the forest reserves were in the high mountains, they were usually able to trade their claims for more-valuable land at lower elevations. Then the speculators would often sell their newly acquired land at high prices to big timber companies, or remove the standing timber and then sell the high-priced "agricultural" land to homesteaders, or simply cut the timber and let the land revert to the state or county without paying taxes. Any way it was done, the frauds amounted to millions of dollars lost by the government.

One attorney justified the fraud: "Why ... nobody is robbed—there is no robbery. The [paid land] locator gets value for the exercise of his right as a citizen; the government gets paid its price for the land, and the land goes to the man or company that sooner or later puts it in use, while the state begins to get annual taxes, which it did not before [as public domain]; so nobody is hurt, but everybody profits, though, ... of course perjury may have been suborned—but that was sanctioned by common usage in land matters." (Pease 1907: 148)

The Organic Act supplied another basis for fraud by providing that lands within the forest reserve boundaries found to be more valuable for agriculture or mineral production should not be retained in the forest reserve system. The land speculators found they could easily defraud the government if they knew where the forestland was to be withdrawn for a new reserve. Employees of the GLO, local townspeople, surveyors, and even elected officials were often bribed to give advance information regarding the establishment of new forest reserves. "Tens of thousands of dollars were paid by middlemen and bribes as high as one or two thousand dollars were paid to surveyors, land office receivers and recorders, inspectors, prospectors, attorney generals, congressmen, and United States Senators." (Robertson 1984: 32) Thus the withdrawal and establishment of forest reserves quickly became a very lucrative business for everyone except the federal government. The GLO's hold on the reserves began to weaken when land-fraud scandals throughout the western states started to receive widespread publicity in 1899.

On July 21, 1901, *The Oregonian* newspaper published an article entitled "Big Timber Graft." The report about the proposed Coquille FR in southwest Oregon stated: "[T]here are many ways in which the Government may lose and private corporations and individuals may make illegitimate gain from such

Land Office rush to obtain timber claims, Lakeview, Oregon, 1907. Author's collection.

reservations as is now proposed for Southern Oregon." Within five days GLO Commissioner Binger Hermann, himself a politician from southern Oregon who was later indicted in Oregon land frauds, suspended all claims in the area under consideration as a forest reserve. Hermann may have chosen to suspend the claims so that GLO investigators would not get "too close to home," as his relatives were directly involved in the initial survey of the land for the reserve (*Morning Oregonian*, July 31, 1901) and illegal land claims by his relatives and friends were a strong possibility.

Stephen A. Douglas Puter, self-proclaimed "king of the Oregon land-fraud ring," related that at least several of the forest reserves in Oregon, Washington, and elsewhere were actually made at the suggestion and encouragement of the land speculators, including himself. While serving time in the Multnomah County jail for land fraud, Puter, in collaboration with Horace Stevens, wrote a book that was titled *Looters of the Public Domain* (1908). This insider account of the defrauding of the U.S. government gave names, dates, and dollar amounts in the coordinated actions of many speculators in the Pacific Northwest, especially the state of Oregon.

It was common practice to state that the claimant had improved the land by removing trees and either farming or grazing animals. Another provision was building a permanent dwelling. Some of the dwellings were small doll houses; the law did not say the dwelling had to be actual house size. The famous "11-7" case concerning fraud in Township 11 South, Range 7 East (in the Marion Forks area of the Detroit Ranger District) of the Cascade Range FR (now Willamette NF) was a classic in the land-fraud trials. Land grabbers in 1900 had convinced 10 people to file 12 homestead claims (two people signed for additional claims by using fictitious names) in Township 11-7 and to swear falsely that they were living on the claims at the time the Cascade Range FR was created. Not one of the 10 claimants had ever seen the land they filed for and none lived within 30 miles of the location.

Washington State did not have the same type of massive land-fraud problems as Oregon; although there were no indictments or trials that would rival the sensational cases in Oregon, there were at least some instances of timber fraud. Puter devoted an entire chapter in his book to the dealings of the Northern Pacific Railroad and the creation of the Mt. Rainier FR and later Mt. Rainier National Park in 1899. The Northern Pacific's land grant was given by the federal government to subsidize the construction costs for the railroad and promote settlement. Approximately one million acres of the Northern Pacific land grant were within the boundaries of the Washington FR and nearly 100,000 acres were within Mt. Rainier National Park. The company was glad to exchange, or as Puter wrote "steal," those lands for more valuable public domain timbered lands in other parts of Washington, Oregon, and Idaho.

An enlargement of the Cascade Range FR in 1901 brought new allegations that GLO officials had taken bribes from speculators who stood to benefit from the expansion. A newspaper account from *The Oregonian* on October 10, 1902, noted that the Captain Salmon B. Ormsby's forest superintendent position was to be eliminated for cost-cutting efforts, or in the official language "superfluous and tending to retard the expedition of public business." Ormsby was allowed to stay on until December 31, 1902, to complete a report on the Blue Mountains FR. He then resigned so that the position could be abolished. All state superintendents, except for Ormsby, were terminated in October and management was given to each forest supervisor on their respective forest reserve.

The land-fraud trials were important to the future of the national forests in the Pacific Northwest as they were instrumental in ending the lieu-land provisions in the Organic Act of 1897, which were repealed on March 3, 1905 (Messing 1966).

Management by the General Land Office 1897-1905

From 1897 until 1905, all forest reserves were under the control of the General Land Office in the Department of the Interior. As we saw earlier in this chapter, the 1897 Organic Act had two important components: one allowed for land speculators to trade land within the forest reserves for public land elsewhere (the so-called lieu land provision), while the other provided for the management of the reserves by the General Land Office (rangers) and the Geological Survey (mappers). The GLO established a comprehensive system of management and control of the reserves and national forests involving state, district, and forest supervisors, district rangers, and technical forestry support.

Many observers were less than impressed. All the GLO state superintendents, forest supervisors, and forest rangers were political appointees with no experience and no training in forestry matters.

The first superintendent of the forest reserves in Oregon and Washington was Salmon B. Ormsby of Salem, Oregon. He was appointed in early July 1897 and, at the same time, another superintendent, Benjamin F. Allen, was appointed to supervise the forest reserves in California and Arizona. In the summer and early fall of 1897, Bernard E. Fernow, chief of the Division of Forestry, made an extensive inspection through the forest reserves of Oregon. In an interview, he said: "The object of my visit was to inform myself of forest conditions in Oregon and to study its forest trees. The first thing needed from a technical point of view is to have reliable, authoritative information as to the quality of Oregon timber. I have for years urged the need of such knowledge for all our woods, that … next to stopping fires, knowledge of our woods is the most practical means of forest protection." (*The Oregonian*, September 29, 1897)

Fernow resigned in May 1898 and was replaced as chief of the Division of Forestry on June 1 by Gifford Pinchot. By the summer of that year, the GLO was able to appoint supervisors for each forest reserve and forest rangers to patrol them. These men often had little experience in forestry matters or even logging operations. Their appointments were generally for the summer fire season, usually July, August, and September. There was not a great deal of confidence in this early forest ranger system. Chief Forester Gifford Pinchot in particular detested the untrained, political appointees who had been hired. He was quoted in a report highlighting some of the troubles:

> The present fault with the system lies in its subordination to politics. A forestry expert who recently returned to Washington from the West, had this to say: "Not over one-fifth of the employees are more than

> apologies for forest rangers. A system has been inaugurated by which salaries assignable to forest rangers have been distributed as chromos [essentially a bribe] to political workers, with little or no pretense of anything else. I found one working as a journeyman carpenter, and yet drawing the regular salary of a forest ranger. I took pains to investigate his work and learned that his district extended down the main road for 12 miles. Another man who had occasion to go over the same road daily and knew every traveler on it, assured me that the carpenter-ranger had not in weeks been more than three miles from the place where he was employed. This, moreover, was not in a commanding position, so as to be of any use in detecting fires throughout the district. About three miles away this ranger had a little fire to which he drove down daily and threw on a bucket of water. A limited area had been so trenched about that there was little danger of serious consequences, and his report as submitted to the department doubtless shows a record of considerable industry. This was in the Cascade region of Washington (*The Oregonian*, October 11, 1898).

The GLO was not equipped to manage the forest reserves, but it did not take the matter lightly. After learning lessons from the first field season, a slightly revised system of management and organization was instituted.

> In the 11 states and territories of Arizona, California, Colorado, Idaho, Montana, New Mexico, Oregon, South Dakota, Utah, Washington and Wyoming there are 36 forest reserves, containing an aggregate area of about 46 million acres. These reserves are divided into nine districts, each having a general officer, known as a forest superintendent, in charge. Each district is divided into supervisors' districts, the number depending on the number of reserves, total area and difficulties of supervision, as affected by topography and liability to fires and depredations of all kinds. For each of these supervisors' districts there is appointed an officer called a forest supervisor, who has direct charge of the reserve or a portion of a reserve forming his district. … Each reserve is divided into patrol districts, the size of each patrol depending upon topography and the liability to fires and depredations; and a forest ranger, whose headquarters shall be at some central point in his subdivision, is appointed for each such district. …Two hundred and fifty rangers for immediate duty were authorized May 2, 1899, to serve until October 15, 1899. One hundred additional rangers were

authorized to enter upon duty July 15, to serve until October 15, 1899. The rangers report to the supervisors and are under their immediate supervision. The supervisors report to the superintendents and the superintendents report to the commissioner of the general land office in Washington …

The supervisor must have his headquarters in or near the reserve of which he is in charge. He must familiarize himself with all the conditions existing in his district, especially in regard to forest fires. … They have immediate supervision of the rangers and are required to be in and through the reserve to see that the time of the rangers is fully occupied in patrolling their districts, clearing up old trails, cutting new trails and performing their duties generally. They make weekly reports of daily service rendered, and monthly reports on the general conditions existing in the reserve. ...[Fire control is an important aspect of the position, where the supervisor must report on the] Total amount expended during the month for tools, the dates of fires, the names and address of the parties responsible for their starting, the origin, the damage done, the probable market value of the timber burned and the effect upon the forest cover and water supply.

A ranger must provide himself with horse and equipment, while the government furnishes him with the various implements necessary to open trails in the dense forest, to construct fire barriers and to extinguish and surround fires. Each ranger is provided with a [round] nickel badge, which is worn as an evidence of his official authority. ...

Oregon.

The three forest reserves in Oregon—the Cascade range reserve, area, 4,492,800 acres; the Bull Run, area 142,080 acres, and the Ashland, area 18,560 acres, or an aggregate area of 4,653,440 acres—constitute a superintendent's district, of which S. B. Ormsby, Salem, Or., is the superintendent. There are three supervisors' districts, the northern division of the Cascade range reserve and the Bull Run being under the supervision of W. H. Dufur, Dufur, Or.; the central division of the Cascade range forming another supervisor's district, under Ralph B. Dixon, of Roseburg; and the southern division of the [Cascade Range Forest Reserve and] Ashland another, in charge of Nat Langell, Jacksonville. Forty rangers were assigned for the entire season—37 for the Cascade range, two for the Bull Run and one for the Ashland. ...

> *Washington.*
>
> The superintendent of the reserves in Washington is D. B. Sheller, Tacoma, Wash. These contain an aggregate area of 8,121,880 acres, comprising that part of the Priest river reserve which lies in this state, areas 103,960 acres; the Washington, 3,594,240 acres; the Olympic, 2,188,800 acres, and the Mount Rainier, 2,234,880 acres. The Washington and Priest river form a supervisor's district, of which Edward Burin, Custer, Wash., is the supervisor; F. C. Mathewson, of Shelton, is the supervisor for the Olympic, and George McCoy, of Napavine, for the Mount Rainier. Twenty-three rangers—eight for the Washington, six for the Olympic and nine for the Mount Rainier—were authorized for immediate duty, and for duty on July 15, 11 more, as rangers-at-large (*The Forester*, reprinted in *The Oregonian*, November 1, 1899).

Grazing was the main management problem for both rangers and forest supervisors. A number of trips were made into the West by congressional members as well as GLO and Bureau of Forestry experts to investigate and discuss the issue. A report from the *Morning Oregonian* on August 6, 1901, noted several important visitors, including Frederick Newell from the U.S. Department of the Interior and Gifford Pinchot from the Bureau of Forestry: "[They] arrived yesterday from Washington, via Utah and Idaho, and leave tomorrow with Representative [Malcolm A.] Moody by private conveyance for a trip through Wasco, Crook, Harney, Malheur and Baker Counties. … Mr. Pinchot joins the party as far as Prineville for the purpose of meeting and conferring with the stockmen interested in pasturing on forest reserves."

During that same summer, Harold D. Langille, then doing USGS mapping in the Cascades, met Pinchot and Newell on the slope of Mt. Hood at Cloud Cap Inn, owned by the Langille family. They discussed a number of questions about the management of the mountains:

> Sheep had been excluded from the northern part of the reserve. "G.P.," as we affectionately knew him, was bent upon determining the effects of sheep grazing. … [The following day, after spending one day hiking around the mountain,] crossed into the sheep country. Soon G.P. gave tongue, for he had found the spoor he sought. Some old mother ewe trailing back to the winter range had paused to peel a cud of juicy, resinous bark from a sapling lodgepole pine. … But here was the *corpus delicti*, proof positive that sheep devour forests, and the proof was in

> the hands of none less than the chief forester himself. A bad day for the woolies (Langille, 1956: 305).

The GLO rangers wore round nickel-plated badges with the words *General Land Office* and *Forest Ranger* engraved on the front. Beginning in 1902 they had dress uniforms based on military examples that dated from the post-Civil War era. In the late fall of 1902, there was a major shakeup of the GLO forest reserve management system. All the GLO state superintendents were fired in October at the end of the fire season, leaving everyday forest management up to the forest supervisors supported by the forest rangers. The notion of state superintendents never came up again, although today there are states in the South and East that have one forest supervisor to manage numerous national forests, as in North Carolina.

U.S. Geological Survey Mapping 1897-1905

The work of mapping the reserves was undertaken by the U.S. Geological Survey, as specified by the Organic Act of 1897. A series of USGS mapping and descriptive work reports were taken on by a small force of USGS employees until 1905, and even later for some areas in the West. The extensive USGS reports often included written descriptions of forest cover, burned areas, water sources, mining activity, grazing, settlements, and potential developments. The Washington FR was the first reserve to be studied in the Pacific Northwest. In this case, as reported by USGS geographer Henry Gannett,

> Three separate and independent parties were organized in this reserve—one in the eastern portion of the reserve, in the vicinity of Lake Chelan; the second in the territory west of the summit of the Cascade Range, operating from Monte Cristo, and a third at Ellensburg, for the purpose of extending reconnaissance triangulations over the entire reserve. … An average force of about 30 men was employed at various times in this reserve [work], and the sum of $13,465 was expended, including the purchase of 20 [pack and riding] animals and various articles of camp equipment. … For the examination of the [eastern portion] … Messrs. W. G. Steel and M. [Martin] W. Gorman [from the Mazamas club], were employed. … For the western portion Mr. H. [Horace] B. Ayres was employed. …
>
> There is no timber cutting within the reserve, excepting a trifling amount for the local needs of the settlers and mines. There is at present very little demand for its timber, the general market being supplied by

> the [private] forests farther west. The facilities for getting timber out of the reserve are very poor. With the exception of the trifling amount of wagon roads [for mining purposes] the only routes would be by streams, and these are extremely rapid and rocky. Much work would be required upon them to make them suitable for driving logs (1898: 25-26, 28).

Archie J. Johnson was hired in 1899 by the USGS to survey and map the forest areas in Oregon, as well as to report on the amount of standing timber, the variety of trees, and the area of forest that had been burned or logged in every section, public or private. (The A.J. Johnson and Gilbert Thompson map of the State of Oregon "Showing the Classification of Lands and Forests" was printed with the U.S.G.S. Professional Paper No. 4 in 1902.) The work proved very difficult, because few roads and only Indian trails traversed the heavily timbered and steep mountain terrain. Much of the time he spent "swamping" his way through the forests as best he could with the instruments at hand and mapping as he went.

> A. J. Johnson … forestry expert, turned up at *The Oregonian* office last evening, suffering from a stupendous cold, contracted in the Bohemia mining district [in the Umpqua NF], where in pursuit of his business he had to ford Row River [a tributary of the Coast Fork Willamette River] in flood, the water coming two feet over the seat in the stage, and it a-swerving on the mountains [wagon roads built by miners]. This river had to be forded 14 times, and about every time the horses had to swim in crossing, the buckboard stage sank completely out of sight, and the passengers had to hold on with both hands to prevent being washed away down into the Willamette river. He does not consider this trip a summer outing, by any means; and … has within the past year traveled 20,000 miles among the mountains and forests of Oregon (*The Oregonian*, December 6, 1899).

A U.S.G.S. forest reserve study by John B. Leiberg was reported in *The Oregonian* October 8, 1900:

> The southern portion of this [Cascade Range Forest] reserve, together with the Ashland Forest Reserve, was examined by Mr. J. B. Leiberg during the past season. Altogether, in the neighborhood of 8000 square miles were examined by him. Upon the west side of the Cascade Mountains, with an ample rainfall, the forests were found dense and

the undergrowth luxuriant. Upon the east, where more arid conditions prevail, the forests are open, with no underbrush … a little less than two-thirds, or perhaps 65 per cent, is forested, and 35 per cent not forested, the nonforested areas lying mainly in the extreme west and in the eastern part of the area examined. The amount of timber found upon the forested area is estimated by him as a little less than 20,000,000,000 feet, an average per acre of the forest land of 6660 [board] feet. …

Fires have widely ravaged the region examined. There is not a single township in which the timber is not more or less fire marked. Without much doubt the present agricultural areas, once grass-covered and carrying scattered stands of oak, were burned over quite as extensively as the timbered tracts [by the Indians and early settlers]. The only tracts that have escaped are the swampy sedge and tule-covered areas bordering the Klamath lakes and marsh, and such spots in the higher elevations where bare lava or pumice fields made the spread of fires impossible. Of the forested area examined, comprising in round numbers 3,000,000 acres [south portion of the Cascade Range FR], a total of 2,975,000 acres, or 99.992 per cent, are fire-marked.

USGS field party at Lake Chelan, Washington, 1902. USFS photograph.

An examination of the Washington Cascades between Mt. Rainier and the Washington FR was undertaken in the spring of 1900 by Harold Langille and Fred Plummer, both USGS mappers, at the request of USGS geographer Henry Gannet, and their report was printed in 1902. Gannett in the spring of 1901 again called upon Langille to make another survey, from the Columbia River to a point south of Mt. Hood. "My need for education as an engineer was immediate. It was urgent. So I hurriedly sent three dollars to the nearest bookseller, received a two-volume set of 'Gillespie's Surveying'… [T]he plane table was set up on the back stoop … A sight at the stovepipe on Ho My's washhouse began my experience as a topographic engineer." (Langille 1956: 304)

GLO Forest Rangers

Forest ranger standards were, on paper, quite rigorous. A circular of instructions to forest rangers was issued in August 1898 by GLO Commissioner Hermann as follows:

> It will be the primary duty of forest rangers to patrol the different forest reserves for the purpose of preventing and extinguishing forest fires, preventing timber trespass, and trespass by stockmen in reserves where grazing is prohibited. … Every forest ranger will be assigned to a reserve, or part of reserve, and will work under the immediate supervision of a forest supervisor. He will make his headquarters at some elevated central point in the reserve overlooking the territory assigned to him. This point will be designated by the supervisor in charge of the reserve. He will be required to provide himself with a saddle-horse, the cost of maintenance of which will not be chargeable to the government; also, where necessary, he will provide, at his own expense, a camp outfit for use while on duty in the reserve.
>
> He will be expected to thoroughly familiarize himself with the topography of the territory to which he is assigned, and with all roads and trails leading through it. Where necessary he will blaze trails from point to point in the reserve, and, where practicable, he will cut fire breaks and burn brush, at dangerous points, to prevent the spread of forest fires. Where the lines of reserves have been surveyed, forest rangers will see that the marks of survey are not obliterated, and keep the lines at all times plainly marked.

The Oregonian of May 9, 1900, recounted the national GLO direction on hiring and qualifications of forest rangers:

> They should be able to ride a horse and take proper care of the animal in range riding, and be practical woodsmen in the sense of possessing a knowledge of woodcraft common to those brought up in a timbered region. …They should know, in a general way … what timber is mature and suitable for cutting; methods of reforestation; should have a clear knowledge of the best methods of fighting forest fires under varying circumstances, particularly of attack in its earliest stages, and of reducing the danger of fire to a minimum; a knowledge of the effect of sheep and cattle grazing upon the forest cover, and its relation to or pollution of water supply, and whether or not such grazing would prove injurious in the locality where a particular ranger might serve, giving reasons for his deductions; the best methods of cutting and clearing trails or roads, making fire breaks, and patrolling his range in the most effectual manner.

The early forest rangers were appointed by the GLO in Washington D.C., often through the auspices of the U.S. senator from the state, and were, as noted earlier, political appointments. In some cases, the qualifications could be flexible, as in the case of C. V. Dodd from McKenzie Bridge. Apparently his only qualification was that he had been a captain in the Civil War. Sometime later, Ray Engles noted that Dodd "carried a pistol with four shells, a large beard, and a round badge." He spent most of his time at the Log Cabin Inn at McKenzie Bridge. Another GLO ranger was Homer Ross in Washington State. He recounted his appointment: "I applied to G.F. Allen, Supervisor of the Rainier Forest Reserve. In due time I was hired at $60 per month for July, August and September of that year [1904]. I furnished two saddle horses, one pack horse. … I liked the work and the men I came in contact with, and also enjoyed wearing that big, round, nickel-plated badge." (Ross 1986: 1).

GLO rangers were not always the best men for the job. For example, Forest Inspector I. A. Moody reported to the commissioner of the GLO on October 1, 1901, that in the southern portion of the Cascade Range FR:

> [O]n October 11th-15th I discovered that Ranger J. R. Wick of said reserve, has not been performing service for the Government but has been performing service to others on his own account, during the time he has been charging the Government therefor [sic]. One J. J. [or G.G.] Crary Keeps a Hotel or Summer resort at Pelican Bay

[along the western shore of Upper Klamath Lake] within the Reserve on patented lands about 50 miles from Ashland, Oregon. Ranger Wicks headquarters are nominally on the Kasson place (script land near Pelican Bay[)] at Crary's urgent and persistent solicitation to the Super[visor] and Sup't [Superintendent], Wick was taken from Ashland Reserve, Or. and stationed on the Kasson place (a cabin & stable being thereon) and Ranger [W.M.G.] Kropke was transferred from Kasson place to Ashland Reserve. Carry has been using Ranger Wick during the Summer to assist him in entertaining his guests, taking them boating on the Lake and sending him on trips to Crater Lake 40 miles distant and also to do chores about the Hotel. … I respectfully recommend that he have deducted from his salary ten days pay for the two trips he is known to have gone to Crater Lake, and that he be allowed no pay for the fifteen days in October he is supposed to have been on duty but which he confessed he was working for Crary for his board, and that he be dismissed from the service.

Chapter 3

Taming the Wild Forests 1905–1910

Department of Agriculture Takes Over

Congress transferred all forest reserves and their management to the Department of Agriculture on February 1, 1905 (the Bureau of Forestry was established from the older Division of Forestry on March 2, 1901, and the Forest Service was created on July 1, 1905, by a simple name change). This occurred in the context of the land-fraud trials in Oregon. After the transfer, Pinchot moved quickly to make federal forest administration above reproach. Employees were appointed through the Civil Service Commission rather than by political appointment. It was a critical way to show the public that these important public forest lands could and would be managed, a process later referred to as "taming the forests."

Yet the forestry experts until 1905 did not have any forests to manage. They provided forestry advice to the GLO and to the private sector when requested. During this period, a number of forest plans were undertaken on large forested private lands, mostly in the South. (The tradition of providing forestry assistance to the private sector continues to this day through the state and private forestry branch of the agency.) For young foresters from the Washington Office of the Bureau, being assigned to work on these plans was a valuable experience.

The American Forestry Association, a group devoted to scientific management of forests, held its American Forest Congress (AFC) from January 2 to 6, 1905, in Washington, D.C. Pinchot, in charge behind the scenes, had convinced the Association to hold the event. His major goal was for the AFC to approve the transfer of the forest reserves from USDI to USDA. A reception was held for the delegates at the White House, where they were greeted by President Roosevelt, who said, "I would like to add one word as to the creation of a national forest service which I have recommended repeatedly in messages to Congress, and especially in my last. I wish to see all the forest work of the Government concentrated in the Department of Agriculture. It is folly to scatter such work, as I have said over and over again." Within days of the end of the AFC, a bill was passed by the Congress and signed into law by Roosevelt. The Transfer Act of 1905 (33 Stat. 628) transferred the forest reserves, just as Pinchot wanted. In a letter probably written by Pinchot or one of his associates, Secretary of Agriculture Wilson on the same day wrote instructions to Pinchot:

> In the administration of the forest reserves it must be clearly borne in mind that all land is to be devoted to its most productive use for the permanent good of the whole people; and not for the temporary benefit of individuals or companies. … the policy of this Department for their protection and use will invariably be guided by this fact, always bearing in mind that the *conservative use* of these resources in no way conflicts with their permanent value. You will see to it that the water, wood, and forage of the reserves are conserved and wisely used for the benefit of the home-builder first of all, upon whom depends the best permanent use of lands and resources alike … and where conflicting interests must be reconciled the question will always be decided from the standpoint of the greatest good of the greatest number in the long run.

When the Forest Service took over in 1905, Chief Forester Gifford Pinchot announced a contest among Washington office employees to design a new badge. One of these Forest Service employees, Edward T. Allen, who was attracted by the type of shield used by the Union Pacific Railroad, traced an outline of the shield (from a Union Pacific timetable) on a sheet of paper. He inserted the large letters U and S halfway from the top to the bottom of the shield, leaving a space between them. William Hodge, looking on, was inspired to sketch a fir tree on a sheet of roll-your-own cigarette paper he took from his pocket. He laid this between the U and the S. The two men quickly wrote FOREST SERVICE across the top and "DEPARTMENT OF AGRICULTURE" across the bottom. The design, often called the "pine tree shield," was approved and remains the official badge over 100 years later with only two slight changes: both the center tree and the letters U and the S were lengthened. The Forest Service shield, as it is called, is printed on all agency publications, signs, and used for the identification of Forest Service vehicles and other equipment. Several months later, on July 1, 1905, the Forest Service was created.

Only a few GLO rangers made it into the new Forest Service. Homer Ross, who was a GLO ranger in 1904, noted that in the spring of 1905 he was appointed a forest guard. "They sent me a big bronze shield badge instead of the nickel-plated one and wrote me that the Forest Reserves were being transferred from the Interior Department to the Department of Agriculture.

... I took the Civil Service examination that fall and passed, and in [the] spring, 1906, I was appointed as assistant Forest Ranger at $75 a month." (Ross 1986: 1) The rangers that were accepted into the Forest Service were men of practical experience; none had college or university training.

> Altogether we have the names of about 90 men in the Region Six who came into the Forest Service from the Department of the Interior in 1905. Many of these old Interior men accepted the conservation ideas and became the solid backbone of the organization. Some dropped out immediately for business or personal reasons. A few lacked ability or temperament and were dropped. A very few were victims of drink (we now call them "alcoholics"). Those that stuck, together with new recruits and the few technically trained men that served as leaven for the whole loaf, soon shook the Forest Service down into an active bureau that adjusted itself to everyday problems. Those first groups contributed much. A few became leaders—more were doers—qualified woodsmen who worked long and faithfully for small pay and with little regard for days or hours of work. All of them deserve high praise and commendation for a job well done (Merritt 2005: 6).

New Forest Reserves Added

Meanwhile, President Roosevelt continued to make proclamations creating new forest reserves and enlarging existing ones. Numerous field investigations, which included descriptions of timbered areas and maps, continued for several years. In both Oregon and Washington, the reports were presented to the Secretary of the Interior for final disposition. Almost all the proposed forest reserves were withdrawn from settlement, then officially proclaimed by the president. In some cases, the withdrawals came quickly, as there was a constant battle with land sharks and timber companies for obtaining, legally or illegally, large forest areas. The Blue Mountains FR was one such "emergency" case. Between 1904 and 1906, 10 new forest reserves were established in the Pacific Northwest, including:

Baker City (OR) 1904
Blue Mountains (OR) 1906
Chesnimnus (OR) 1905
Fremont (OR) 1906
Goose Lake (OR) 1906
Heppner (OR) 1906
Maury Mountain (OR) 1905

Siskiyou (OR) 1906
Wallowa (OR) 1905
Wenaha (OR & WA) 1905

On January 7, 1907, President Theodore Roosevelt proclaimed thousands of acres of timberland for a new reserve in the Whatcom County area of Washington State, near the Canadian border. As a howl of protests was heard from around the state, including the *Seattle Times*, Senator Samuel H. Piles threatened federal legislation to remove the withdrawal. Gifford Pinchot and the president yielded to intense lobbying and restored the land to public domain, claiming it was a "clerical error."

Soon thereafter, Senator Charles W. Fulton of Oregon, who had been implicated in the land frauds in the state, introduced an amendment to the annual agricultural appropriation bill denying presidential authority, authorized under the Creative or Forest Reserve Act of 1891, to establish national forests in the six western states of Washington, Oregon, Idaho, Montana, Wyoming, and Colorado. Afterwards, only Congress was allowed to establish these reserves. The Fulton amendment, at the suggestion of Gifford Pinchot, also changed the name to national forests in order to make it clear that the forests were to be used, not preserved.

With less than a month before the act had to be signed into law, Pinchot and his assistants drew new forest reserves on a map of the United States. As soon as the maps were finished and a proclamation written, the documents were carried to President Roosevelt, who signed each proclamation to establish another new reserve. In a matter of a few days, Roosevelt proclaimed over 16 million acres of new forest reserves, since referred to as the "Midnight Reserves." Opponents were furious, but they had been outflanked and the new reserves/national forests stayed, though the president's power to establish forest reserves ended on March 4, 1907, when the annual agricultural appropriations bill became law. In the Pacific Northwest six new national forests were proclaimed, all but one in Oregon:

Blue Mountains NF—added the Maury FR
Colville NF—in Washington State
Coquille NF
Imnaha NF—created from Wallowa and Chesnimnus FRs
Tillamook NF
Umpqua NF—Coast Range

The creation of these new national forests essentially set the basis for the federal forest system in the Pacific Northwest as we presently know it. There have been many changes in the names and boundaries of the national forests in the years that followed, but the basic land areas that were set aside between 1892 and 1907 are still with us today (see Appendix B for a list of the forest reserves and national forests from the earliest days to the present).

Forest Homestead Act of 1906

Despite the repeal of the lieu-land provisions of the Organic Act of 1897, there were still many critics of the forest reserve program in Congress and the various western states. One of the constant concerns was that when the reserve boundaries were drawn, they often contained small farms or land that could be farmed. Settlers and residents of the West were no longer eligible for homesteads under the earlier homestead acts from the mid-1800s or the Timber and Stone Act of 1878. Mining claims were another option, but the acreage allowed was small, usually 80 acres, and there had to be proven mineral deposits on the land. There was still pressure to open up more federal and state land to homesteading. Potential farmland inside the national forest boundaries became the target of timber companies, grazing interests, and legal homesteaders. As the intent of the reserves was to protect the timbered lands and watersheds, opponents of the Forest Service gleefully expressed their opposition to including farmlands within the reserves.

The Forest Homestead Act was passed and signed into law by President Roosevelt on June 11, 1906. For many years this law would be referred to as the "June 11th Act." The act allowed some limited homesteading within the national forests, but specified that certain conditions had to be met by the homesteaders, including that no more than 160 acres could be claimed and the land must be more valuable for agricultural use than for timber production (Woodruff 1906). C. J. Buck, who later became the North Pacific district forester, noted: "However, the natural result under loose instructions, lack of experience [on part of the Forest Service employees], and haste was the opening among the agricultural lands of tracts having considerable [timber] value." (Buck 1918: 808)

Like the older lieu-land and homestead laws, this new act would provide plenty of opportunities for fraud. However, there were no large-scale conspiracies such as had been common in earlier years. The Forest Service was very active in examining the forest homestead applications; forest examiners were very thorough relying on field notes, signed depositions, maps, and photographs of each claim. A few cases reached the courts but fewer still were allowed to go to patent. Those that were patented clearly fit all the requirements in the act.

The Inspectors Office Established in 1907

From August 1907 to November 1908, Edward Tyson (E. T.) Allen was the chief inspector for the Forest Service in the Pacific Northwest; more details of his life are given below. The Inspectors Office headquarters were the Tilford Building at the corner of S.W. 11th Avenue and Morrison in downtown Portland. Allen's Inspection District consisted of Oregon, Washington, and Alaska. He was in charge of inspectors M. L. Erickson, W. H. B. Kent, and Fred Ames, while Shirley Buck was the assistant inspector. According to Secretary of Agriculture James Wilson, these inspectors were placed in Portland, and other locations around the country, to have "local officers [of the national forests] under the eyes of thoroughly trained and competent inspectors." Probably most important was providing fiscal oversight for the forest supervisors, as well as proper hiring and firing of rangers.

A ranger meeting was held in 1907 in Roseburg to discuss common methods of managing the national forests, as well as solving and learning from the mistakes and successes of forest supervisors and rangers. "From all over the Oregon country and from northern California the trails led to Roseburg, Oregon, by horseback, railroad, and stage." Scott Leavitt, a new Forest Service guard employee, told of that first meeting:

> The National Forests were new then … and many of them were just being put under administration. At this ranger meeting we got instructions from the inspectors, discussed protective measures and swapped notes on our problems, particularly the questions of fire protection and grazing. For the first time in history, restrictions were being put on logging and grazing, and, believe me, those rangers were self-reliant and resourceful men. One of their principal jobs was to create public sentiment in favor of forest administration and conservation. They had to be able to go right in to a sheep or cow camp, sit down by the fire and preach the gospel of conservation to men whose ears were not always sympathetic, and convince them of the value to themselves and others of making the forest wealth permanent. To accomplish that a ranger naturally had to be completely sold on the idea himself. To put the idea across he almost had to be fitted for a place in the diplomatic corps (Chesnut 1942: 149, 190).

The North Pacific District Office Established in 1908

A major organizational change occurred in July 1908, when six District Offices (DOs), which have been called Regional Offices since 1931, were layered between the national headquarters (Washington Office) and the more than 130 national forests. Generally, the new District Offices followed the territory that previously had been Inspection Districts. Prior to this time, each national forest supervisor had communicated directly with the Washington Office. Gifford Pinchot decided that the Washington Office decision making should be decentralized for better administration.

As a result, in November 1908, 377 foresters, clerks, and stenographers were transferred to the new District Offices scattered throughout the West. Portland became the location of the new North Pacific District Office in early December. The North Pacific District (D-6) at that time included all national forests in Oregon, Washington, and the Territory of Alaska. From Washington DC came a transfer of a large number of the Washington Office staff to the new Regional headquarters. Will C. Barnes commemorated this transfer in verse (1929: 44):

> Oh, they're whispering in the corners and talking in the hall
> They are scheming and a-planning where to migrate in the fall,
> They are telling one another of the places they like best;
> Oh, the whole blamed outfit's "locoed" 'cause we're going out West.
>
> "Have you ever lived in Portland?" "Is it wet or is it dry?"
> "Do you think you'd like Missoula?" "If you do, please tell me why."
> "Is the living high in Denver?" "Are the ladies there well dressed?"
> Oh, these are the burning questions, 'cause we're going out West.
>
> "Now I want to go to Frisco even though the earth does quake."
> "Well, I'm wild to see a Mormon so I'd much prefer Salt Lake."
> "Do you think that I'll get homesick?" "Are the Frisco fleas a pest?"
> What a turmoil has been started 'cause we're going out West.
>
> "Oh, they say that board's expensive in the town of Albuquerque."
> "But you needn't take a streetcar for to reach your daily work."
> "Well, I've heard the living's awful (Now please don't think me silly)
> But really, do they live out there on only beans and chilli?"
> Oh, such like doubts and troubles daily agitate the breast
> Of each one in the Service 'cause we're going out West.

The first North Pacific DO was located in the Beck Building at the corner of Broadway and Oak in downtown Portland. Edward Tyson Allen was the district forester, George H. Cecil the assistant district forester. The main divisions in the DO were: Charles H. Flory was in charge of operations (which included engineering, occupancy, accounts, and maintenance sections); Fred Ames was in charge of silviculture (including tree planting and silvics sections); and Howard K. O'Brien was in charge of grazing. An office of lands was soon established, with the offices of public relations and fire control coming later. By June 30, 1909, there were 65 employees at the DO and 299 Forest Service employees on the national forests and ranger districts in Oregon, Washington, and Alaska. From that time on forest administration was decentralized, with the bulk of the forest work handled by the DO and by the various forests within the district. Only questions of large importance or on national policy were referred to the Washington Office.

Women played a professional part in Forest Service work after the establishment of DOs in 1908. During this period they were mostly hired as stenographers, file clerks, and for drafting work.

After 1913 there was occasional employment of women as lookouts and for fire patrol, especially in World War I and afterward. Women were not allowed to fight fires or do other dangerous, outside work in the national forests. It would take decades for this change to occur. On-the-ground foresters were white males, many who were forestry school graduates.

District Foresters

Edward T. Allen was the first district forester for District 6 (D-6). Allen's family had homesteaded on the upper Nisqually River near the Pacific FR, that later became Mt. Rainier National Park. Allen had no formal schooling but was tutored at home by his parents. In 1898, Pinchot met Allen, who was then working as a reporter on a Tacoma newspaper, and persuaded him to join the Division of Forestry. He moved to the Forest Service in 1906 when the forest reserves were transferred to the Department of Agriculture, coming to the Pacific Northwest as inspector and becoming district forester in 1908. He left federal employment in 1909 to work for the Western Forestry and Conservation Association. His major accomplishments during his short time in the District were to upgrade field personnel and work with lumbermen and state officials against the common enemy, fire.

Charles S. Chapman succeeded Allen in 1909 as district forester. Chapman was a member of the first (1902) class to graduate with a masters of forestry degree from Yale. He worked several years for the Bureau of Forestry, left public service to work as a private forester, returning to the Forest Service in 1907. Like Allen, he spent most of his time as district forester trying to build good relations and cooperative efforts with the lumber industry, especially over forest fires, as well as to educate the public about fire. Chapman resigned in 1911 to become the secretary and manager of the Oregon Forest Fire Association.

Creating Ranger Districts and Ranger Stations

The lowest level (sometimes referred to as the ground level) of the Forest Service is the ranger district. From the time when the forest reserves were established, roaming rangers checked on grazing, fires, and trespass cases, though the first ranger district did not come into existence around 1908, about the same time as District Offices were first established. These early ranger districts were organized around grazing units; grazing remained the main administrative challenge for decades. By the 1920s, many old grazing units had been combined into one or more ranger districts.

Many national forest sites were considered for ranger stations. Most were adjacent to water supplies, especially rivers, and were often sought by both homestead claimants and water-power companies. The DO sent crews out to find the best management sites, and to survey, map, and record each site. After the surveys were completed, the most desirable sites were formally withdrawn from homesteading, mining, and power companies, though not all were used as ranger stations. For example, the Willamette NF surveyed more than 75 possible ranger station sites, of which fewer than 20 were ever used for administrative purposes.

Constructing the Diamond Lake Ranger Station, Umpqua National Forest, Oregon, c. 1907. Author's collection.

Beginning in the 1905 to 1910 era, construction began on hundreds of ranger stations in the Northwest. They were usually located along main trails or adjacent to lakes and rivers. Early ranger stations were considered as sites for the ranger to get out of inclement weather to keep warm and dry. Generally, they were placed about a day's horseback ride apart. If rangers were fortunate, they were able to take over abandoned homestead cabins, but most early stations were nothing more than shelters without windows; some were without doors or floors. For years, through the 1910s and part of the 1920s, a battle raged with Congress about the amount of money to be spent on building ranger stations. Restrictions on expenditures for buildings kept many from being built or improved until the Civilian Conservation Corps projects of the 1930s.

In the summer of 1905, with the start-up of the Forest Service, the old GLO system of political appointment of forest rangers was dropped. An examination of candidates under Civil Service Commission rules became standard. One newspaper announcement (in part) indicated that changes were coming:

> The United States Civil Service Commission will hold an examination … to secure eligibles from which to make certification to fill vacancies as they may occur in the position of Forest Ranger. …The examination will consist in announcing practical questions in writing and in a field examination to test the ability of the applicants to actually perform the duties of a forest ranger. In the field examination applicants, among

Forest Ranger Cy Bingham, Cascade National Forest, Oregon, 1903. USFS photograph.

GLO ranger camp at Davis Lake, Cascade Range Forest Reserve, Oregon, c. 1904. Author's collection.

> other things, will be required to shoot at a target with a rifle and pistol, and should bring with them these arms for the test. The age limits will be, from 21 to 40 years on the date of the examination (Brown 1960: 52).

Probably the most difficult part of the ranger exam was properly tying a diamond hitch to hold boxes on a horse or mule. In some exams, the candidate was required to cook a meal over an open fire, then eat it. Within a few years, new categories of forest rangers were created, including assistant rangers, forest guards, forest assistants, forest examiners, forest clerks, protective assistants, junior forest rangers, and district rangers. Over the decades that followed, more categories were created as management needs changed.

Upon passing the exam, the new rangers were issued uniforms and probably most importantly a bronze badge (see Chapter 2) and the *Use Book*. The dark green, hardback *Use Book* was first issued to all rangers in 1905. It contained all the important laws and regulations that applied to the management of the national forests. It was 142 pages long, approximately 4 inches by 6 ½ inches; designed to fit in the pocket of the ranger's shirt or jacket. It was constantly referred to by rangers and the public. In addition, each field ranger had to write in a special blank diary book everything they did every day and everyone they met. This was a way to ensure that, upon inspection, the rangers were actually doing the work they were hired to do. The *Use Book* was updated yearly until 1908, then periodically after that, evolving into the *Forest Service Manual* by the 1930s. A 42-page reader-friendly version of the *Use Book* called *The Use of the National Forests* was printed for public consumption in 1907 and handed out to anyone who wanted to know about the purposes of the national forests and their management.

In the early days, the forest rangers earned somewhere between $900 and $1,500 per year. Rangers in the field, as most were then, had to provide most of their own equipment at their own expense, including horses, pack animals, saddles, shovels, axes, food, and bedding.

Experiment Stations Established

Pinchot established a section of special investigations that by 1908 had division status with experiment stations set up in the West. In Region 6, a new division of research was organized, which was part of the division of silvics, headed by Thornton T. Munger (silvics was under silviculture, headed by Fred Ames). In 1915 a separate division of research was set up in the Washington Office under Earle Clapp. Forest and range experiment stations were established that resembled agricultural experiment stations. These experiment stations, like the Pacific Northwest Forest and Range Experiment Station, were independent of the DO, reporting directly to the Washington Office. Much of their work was applied to forest and range management; there was close cooperation between research and management since the majority of research work was carried out on the national forests.

Range Management

Range management remained a major effort after the Forest Service took over forest reserves/national forests. Probably the most significant early problem was charging of grazing fees for each head of cattle, sheep, horses, or goats that grazed upon the national forests. Previously, no fees had been levied against sheep and cattle owners on the public ranges of the West. Beginning in 1906, fees were charged for each animal grazing on national forests. Although the new fees charged were nominal, grazing interests have protested each fee increase to the present day. Over time, there has been a reduction in number of grazing animals allowed on the national forests. Newspaper accounts in the early 1890s, for example, reported more than 600,000 sheep were grazing in the Cascades of Oregon, and perhaps as many as 300,000 in the Cascades of Washington.

As noted in Chapter 2, grazing studies in 1896 and 1897 found overgrazing, regulation of grazing in the national forests was settled in the Pacific Northwest earlier than in any other Region. The Pacific Northwest Forest and Range Experiment Station continued grazing studies for decades, often including photography comparing grazed areas with exclusion areas, showing that many areas were severely overgrazed. Reductions in the number of grazing animals ensued, much to the dismay of the sheep and cattle owners.

John Muir visited Portland during the summer of 1896 and again as part of the Harriman expedition to Alaska in 1899. Muir has often been portrayed as in opposing Pinchot, the Forest Service, and the national forests, but the truth is somewhat different (Miller 2000). He supported the establishment of forest reserves, including logging within their boundaries, recognizing that they were one step removed from national parks and had different purposes for management. But he was strongly opposed to sheep grazing in the forest reserves:

> John Muir, the well-known California geologist and explorer, who is a member of the Harriman expedition to Alaska, is an enthusiastic supporter of the forest reserves. He has no patience with those who want to let herds and flocks graze on those lands. "Better bring the bands of sheep right into the city of Portland and let them denude every lawn and destroy every flower garden than permit them on the forest reserves," said he, in speaking to an *Oregonian* reporter. "The damage they would do here would be infinitesimal compared with what they would do on the reserves." (*The Oregonian*, May 31, 1899)

The Pinchot policies of charging for grazing and research studies were strongly endorsed by the National Woolgrowers Association in their annual meeting held in Portland on January 11-15, 1904. Nevertheless, grazing problems arose in central and eastern Oregon and Washington, including conflicts between local stockmen and tramp sheepmen (that is, sheep owners without farms), cattle and sheep owners' conflicts, and conflicts between grazers and homesteaders. For the most part, such disagreements did not arise on westside forests as there was little grazing activity.

The Forest Service was able to settle local problems on local grounds. Sometimes the solution involved a ranger showing a pistol and threatening to use it if conflicts were not solved. Flock owners from distant ranches had little, if any, interest in local conditions. Forest Service grazing policy evolved to give preference first to stockmen residing within the boundaries of the national forests; second, to those with land holdings in the forest but who did not live there; third, to stockmen living near the forest; and fourth, to those living at some distance based on prior use of the forest range. On the Colville NF, there was a story about a non-local sheep owner: "Sheepmen tended to avoid the Colville ranges in the early years also because of the country's bad reputation for "death camas," which was, and still is, a deadly poison to sheep and man alike. One stock owner, against the forest supervisor's advice, herded his hungry flocks from the [Great Northern] train through a driveway lined with the camas and overnight lost 1,300 of his 1,960 sheep." (Holstine 1987: 50)

Timber Management

The management of the timber resource was essentially a low priority in the early decades. There were small sales for firewood and mining, but large-scale timbering would have to wait until the late 1910s and 1920s. One problem was that the national forests were seldom near population centers and railroad lines so the trees were inaccessible for easy, economical logging. The prevailing method of log extraction was rolling logs into nearby rivers where they could be floated or "splashed" to sawmills downstream. In some cases, long log flumes were used for the same purpose.

In addition, the fledgling timber industry was not especially interested in national forest timber because they were worried that, if the huge storehouse of federal trees were cut and "dumped" on the market, lumber prices would be depressed. Some smaller operators were interested, built sawmills near national forests, and acquired some limited market access for lumber. For the most part such operations, especially those relying on railroad construction to bring the big logs out of the woods, were in short supply.

In 1909, 158 timber sales were made, with 32 million board feet cut at an average value of $1.69 per thousand board feet. Probably the largest timber sale during this period was on the Crater NF in 1909 on the northwest shore of Upper Klamath Lake at Pelican Bay and Odessa. It comprised three sales amounting to about 17 million board feet and another two smaller sales totaling almost 808 thousand board feet.

Engineering

Engineering was, for the most part, related to surveying, mapping, and administrative structures. Survey work was important; nobody knew exactly where the forest boundaries were located. The edges of the forests looked fine on a map, but rangers needed to spend time establishing those boundaries. Once the boundaries were known and marked with signs, work was needed to find township and section corners. Maps, often printed in the District Office, were of growing importance, not only for employees but increasingly for the public. Other than USGS topographic maps, many out of date, the early rangers had to rely on forest atlases. These were made for most of the forests in the District beginning in 1909. Ray Grefe pointed out that in 1910 "maps [were] extremely questionable as to coverage and accuracy." (1990: 275)

One interesting question arose from early reports and maps: should the boundary of the national forests be along a ridge separating one drainage from another or should they meet on the river with one side in one forest, the other in another? Around 1909-10 the issue was settled—ridge tops would divide the national forests. Many thousands of hours of work were spent surveying forest boundaries, mapping them, and posting USFS property signs along the boundaries. On the eastside forests this was easier than on westside forests that were heavy with timber and underbrush. In some cases, the rangers could rely on old survey maps and notes, but in more remote locations the job was difficult as there were no adjacent land surveys, much less township, range, and section lines and marked corners.

Fire Management

Fires have plagued the Pacific Northwest for thousands of years. American Indian tribes and bands commonly set fires, and settlers did so either purposely or accidently. Railroad engines spewing fire brands from their smokestacks caused many fires. Lightning often struck the high Cascades in hot summer months, but there are only a few recorded instances of fires started by lightning until the early 1900s. The Yacolt Fire of 1902 on the Columbia NF burned 238,920 acres, killed 38 people and destroyed more than 20 small settlements. It was a wakeup call to the Forest Service.

Mineral Management

Mining even to this day derives from the General Mining Law of 1872. This important act was passed to affirm the principle that gold, silver, and other minerals belong to the person who finds them on public land, including national forests. At this time, a miner would stake a placer claim usually along a creek or river. Such a claim was generally 1,600 feet by 500 feet, or about 20 acres and required a minimum of paper work. Lode or hard-rock claims that needed tunnels to access the mineral ores could measure 1,500 feet in length, with 300 feet on either side of the middle vein, all in parallel lines. These claims were generally about 160 acres. After meeting minimal requirements under the act, then paying $2.50 per acre for a placer claim, or $5 per acre for a lode claim, the claimant could become the legal owner of the minerals both on and under the surface, as well as the land itself, with no obligation to pay royalties or fees on production or value of the minerals found. In many cases, the prospector did not need to hurry to mine or seek patent, as the

Blue River Mining District, Lucky Boy Mine, Cascade National Forest, Oregon, c. 1910. USFS photograph.

claim could be held indefinitely by performing annual assessment work that in many cases involved only a minimal clearing of land, digging underground or in pits (adits), or improving their living quarters—and this is true even today. Mines on the national forests required cooperation between the General Land Office, which gave title to the mineral land, and the Forest Service. Generally, the GLO allowed patents on every claim unless the Forest Service disputed the claim. Every national forest had at least some mining operations. Probably the most mines or claims in the Northwest were on the Mt. Baker, Snoqualmie, Willamette, Rogue, Siskiyou, Wallowa, and Whitman NFs, but that varied by year and the price of minerals, especially gold.

Wildlife and Fish Management

Wildlife and fish in national forests represented one of the most widely utilized resources. Hunting for subsistence was—and still is—a tradition in many rural communities. Nationwide, rural dwellers, farmers, loggers, and miners geared up at the beginning of hunting season to get their deer or elk. Trapping fur-bearing animals for pelts and predators for bounties was also a way of life in remote settlements and small communities. State and county bounties in this period were often around $2 for bobcats; coyotes $2.50; bears $10; cougars $30; and wolves $50. Each year an increasing number of anglers entered national forests, especially with the use of automobiles. Unfortunately, illegal hunting and fishing (poaching) also increased. Penalties for overfishing and hunting out of season were prescribed by the two states in the early 1900s.

Wildlife management involved a complex interaction between federal agencies, the state, and individuals. The Forest Service managed the land within the national forests. The USDA Biological Survey, predecessor to the U.S. Fish & Wildlife Service, carried on wildlife research and some predator control. The state administered and enforced game laws. The Forest Service worked with the state in having its employees, usually the district rangers, act as deputy game wardens. Individuals varied in their cooperation with the government agencies. Local trappers and predator hunters often resented the government encroaching on their livelihood. In the Crater NF, for example, a USDA Biological Survey hunter named Hammersley suffered some indignities. As George Cecil (as quoted in Rakestraw and Rakestraw 1991: 37) wrote, "This office is rather reluctant about sending a hunter there again because it appears that some one who has not much use for the Forest Service killed Hammersley's horse and robbed his camp of all his supplies, etc."

Recreation Management

Recreation—especially mountain climbing and hiking—in what became the Northwest's national forests started in the late 1800s. The first western mountaineering club, the Oregon Alpine Club, organized in 1887 with Will Steel as president. He became concerned that membership rules were not tight enough, since many hikers and non-climbers had become members, and the club's leadership was not active on the political front. Steel organized a new group, the Mazamas, emphasizing the club's alpine character and membership and its aim to be politically active in forest reserve and national park issues (Williams and Mark 1995). Later, the Mazamas would break into two groups, the Mazamas serving Oregon and the Mountaineers serving Washington. Both clubs have been active in national forest matters in the region for over 100 years.

Hot springs and mineral springs were among the earliest places in the Pacific Northwest where summer resorts were established. In general, they had a common history: establishment of the resort, an initial period of prosperity and growth, and a period of decline. Most hot springs resorts were established between 1870 and 1900, along main roads and trails. Many have been in private ownership for years and most predate the establishment of the national forests (e.g., Belknap Hot Springs, Carson Hot Springs, Sol Duc Hot Springs). Yet many hot springs have never passed into private hands and are still part of the management of the national forests, like Terwilliger Hot Springs on the Willamette NF and the Bagby Hot Springs on the Mt. Hood NF.

Foley Springs Resort near McKenzie Bridge, Willamette National Forest, Oregon, c. 1935. Author's collection.

The years from 1900 to 1940 were the heyday of summer resort hotels, many associated with various hot springs scattered throughout the Cascades and Olympics. It was an era when the pace of recreation was more leisured than today. Visitors would arrive at a summer hotel with their trunks, fishing gear, and hiking boots to stay for a week or a month at a time. The hotels, usually large frame structures with wide verandas, were noted for good cooking and pleasant living. Visitors hiked, picked berries, and fished. Many resorts had croquet lawns or tennis courts for social games. The automobile made the resorts accessible, but in the end ruined them. People become interested in covering miles rather than in staying in one place and motels sprang up to cater to temporary auto visitors.

Research Work on the National Forests

James T. Jardine carried out groundbreaking grazing studies on the Billy Meadows experimental grazing plot in the Wallowa NF where he enclosed 2,500 acres for the work. The results from these studies helped develop techniques for range reconnaissance as well as standards for deferred grazing and rotation. Jardine was placed in charge of grazing studies in the Washington Office in 1910.

National Monuments and National Parks

The first national park in the United States was established in 1872: Yellowstone National Park. The park remains the premier national park in the system today. The first public land withdrawal for a possible national park in the Pacific Northwest was around Crater Lake in the Oregon Cascades; however, it was not officially designated as a national park until 1902 (see Chapter 2). Mount Rainier National Park was carved from the old Rainier FR in 1899.

Both these parks had widespread support from people in the two states, as well as their respective congressional delegations.

One particularly contentious effort to establish a national park involved the Olympic NF. As early as 1904, Francis W. Cushman, U.S. representative from Tacoma, introduced a bill in Congress to establish an Elk National Park on the Olympic Peninsula within the Olympic FR. Cushman and others were concerned about the slaughter of the Olympic or Roosevelt elk, which were being killed by poachers in great numbers for their teeth, which were used on the Fraternal Order of the Elks watchfob ornaments. His bill did not pass Congress. In the summer of 1907, Mountaineers hiked into the wild and remote Olympics. After spending two weeks in the wilderness, they returned to Seattle with several ideas on how to preserve and protect the wild area. By early 1909, the group sought help from the Washington delegation to Congress to consider national park status for the area. The USDA Biological Survey, later a major part of the USDI Fish &Wildlife Service, became involved in the matter as it:

> recommended a national monument withdrawal to protect the breeding grounds of the Olympic elk from roads and from mining and logging; this recommendation was opposed by the Forest Service chief law officer. On March 2, 1909, Representative W. E. Humphrey of Tacoma drafted a proclamation for Theodore Roosevelt's signature which would withdraw the Mount Olympus National Monument. The presidential proclamation was issued the next day, and soon thereafter settlers and miners began to object to the 600,000-acre national monument. Forest Service [Assistant] District Forester George Cecil ... requested that the monument be abolished. However, Acting Chief Forester C.S. Chapman rebuffed him in deference to the Biological Survey, which "would very decidedly disapprove." The district forester pressed the issue again in the following year, hoping that the Biological Survey would accept establishment of a state game refuge in lieu of the monument (Twight 1983: 34).

The Antiquities Act of 1906 established penalties for destroying, injuring, or excavating any historic or prehistoric ruin or object of antiquity located on federal lands and authorized the president to set aside national monuments by proclamation. Over the years this has been broadly interpreted to include elk protection in the Olympic Mountains of Washington, geological features found near Yellowstone National Park, and glaciers in Alaska. (Rothman 1994, Williams 2000d)

But national monuments are something of a rarity in the Pacific Northwest (see Chapter 6). There is only one early national monument, which dates from 1909: the 480-acre Oregon Caves, located on the Siskiyou NF in southwest Oregon. Its establishment as a national monument was recounted by ranger M. M. Lewis: "During the year of 1907, I learned that parties living in the Illinois valley were inquiring as to the possibilities of securing a title to the Oregon Caves by filing a mining claim on them. I reported the matter to M. J. Anderson who was Supervisor at the time. He initiated the action which led up to the withdrawal of the Oregon Caves as a National Monument." (*Six Twenty-Six* August 1935: 13)

An Era Ends—Gifford Pinchot Fired in 1910

President Roosevelt left office in early 1909, replaced by his hand-chosen successor William Howard Taft. Pinchot remained as chief forester. Pinchot was an inspiring leader of the Forest Service. But he soon came under fire for challenging Secretary of the Interior Richard Ballinger over leasing coal lands in Alaska to national syndicates. After only a few months, President Taft could no longer take the backbiting that was going on between a bureau chief and the secretary of the Interior. Pinchot was fired for insubordination on January 7, 1910, along with Associate Forester Overton Price and Assistant

Gifford Pinchot at his desk in Washington, DC, c. 1905. USFS photograph.

Law Officer Alexander Shaw. The letter from Taft (in the Pinchot collection at the Library of Congress), in part, ends with the following statement:

> By your own conduct you have destroyed your usefulness as a helpful subordinate of the Government, and it therefore now becomes my duty to direct the Secretary of Agriculture to remove you from your office as The Forester.
>
> Very sincerely yours,
> WM. H. TAFT

Morale in the agency plummeted as rangers who had grown up (though only for five years) with Pinchot no longer had this charging stallion to lead the way, fight the battles with special interest groups, or lead Congress to write the best conservation legislation (Miller 2001). Pinchot was replaced by his old friend Harry Graves, then head of the Yale School of Forestry. Graves had to move the agency away from its almost cult-like fascination with Pinchot and in a new direction. It should be noted that many in the Forest Service over 100 years later still venerate Pinchot as the protector of the U.S. Forest Service and the national forests.

Chapter 4

Building the Management System 1910–1929

Management of the North Pacific District underwent a number of major changes during this period. The Forest Service, steeped in Pinchot's principles of scientific management and conservative use of the national forests, found itself in the perfect position to translate these principles into on-the-ground management. The agency grew from providing simple fire protection with some recreation and limited timber production to a fully functional organization well integrated into forest management and protection. Control of forest fires became the primary concern in this era. World War I and its aftermath proved challenging. In another development, Congress in 1920 imposed a limit of $800 for the building of ranger stations, which effectively halted construction of many remote stations, but a little later in the decade construction began on improving the ranger stations. Roads began to replace trails. Forest Service rangers were entering the "white hat" era of forest protection and development. The Stock Market Crash in late 1929 challenged the nation and the Forest Service to respond to changing societal conditions and needs.

District Foresters

District Forester Charles Chapman was succeeded by George H. Cecil. As a young man, Cecil had read an article on forestry by Gifford Pinchot in the *Youth's Companion* and became thrilled at the conservation message. In 1902, Cecil enrolled in the Biltmore Forest School. He graduated a year later and joined the Bureau of Forestry. In 1906, he went to work in the Yellowstone Forest Reserve. Then he spent a year in District 1 (Missoula) before transferring to Portland to work under E. T. Allen as associate district forester, a job he held until April 1911, when he took over as district forester. A warm individual with a capacity for making friendships, he did much to keep up the morale of the Forest Service in the District. Cecil, like Allen and Chapman before him, worked with the timber industry and did much to create an efficient fire control administration on both public and private lands. Cecil left Portland on December 31, 1924, after Forest Service Chief William B. Greeley personally chose him to head the Angeles NF in southern California, which had major fire problems, with the expanding urban population. Cecil's service with the Angeles NF was so successful that in 1930 he was offered and accepted the position of manager of the Conservation Department of the Los Angeles Chamber of Commerce and executive secretary of the Conservation Association of Southern California.

He was succeeded as district forester in early 1925 by Christopher M. Granger. Granger was a 1907 graduate of Michigan State College. He spent most of his career in the Rocky Mountain District (Region) on the Medicine Bow and Pike NFs, as well as in the District Office there and in the Washington Office. While in the District Office in Portland, he was instrumental in the development of social amenities for the Forest Service, such as minimizing timber cutting along roads, developing campgrounds and summer home construction, and increasing wildlife management. He served until 1930, when he was transferred to the Washington Office to develop the national timber survey (Forest Resource Survey) called for by the McSweeney-McNary Act of 1928. A few years later, he became active with the Forest Service's part in the Civilian Conservation Corps program.

National Changes in the Forest Service

Gifford Pinchot had personally established a new federal agency; the transition from his agency to one that would run smoothly fell to Henry Graves and his successors. During Graves' tenure the major distraction from national forest management was World War I. Graves was in charge of the 10th and 20th Engineer (Forestry) Regiments, which were assigned to France, at the rank of colonel, a title that he retained for years afterward. Graves was replaced as chief forester by William B. Greeley, who served until 1928. Greeley had also served in the 20th Engineers at the same rank as Graves, and he preferred to be called Colonel Greeley. Robert Y. Stuart served as the fourth chief forester, from 1928 until 1933. He had served under Commissioner of Forestry Gifford Pinchot in Pennsylvania from 1920 to 1927. His time as chief forester was marked by the onset of the Great Depression in the fall of 1929. He died after a tragic fall from the seventh floor of the Washington Office, suspected as suicide, which many attributed to overwork.

During this period, the dress uniforms worn by forest officers changed. The bronze badge was decreased in size by roughly half, but it was still prominent on the dress uniforms and field shirts. By 1922, an official badge half again as large was made for the dress uniform (it was later referred to, incorrectly, as the women's badge).

Developments in the District

Staffing increased in the North Pacific District and the District Office moved from the Beck Building to the new Post Office Building, at Broadway and Glisan, on January 12, 1919. The Alaska forests were designated as District 8 on January 1, 1921 (the name would be changed to Region 10 in 1934). District 6 sent a number of DO employees to staff the new District Office in Juneau.

In June of 1924, Congress passed a bill that authorized the establishment of the Pacific Northwest Forest Experiment Station for Oregon, Washington, and Alaska and Thornton T. Munger was appointed the first director. By 1926, the North Pacific District D-6 office had grown to 84 employees, while the field forces in Oregon and Washington consisted of 300 employees.

Beginning in 1910, the DO offered a special course for rangers at the University of Washington and other forestry schools in the West. This ranger course was an attempt to train non-university rangers in the needs of management. The district forester's report for 1911 recounted the first year of operation:

> During the past winter 25 men, including rangers and guards, attended the ranger school at the University of Washington. Lectures were given on the various lines of National Forest work in this school by members of the District office in addition to the regular teaching force [from the university]. The lectures given by District officers were mimeographed and sent to every ranger in the District, thereby presenting to every Forest officer a more complete compilation of instructions and a larger discussion of National Forest work than has been presented heretofore.

Yet a problem quickly arose when the fiscal people in the DO ruled that men attending ranger school could not be paid their government salary, and thus their travel expenses could not be reimbursed. Within a couple of years, the ranger course was no more. It took many years for the rules governing these types of expenditures to change.

Ranger and guard stations, as well as lookout houses and towers, were built and staffed in hundreds of locations throughout the Region in this era. The remoteness of these stations didn't seem to bother the rangers or their wives, though conditions were, in most cases, primitive at best. Wives were not employed by the Forest Service at that time, but they had many responsibilities at the stations.

Black Rock lookout house with Mrs. Curtis, Umpqua National Forest, Oregon, 1923. USFS photograph.

The publication *Six Twenty-Six* (indicating District 6 and 26 national forests) began in 1916 as a monthly newsletter for employees. Over the years the name changed to the *R-6 Administrative Digest*, then to *Greensheet*, and it now is sent electronically. Another change, little noticed by many, occurred in 1920 when, at the suggestion of District 6, the Forest Service began printing the USFS shield on all official stationery. In a somewhat optimistic view of what this would achieve, the District Office stated that "in the advertising sense we are 'selling' the National Forests. If we can get this one picture of 'the green tree' over to the majority of the American public, we shall have stabilized forestry and forest protection in the United States." (*Six Twenty-Six*, July & August 1920: 48-49)

The District Office reported on a district-wide supervisors' meeting in Portland in early 1917. A story in the *Six Twenty-Six* of April 1, 1917, related that "the features of the meeting were Mr. [Chief Henry] Graves' visit and his inspiring talks to the men. The meeting was a decided success and the men returned to their stations with a zeal for public service renewed."

The District gained "ranger" boats for administrative purposes in several lakes and estuaries. In 1917, there was an announcement that two Columbia River fishing boats destined for the Siuslaw NF had been completed by the Wilson Brothers from Astoria, Oregon. One of the boats "will be used on Umpqua Bay and tributaries, the other on Siuslaw Bay and tributaries. An additional ranger boat for Alaska is nearing completion in Seattle. These boats will be used for carrying men and equipment, and in other administrative business." (*Six Twenty-Six* August 1917: 9) In 1923, a new ranger boat was launched on Lake Chelan and was named the *Forester* by boat captain Southwick.

The District Office was reorganized in early 1920 to meet new forest inspection standards. Five forest supervisors were reassigned from their forest responsibilities into a new section of timber working plans in the District Office. "Forest Supervisors Sherrard, Weigle, Brundage, Fromme and Evans have been appointed as District Forest Inspectors effective January 16, as a result of a re-organization plan worked out by the Secretary [of Agriculture] and Colonel [Chief Henry] Graves. This change is to meet the growing needs of the service and partly offset by more frequent inspections the loss in efficiency caused by the taking on so many new men to fill vacancies caused by resignations." (*Six Twenty-Six*, February 5, 1920: 19) Each of the five men was assigned specific duties in the new District Office section. Thomas H. Sherrard and Fred H. Brundage were assigned to review fire protection plans; W. G. Weigle examined data on the pulp resources in the Puget Sound area as well as making a working plan for the area; Robie M. Evans assisted eastside forests with working plans for ponderosa pine management; and Rudo L. Fromme assisted with policies regarding grazing and recreation, as well as fish and game propagation and protection.

DIGNITARIES VISIT THE DISTRICT

In 1915, a three-person committee of Shirley Buck, T. T. Munger, and Albert Cousins was appointed by Regional Forester Cecil to respond to Chief Graves' concerns as to why the costs of administering the national forests were increasing. The committee in their report found that employees and forest supervisors spent "an unwarranted amount of their time … accompanying inspecting officers; much of this must be considered unprofitably spent. As nearly as can be remembered, during the past six months 22 different men from the Washington office visited Portland on official visits." (District 6 Committee 1915: 6)

Congressional, dignitary, national headquarters, and chief visits have been an ongoing part of management from the beginning of the Forest Service. During the summer of 1920, Chief Greeley and Representative Gilbert Haugen visited the Northwest on a two-week trip through Districts 6, 4, and 2. They visited briefly the District Office, then accompanied by District Forester Cecil and T. T. Munger drove up the Columbia River Highway to Eagle Creek. They crossed the Columbia River at Cascade Locks to visit a private logging Camp 8 in the upper Wind River Valley. They visited the Wind River Experiment Station and nursery on the Columbia NF. The following day, after Cecil and Munger returned to Portland, the party took a train to Baker City. They toured part of the Whitman NF, viewing a railroad logging operation, and then rode on horseback to the summit of Mt. Ireland. At dinner, they discussed the difficulties of grazing management. Later that night, Greeley and Haugen left on the train for Ogden, Utah. The District Office summarized benefits of the visit:

> We feel that this visit has been of the greatest possible benefit to the Forest Service because of the opportunity afforded Congressman Haugen to see the Forest Service actually at work on its many activities, to learn first hand of the varied and numerous difficulties that must be over-come, to obtain an appreciation of the trials and tribulations of the western stockmen and lastly, to find an agreeable and amicable under-standing existing between the representatives of the Government and the users of the Forest (*Six Twenty-Six*, August & Sept 1920: n.p.).

Two years later, Dol Smith, head of Washington Office public relations, visited the District. He spent two days on the Siskiyou NF visiting the Oregon Caves National Monument and spoke before the Grants Pass Chamber of Commerce. He visited the Wind River Experiment Station and Eagle Creek Forest Camp. He spent a day in Tacoma and another in Seattle before returning to the East.

Chief Greeley again visited the District, arriving on May 21, 1923. He was shown sites on the Fremont, Crater, Umpqua, Cascade, Wallowa, and Whitman NFs, as well as the District Office. In early July, Greeley returned with President and Mrs. Harding along with Secretary of Agriculture Henry Wallace. The party arrived in Spokane, and visited briefly the Meacham area and the nearby Umatilla NF. In Meacham, President Harding talked with Charles Becker, an old Pony Express rider, who had carried the news of the election of Abraham Lincoln in 1860. The party went on to Portland where Secretary Wallace spoke to department employees at the Public Library. The secretary was taken after dinner for a ride along part of the Columbia River Highway. The presidential party left that same night for Tacoma and Alaska.

In the 1920s, the District Office became enamored with presenting the Forest Service to the public through an unconventional method—entering floats in parades, including the Portland Rose Parade, and encouraging individual forests to enter floats in local parades. Many of these floats focused on fire prevention. The District Office also took the lead in working with motion picture companies to shoot film on various national forests, produce magic lantern slides for public speeches and informational talks, and produce forest maps for free public distribution.

Talks by Forest Service employees, illustrated by magic lantern slide shows and movies, were very popular in remote locations where there was little interaction with "big city folks." Schools in all parts of the two states participated in the fire prevention effort. Mt. Hood NF ranger Albert Wiesendanger was often sent to give illustrated talks to schools during the winter months. Even after his retirement in the 1948, the District Office gave him a set of lantern slides to continue his good work through the Keep Oregon Green Association, where he worked as executive secretary for the next 32 years!

World War I

The United States entered World War I in the spring of 1917. While the war effort had little effect on the national forests, almost immediately after war was declared against Germany and her allies, there was a recruiting campaign for experienced foresters and woods workers to join the war effort through the 10th Engineers (Forestry) Regiment. Another forestry regiment was formed in 1918—the 20th. By the end of the war, both regiments had been combined into the 20th. The major functions of the two regiments were to find suitable timber, cut trees, and build sawmills to provide wood for the railroad ties and for the trenches in France. In addition, specialized wood was cut for airplanes and other necessary war products. Many Forest Service men, including, as we have seen, Graves and Greeley, joined the forestry regiments, including a number from the Pacific Northwest.

While many Forest Service men were off fighting the war, there was a lack of personnel to protect the forests from fires. As a result, in a number of cases, women were hired to serve as fire lookouts in the national forests. A few, like Hallie Daggett on the Klamath NF, had actually been hired prior to the war, but most women were employed during the fire seasons from 1917 to 1920. In 1918, women lookouts were stationed at Cleveland Mountain on the Snoqualmie NF (Mabel McBain, a member of the Mountaineers); Columbia Mountain on the Colville NF (Gladys Murray); on the Cascade NF for the upper McKenzie River country (Helen McCormick); and Sanger Peak on the Siskiyou NF (E. Cornell) to name a few. Several women continued to work as lookouts well into the 1930s.

Beginning in 1917, there was a strike of woods and mill workers in the Pacific Northwest, led by the American Federation of Labor (AFL) and the Industrial Workers of the World (IWW or Wobblies). Wood production from Northwest forests basically halted during the summer and fall of that year. The Army was concerned enough that Colonel Brice P. Disque was brought in to study the situation. He recommended that the Army form a

Breaking a log jam in Trout Creek, Columbia National Forest, Washington, c. 1905. USFS photograph.

new, patriotic labor-management union named the Loyal Legion of Loggers and Lumbermen (LLLL or 4L) to counteract the AFL and IWW unions. He also recommended a new Army unit be established—the Spruce Production Division (SPD)—to cut Sitka spruce for airplanes and Douglas-fir for ships. Both these recommendations were implemented in early 1918. Almost all the LLLL members came from woods and mill workers, as well as company managers. The LLLL essentially stopped the AFL and IWW strike and lumber production resumed (Hyman 1963). The Forest Service District Office helped the SPD find large stands of spruce, but only a few were found to be accessible by railroads. The entire District Office timber survey staff, along with 11 temporary assistants, worked from May 8, 1918, until October to map the west side of the Olympics for Sitka spruce and other evergreen stands. Other studies of spruce began in the summer of 1919 in Clatsop County, Oregon, and in Clallam County, Washington. SPD operations extended into the Olympic NF (and Mt. Olympus National Monument) and the Siuslaw NF.

The SPD started operations in early 1918, establishing 13 railroads and numerous sawmills along the coasts of Washington and Oregon. A few Forest Service men joined the SPD, but most were already involved with the 10th forestry regiment. All told, over 30,000 SPD men were assigned to the Pacific Northwest for the duration of the war and into early 1919. In July of 1918, some 150 SPD men were sent from Vancouver Barracks to fight the Cispus fire on the Rainier NF before it reached the Columbia NF. Other SPD soldiers were sent to private sawmills to stop any sabotage of machinery by the unions to slow lumber production.

Spruce Production Division truck on plank road, Washington, 1918. Author's collection.

After the war, in 1919, huge amounts of railroad, sawmill, and logging equipment, as well as a very large truck fleet, were sold to private lumber interests for pennies on the dollar. The equipment helped to jump start several private logging operations along the Pacific coast and Puget Sound (Evans and Williams 1985).

Homesteaders and Trespass

During the 1920s there were a number of attempts to illegally homestead on the national forests in the Region. Pressure was being applied by hundreds of would-be settlers along the edges of the forests and sometimes inside them. The would-be homesteaders pressed on regardless of whether or not there were laws that allowed homesteading on the national forests, but the harsh winters often drove them away after a couple of years. They found it impossible to grow crops and vegetables at the higher elevations where frost was common every month of the year. Some, however, continued to believe that they were entitled to homestead on any federal land, including national forests. One such case involved the Fish Lake Burn area on the Umpqua NF where a half-dozen well-intentioned homesteaders were squatting on national forest land. They were removed at gunpoint by forest rangers, and tried in federal court. Another instance was recorded on the Deschutes NF in the 1920s:

> One summer during the reign of Supervisor Hastings, the Crane Prairie ranger was amazed to find a small traveling caravan of several families was settling at Crane Prairie. They had traveled from some far place with horses, wagons, household equipment, dogs, chickens and a cow or two … [After investigating,] [t]he Supervisor hurried to a phone and called the District Forester Geo. Cecil at Portland, who referred

> the matter to C. J. Buck in charges [sic] of Lands; while he, meanwhile, held the phone … Mr. Cecil was engaged in a conference at the time and merely shouted into the phone: "Arrest them right now for cutting trees and take them to the United States Commissioner" and hung up (Baker 1950: 16-17).

It turned out that the commissioner himself had advised the trespassers to settle there. After talking to a federal law clerk, the commissioner advised the families that they should go back to Crane Prairie, remove brush and trees that had been illegally cut, then leave the area immediately. The defendants said they had no transportation back to Crane Prairie from Bend, but the commissioner "[s]ternly bade them to walk and not to dally along the road, to get the work done and to leave the country. All of which they did, and no settlers ever thereafter bothered the Forest Service at Crane Prairie." (Baker 1950: 17)

An interesting series of related events in 1923 occurred on private property at Spirit Lake on the Columbia NF. The site was the Land cabin that was probably on land purchased from the old Northern Pacific Railroad land grant. It seems that 40 boxes of dynamite had been stored in the old cabin since before World War I, but since it was on private property, there was little that the Columbia NF could do about the hazard until a deal had been worked out with the county concerning liability. The story goes that Forest Supervisor A. H. Wright

> had the powder removed and burned last winter, while it was frozen, after having obtained from the Skamania County officials assurance that the County would assume responsibility in the event of a claim for damage by the absent owners [but the cabin remained] … In October, two men slept in the cabin over night—in the morning one lighted a cigarette, and carelessly tossed the match on the floor. The glycerine-soaked floor caught fire. The men grabbed their belongings and ran; just as they reached the door, the glycerine … exploded, knocking one of the men down, and throwing the other several feet out doors. A perfectly good mountain cabin is now minus a roof and a floor (*Six Twenty-Six*, December 1923: 14).

Jurisdiction of the O&C Lands

Around 1900, the Southern Pacific Railroad stopped selling the lands that had been granted to the company by Congress. The sale of the land to homesteaders at $2.50 per acre was one of the conditions of the land grant. When the land sales stopped, the company was technically in violation of the grant. The railroad also illegally sold hundreds of thousands of acres of timber land in the foothills of the Coast Range to the Coos Bay Lumber Company and more timber land in the Mohawk Valley northeast of Eugene to Booth-Kelly Lumber Company. The state and federal government brought suit against the railroad on those grounds.

After a 1915 Supreme Court decision against the railroad, the Departments of Agriculture and Interior saw that the O&C lands in western Oregon would be returned to the government and that a decision would need to be made as to which was the most appropriate agency to manage them. A quarrel developed between the two departments, as both wanted to claim the increasingly valuable timber land that was located along the easily accessible valley floors from Salem south to the California border. Both agencies had logical arguments and strong support for their respective positions. The Department of the Interior argued that the O&C lands should be managed as one block of land by the GLO in order to simplify management, especially concerning the sale of the land to homesteaders. The Forest Service argued that thousands of acres of O&C land were adjacent to and intermingled with national forest land and considered the O&C lands within the national forest boundaries part of those national forests. In addition, the Forest Service believed that management by the GLO would result in serious land management problems.

A state conference regarding the disposition of the O&C lands was held in Salem in September 1915. There were many suggestions, including one proposal by Governor James Withycombe to transfer the unsold O&C land to the State of Oregon. Testimony from well-informed and well-intentioned citizens, including ex-Forester Gifford Pinchot, was received to incorporate the O&C lands into the Forest Service. Local officials and timber interests backed the Interior Department, as they felt it to be more willing to open the timbered lands to homesteading or timber production, either of which would increase revenue to the state or county, as well as opening large tracts to logging. The counties were fearful of losing more timber land to the Forest Service. The problem for the counties was that the national forests, at that time, were not producing revenues in large quantity, and thus were seen as huge burdens to state and county treasuries (Richardson 1980). In August 1915, *The Tax Liberator*, a Roseburg newsletter, expressed a minority of Oregon sentiment that was adamantly opposed to the Forest Service gaining control over the O&C lands:

> The United States Forest Service is making a determined effort to secure control of the Oregon and California Land Grant. Success in this effort would be a positive calamity to the State of Oregon. Bureaucratic despotism administered by impractical theorists has been an insurmountable bar to the development of our state. ... THE FEDERAL FOREST SERVICE HAS NO MORE RIGHT TO ATTEMPT TO GAIN POSSESSION OF THE OREGON AND CALIFORNIA RAILROAD LAND GRANT THAN IT HAS TO ATTEMPT TO GAIN POSSESSION OF PRIVATELY OWNED LAND IN THE HEART OF THE CITY OF PORTLAND. [Emphasis in the original.]

The 1915 conference reached no clear decision, but one of the three conference resolutions to Congress was "that we are unalterably opposed to any further increase of forest reserves in the state of Oregon." (*Daily Journal*, September 18, 1915) The situation remained unresolved until Congress settled the question in the following year when the Chamberlain-Ferris Act of June 9, 1916, was passed. This act determined that the railroad should forfeit the land grant and that the unsold land should revest (return) to the government. The government therefore received about 2,900,000 acres of land that were put under the jurisdiction of the Department of the Interior, General Land Office (Richardson 1980). However, O&C lands within the national forest boundaries would continue to be managed by the Forest Service, as they are still today. (Similarly, three years later, the Roseburg-Coos Bay Wagon Road Grant was also revested.) The 1916 revestment law was felt to be disastrous to the counties in western Oregon, in spite of the fact that 50 percent of the revenues were to be divided between the state school fund and the counties in lieu of taxes. (In California, the Southern Pacific Railroad also held millions of acres of O&C lands. This land was not part of the law suit or the revestment act.)

Northern Pacific Railroad

While Washington State did not have the same problems with the railroad grant lands as Oregon did, there was at least one lawsuit against the Northern Pacific Railroad. In 1921, the Supreme Court ruled that railroad grant land within national forest boundaries in Washington was still the property of the Northern Pacific Railroad. In March 1924, as reported in a Forest Service newsletter, there was an effort to investigate the same railroad grant. Agriculture and Interior Secretaries Henry Wallace and Hubert Work asked a committee of Congress to make "an investigation ... of some 3,000,000 acres of N.P. [Northern Pacific] land in Washington, Idaho, and Montana." (*Six Twenty-Six*, March 12, 1924: 1) Nothing came of the investigation and the land remained with the railroad. Interestingly, when Mount St. Helens exploded in 1980, the land on much of the mountain was still in private ownership, a reminder of the railroad legacy that helped form the Pacific Northwest.

National Forest Changes

Representative William Humphrey on July 15, 1911, introduced a bill in Congress to establish a national park in the Olympics to take the place of the Mt. Olympus National Monument in the interior of the Olympic NF. District Forester George Cecil and the new Olympic NF Supervisor P. S. Lovejoy were both adamantly opposed to the bill. The Mountaineers, however, pressed on with their campaign for national park legislation but without success. In the summer of 1915, Chief Henry Graves visited the monument, recommending that the Forest Service could and would adequately protect the elk and the monument. The monument was reduced by one-half, in part based on Forest Service recommendations, on May 11, 1915, by presidential proclamation. A new park bill was introduced in 1916, at the same time Congress was discussing the question of establishing a national park service. The bill died in committee. By early 1917, there was growing concern about the availability of Sitka spruce timber within the Olympic NF, which was in great demand for the war production effort. Sitka spruce, western red cedar, Douglas-fir, and western hemlock grow to enormous size within the boundaries of the forest and monument. As a result, some of the forest area was eliminated from protection to increase the availability of this much-needed war supply. After the war ended, efforts continued for the next two decades to establish an Olympic National Park in the heart of the Olympic Peninsula (see Chapter 5).

The Wenaha NF was combined in 1921 with the Umatilla NF because of the potential for cost savings in administering the one unified forest.

Timber Management

Timber management during the era was marked by a few large sales, mostly long-term railroad logging operations. The notion was that if the timber on an entire watershed was sold for development, the timber companies would begin building railroads to access the lower-elevation timber, then take decades to reach the top of the watershed. By that time, it was hoped, the lower elevations would once again be ready for harvesting, producing a sustainable operation over many years. Several of these large sales were made in the Region, including a large Bear Valley timber sale in 1922 on the Malheur NF that was enormous—550,000 acres. The initial timber offering was for 890 million board feet of mixed ponderosa, Douglas-fir, western larch, white fir, and lodgepole pine. It was at the time the largest body of timber offered for sale in the Pacific Northwest. The operation required 80 miles of railroad lines, along with several five- to 10-mile branch lines to access the timber. The timber sale was made to Fred Herrick, but the company defaulted and the sale was readvertised in 1928. The Edward Hines Lumber Company bought the timber. On January 22, 1930, USDA Secretary of Agriculture Arthur Hyde touched a button in Washington, D.C., that started the Hines, Oregon, mill in operation.

Another large timber sale was in 1923 on the North Fork of the Willamette River (Willamette NF); it was the largest sale to date in the westside Douglas-fir region. The area was benchland accessible by a spur line from the main Southern Pacific line. The sale, made to the Westfir Lumber Company, consisted of 685 million board feet of timber, 90 percent of which was mostly old-growth Douglas-fir. A company mill town, Westfir, was established at the base of the project area. The timber was to be cut at the rate of 50 million board feet per year. About a year later another large sale was made to Booth-Kelly interests on Salmon Creek just out of Oakridge. The Southern Pacific Railroad brought the Oakridge lumber industry to life. The national lumber market peaked in 1926; two years later it was in serious trouble.

In 1918 and 1919, an unusual timber sale was made on the Siuslaw NF, consisting of cascara or chittim bark stumpage—the bark was used as a natural laxative. During 1918, around 150 pounds of the bark were sold from the Siuslaw NF. The sale in 1919, awarded to A.C. Lake of Noti, Oregon, was expected to total over 10 tons of cascara.

Timber harvest levels in the District increased during this period (see Appendix E). For example in 1910 there were 238 timber sales, with the cut volume at 49 million board feet at a value of $1.95 per thousand board feet. During the following year there were 315 timber sales that resulted in 48 million board feet being harvested. By 1915 the sales more than doubled to 866 and harvest increased to 134 million board feet. Five year later the sales almost doubled again to 1,339 and harvesting was at 200 million board feet. By the end of the period in 1929, just at the start of the Great Depression, there were 1,344 sales with 514 million board feet of timber harvested.

An infestation of pine beetle killed millions of acres of ponderosa and lodgepole pine in southern Oregon and northern California in this period. In an attempt to control the infestation, pine beetle crews felled dead or dying trees, peeled the bark from the infested trees, and burned the bark and felled trees. The beetle control project covered 1.3 million acres on the Crater, Fremont, and Modoc NFs; Klamath Indian Reservation; Crater Lake National Park; and other public and private lands.

Standing dead trees, killed by fire, insects, or disease, are referred to as snags. They were considered to be deadly to forest workers if the old tree trunks toppled as well as serving as fire sources that could spew burning material to nearby locations. The District and the forests made it a priority to cut the snags down—only years later would the snags be found useful for cavity-nesting birds and other creatures. The usual method was to simply send out a crew to cut the snags down, but there was at least one experiment on the Mt. Hood NF in 1927-28 where explosives were tested: "promising results were obtained by using high explosives in holes bored in the snags with a portable automatic [electric powered] auger…a very tall snag…was felled in 39 minutes by boring and shooting. Experienced fellers estimated that to fell this snag with the saw would have required from 3 hours to half a day." (*Forest Worker* 1928: 12)

Recreation Management

Fred W. Cleator, who served as the regional forest examiner for 20 years, had the responsibility for the planning and development of recreational areas across the Pacific Northwest. Cleator's contributions were widespread, ranging from protecting the rain forest of the Olympic NF and the alpine peaks of Goat Rocks, to the rivers and mountain valleys of the Willamette and the Rogue River-Siskiyou NFs. Cleator's policies for the Region were well adapted to the existing patterns of outdoor enjoyment typical of the Pacific Northwest from the 1890s to 1945.

Relations with mountaineering and hiking clubs were cordial during this era. Forest Service personnel joined the organizations, often helped to plan trips, and took part in climbs and festivities. Cabin sites were leased for field club headquarters. The clubs, in turn, aided the Forest Service in many ways, particularly by giving aid in mountain rescue work, improving trails, and serving as advisory groups for recreational planning. Between 1915 and 1929, there came about a stronger emphasis on using the high country for recreation; and there was a growing recognition within the District that recreational use of the peaks, passes, glaciers, alpine meadows, lakes, and tarns should be dominant with grazing a secondary use. This decision coincided with a decrease in the use of the alpine areas for grazing, so a minimum of friction was created. Within a few years, this interest in the high country would mature into the wilderness movement.

In the early period when there were few visitors, campground building had not been of great importance. The first Forest Service campground was established at Eagle Creek along the Columbia River Highway in the summer of 1916. Albert Wiesendanger was the first "recreation ranger" in the Forest Service, as he was in charge of the Eagle Creek campground, building trails, leveling tent sites, and constructing crude furniture for campers. Wiesendanger noted that "because the area was already so readily accessible and popular, the new camping area could not be merely another undeveloped site set aside for the use of campers ... At Eagle Creek, for the first time, the Forest Service undertook the construction of a public campground in the modern sense." (Wiesendanger 2005: 95-96) Designated funds from the District Office for improving campgrounds were not available until 1923, though there is some evidence that other monies were diverted for this purpose.

Lease of sites for recreational use on public lands had begun as early as 1898 with the authorization of leases for hot springs. Special use permits for recreational purposes were issued before 1915 but they were revocable; this practice discouraged permittees from making permanent developments. Congress through the Term Occupancy Act of March 4, 1915, authorized the Secretary of Agriculture to lease land for hotels, resorts, stores, summer homes, or other structures needed for public convenience or recreation. The lease could not be over five acres nor extend for more than 30 years. The first summer homes in the Region were built in 1916 on the west side of the Metolius River across the river from Camp Sherman. Five years later, Fred Cleator, C. J. Buck, and Perry South completed a land use survey of the Metolius River summer home tract. Summer homes, encouraged by the Forest Service to get support from people recreating on the national forests, proved popular. The Washington Office believed that their popularity showed that the national forests, rather than national parks, should be the prime recreation lands for the people.

A summer resort, Mt. Baker Lodge, with stunning views of Mt. Shuksan across Sunrise Lake, was constructed in 1927 on the Mt. Baker NF. Winter snow skiing in the mountain area near the lodge was tried in 1929. The lodge burned to the ground in 1931.

Another recreational strategy of the District was protection and development of scenery, including protecting scenic strips along lakes, streams, and roads. As time went on, recreation area planning became standard. Establishment

Eagle Creek Campground special breakfast, Oregon National Forest, c. 1920. USFS photograph.

of summer homes, campgrounds, and resorts had in the past been provided according to demand, but it became routine to treat recreation areas such as a stretch of highway, a resort complex, or a summer home area as whole units. Planning these areas became paramount, before they would be overrun by recreationists and destroyed by the very people that had come to enjoy the great outdoors.

MT. HOOD RECREATIONAL DEVELOPMENTS

The Oregon NF included not only Mt. Hood and its surrounding terrain but also the southern wall of the Columbia River Gorge, which, with its spectacular forested cliffs and magnificent waterfalls, had long been recognized as one of the prime scenic resources of the Pacific Northwest. Forester Henry Graves recommended an area 22 miles long and four to six miles wide (13,873 acres) along the Oregon bank of the gorge be designated as the Columbia Gorge Park division of the Oregon NF. Secretary David Houston's order to that effect, dated December 24, 1915, appears to mark the first time the Forest Service dedicated an extended area to purely recreational use. The order prohibited timber sales and the distribution of permits for home sites.

There was a proposal to establish a Mt. Hood National Park, as well as a Yakima National Park (at Mt. Adams), in the 1920s, but these efforts were not successful. The Mt. Hood hotel and tramway controversy of 1927-31 demonstrated the increasing complexity of the recreation problems facing the Forest Service. In 1926, Fred Cleator from the DO and Francis Williamson, Jr., a ranger from the Mt. Hood NF, investigated possible sites on the upper south slope of Mt. Hood and Williamson drew up a recreational plan for the site, including a lodge. The Cascade Development Company of Portland submitted a proposal for a lodge and, most controversial, a tramway to the top of Mt. Hood. The tramway proposal drew considerable opposition. In the spring of 1927 Chief Forester Greeley chaired a public hearing about the tramway in Portland and heard from both sides. He rejected the special use permit; the company appealed his decision to the Secretary of Agriculture.

Secretary William M. Jardine sensed the complexity of the issues, and attempted to cool rising passions on both sides by calling for a study of the entire recreation situation at Mt. Hood. The resulting Mt. Hood committee, composed of a wide variety of local citizens, made its report to the Secretary in August 1928, calling for the construction of the disputed tramway as well as the development of nearby trails, campgrounds, and shelters. Secretary Jardine again delayed his decision by initiating a second study. He called upon Frank Waugh, professor of landscape architecture; Frederick Law Olmsted, the prominent landscape architect; and John C. Merriam, president of the Carnegie Institution of Washington. The three men met in August 1929, and conducted their own on-the-ground study. They could not agree on the tramway; Waugh supported the project while Olmsted and Merriam opposed.

INDIANS AND HUCKLEBERRIES

The federal government began to protect areas around existing huckleberry patches in 1897. A number of huckleberry patches were located in the Rainier and Columbia NFs in Washington, and in the Cascade, Santiam, and Crater NFs in Oregon. Problems developed involving American Indians and the management of the patches. Indians were among the main harvesters of huckleberries and had access to the patches by tradition and treaty rights. In the Twin Buttes area on the Columbia NF, the berry-picking activities were described by Forest Supervisor John Bruckart:

> Each summer as the huckleberries ripened, hundreds of Indians—men, women and children—migrated to the berry fields using all modes of transportation including pack and saddle horses, wagons and automobiles. They put up their teepees in the meadows along the cold mountain streams. Most of the berry picking was done by women and children while the men gathered wood for the camp fires, looked after stock and did some poaching when they could get away with it by avoiding the county game warden ...The Indian women dryed [sic] most of the berries they picked before slow fires except for some fresh berries that were sold after they left the area (Bruckart 1980: 34).

Between 1908 and 1921, protests grew over the fact that Indians had free grazing for stock while white stockmen had to pay a fee. The problem was most acute in the Columbia NF, where Yakima (now spelled Yakama) Indians from the reservation drove their saddle stock for grazing. From the Washington Office grazing section, C. H. Rachford wrote that this was a matter for the Bureau of Indian Affairs and that no restrictions should be placed on saddle and pack horses for other recreationists.

Indian woman huckleberry picker at Mount Adams berry fields, Columbia National Forest, Washington, 1933. USFS photograph.

SKYLINE TRAIL

The High Cascades were traversed by a network of trails, some of them made by Indians crossing from the east side of the mountains to the west; some built by pioneers, including Judge John Waldo, and prospectors; some constructed by early Forest Service men like Cy Bingham; and some of more recent construction. Both west and east of the crest of the Cascades, a network of crude roads had been built south of Mt. Hood making the peaks of the middle Cascades more accessible. The first suggestion for a wagon route along the crest of the Cascade Range was apparently made by Lewis A. McArthur in an article appearing in the *Oregon Journal* on December 27, 1915. He stated, in part:

> The time is not far distant when serious thought will be given to the proposition of constructing a passable wagon road on or near the summit of the Cascade range from the Columbia river to the California line. Such a road will open an almost incomparable storehouse of magnificent scenery of the most varied type. The most important feature of the plan is that the road is feasible from an engineering standpoint and it can be built and used, section at a time. ... It will require effort to bring the plan to its accomplishment but it can be done in time. Such a road as outlined above would have considerable value from other than scenic features. It would be of tremendous value to the forest service in fire fighting, for at the present time men and material have to be taken into most of the territory by pack horses, when fires are off the main roads. It has been suggested that the road also would have a military value.

In 1919, the Forest Service began to consider plans to build the skyline road or trail from the Columbia River to Crater Lake and thence to the California border, in part using existing trails and in part, pioneering new ones. The south portion of the proposed road, from Crater Lake northward to Minto Trail near Three Fingered Jack, was examined between July 10 and October 10, 1920, by Fred Cleator, F. B. Lenzie, an engineer, and E. R. Johnson from the Forest Service. They took a pack string of nine horses.

Carrier pigeons were used to communicate to the supervisor's offices the progress of the party along the crest of the Cascades (an attempt was later made to use carrier pigeons to communicate fire locations from lookouts). Cleator took with him a large number of trail markers and signs and the Oregon Skyline Trail was officially born. The party blazed for the first 40 miles, using the "skyline blaze," which is a long tree blaze with a notch in the center.

Cleator estimated that 99 percent of the road could be built with a grade of 5 percent or less with the rest on a gradient of not over 6 percent. The north portion was examined by the Bureau of Public Roads. The cost estimate for the entire road was around $1.5 million, which at the time effectively killed the project. In 1920, it was very expensive for a road over 300 miles long!

If the road had been built, it would have completely changed the nature of the wilderness battle of the 1950s and 1960s in Oregon as the proposed route would have traversed the length of the High Cascades through portions of the Mt. Hood, Deschutes, Willamette, and Umpqua NFs (see Chapters 7 and 8). Portions of the proposed road were constructed by the late 1920s. These included a section from Crater Lake to Crescent Lake, part of Century Drive along the eastern side of the crest, and a portion between Mt. Hood and Lost Lake on the northern end of the proposed road. The project attracted a great deal of public attention and support in newspapers, but the hard times and lack of money during the Great Depression put an end to the project, as well as to similar plans for a connecting trail built across the river in the vicinity of Underwood, Washington, to the Canadian border.

Engineering

Engineering in the Regional Office was coming of age during this period. Mapping and roads became the prime management concerns of the division. Just before World War I, Lage Wernstedt used panoramic photographs along with triangulation by transits to map over 700 square miles of the Columbia NF. The method was extended to the Cascade, Umpqua, Chelan, and Mt. Baker NFs by the early 1920s.

Ray Grefe described the engineering division in the 1910s:

> The April 1913 Field Program [a USFS program and national employee finding guide] lists Phil Dater as District Engineer under the District Forester, and in January 1917, with George Cecil as District Forester, Engineering was finally given divisional status. In the same year, Allen Hodgson took over Surveys and maps, R.H. Robertson handled waterpower. In 1918, Cecil Lord was added to the staff in charge of surveys. It can well be assumed, then, that with the expansion 1913 to 1918, engineering began to get its feet on the ground. Entry and timber surveys provided data for appreciably improving forest maps. Road and trail construction standards came into being though in very sketchy form (1990:275).

During the summer of 1922, the North Pacific DO sent parties into the national forests to revise or correct the forest base maps. The operation was led by Lage Wernstedt from the maps and surveys section. The first party was sent to the Deschutes, Santiam, and Cascade NFs and covered approximately 900,000 acres during one field season. The mapping work to cover all forests in the District was expected to take another 15 years. The first party found many mapping errors:

> The first among these is the fact that the work revealed a large number of lakes that have not been plotted … the greatest number of these were located between Sisters and Diamond Peak along the summit of the Cascades. A rough count indicates that at least 120 lakes were located … In a number of cases the party discovered that principal mountains, as shown on the existing maps, were more than a mile in error … It was also discovered that a number of mountains … do not exist at all. Whole stream drainages have been shifted and changed … The party was able to definitely locate many streams that were not shown on former maps (*Six Twenty-Six*, November 1922: 14).

Major roads and forest highways were under the supervision of the Engineering Division. Construction of forest roads, as well as use of trucks in the late 1910s through the 1920s made many of the remote stations obsolete and unnecessary. Several funds for road construction and maintenance were established.

This was an era in which scenic mountain highways were becoming popular. The Columbia River Highway flanking the northern edge of the Mt. Hood NF and highways to Mt. Rainier built through the Snoqualmie NF were the first in the Northwest. Outdoor recreation was growing fast sparked by the ownership of private automobiles. There were regional and national efforts to provide areas for summer homes, resorts, forest campgrounds, and downhill-ski areas. Roads leading to and through the mountains brought crowds of summer and winter travelers: existing roads were improved and several important new roads were constructed in the Northwest.

In 1910, the Agricultural Appropriations Act provided that 10 percent of all receipts from national forests be made available for construction of roads and trails in the forests from which the receipts were derived. Three additional laws passed over the next decade addressed the need for more funding of roads and trails: Federal Road Act of 1916 (section 8), the federal forest road construction fund of 1919, and the Federal Highway Act of 1921.

McKenzie River Highway near Blue River, Cascade National Forest, Oregon, 1920. USFS photograph.

> The first general Forest Service engineering conference was held in Denver, Colorado, on February 9 to 15, 1920. Regional Office attendees included George Cecil, Phil Dater and D. V. Stroop. "It was the general opinion … that special attention should be given to clearing [the roadways] in order to preserve the scenic value of Forest roads … thus avoiding any unnecessary cutting … a pleasing alignment, avoiding long tangents. Moss green on an ivory background was selected as the official Forest Service color for road signs." (*Six Twenty-Six* 1920: 2)

The Forest Service began working with the Bureau of Public Roads (BPR) in the early 1920s to improve highways across the national forests in the Region. Dollars were allocated to road access and to improving the existing roads by widening and straightening. Bridges and culverts were paid for with BPR funds. In 1922, the District recommended that the following road projects be approved by the Secretary of Agriculture for funding; apparently they were all approved and work started that summer:

Oregon—Alsea River, Coast Highway, Crescent-Klamath Falls, Eugene-Florence, McKenzie Highway, Medford-Klamath Falls, Anna Creek section near Crater Lake, Mt. Hood-Wapinitia, Prairie City-Unity, and Trail-Prospect.

Washington—Blewett Pass, Inland Empire, Mt. Baker, Olympic Highway-Forks to Bogachiel, Republic-Wauconda, and White Salmon.

The Regional road system was greatly improved during this period. More roads enabled the Forest Service to tie national forest roads to the major highways, setting up the system in effect today. But there were unintended consequences. More and better-quality roads meant more visitors. Recreation demand soared. Trails were impacted from overuse. Resorts, summer homes, campgrounds, and recreation sites (lakes, rivers, ski areas) became more easily accessible year around. Easy access to the forests also increased road hunting for deer and elk; poaching of big game animals as well as overfishing became more common.

Many of the early trails were eventually replaced by roads. In 1920, there was internal discussion about trails—should they be named or not? Naming of the major trails won out, but both trails and roads still have numbers (see Chapters 7 and 8).

Range Management

In 1911, Charles Flemming established an office of grazing studies in District 6 and J. M. Peterson took over the office in 1913. Major studies were undertaken, some on the range plants in the Region and others dealing with predator control. Much of the grazing was carried out on the central and eastern parts of the Region, although there continued to be grazing in the high meadows of the Cascades and Coast Range. From 1911 to 1933 a series of grazing reconnaissances were carried out to locate and evaluate available rangeland. These studies "were the stepping stones to the adoption by the Forest Service of the open method of herding sheep. Since that time this method has been uniformly recognized and practiced by most sheepmen on public and National Forest ranges." (*Six Twenty-Six*, May 5, 1920: 20)

It was very expensive and time consuming to support the grazing program on most forests. George Bright, in a report on grazing, wrote that

> On the Wenaha [Umatilla NF today] … practically the entire personnel is consumed by grazing matters, or by projects whose development is made necessary or desirable through grazing. Cabins are built, springs developed, trails, telephone lines, and bridges constructed, with the primary object of being thereby better able to handle the grazing business … Grazing is the greatest resource provided by the Wenaha. In round numbers one hundred and one thousand sheep and 9,000 cattle and horses were permitted during the summer season on this Forest in 1913, not including the natural increase, which would bring the total number of stock a third higher. In spite of the scarcely more than nominal grazing fees, this stock yielded a return to the government of $9,200 (Bright 1913: 33-34).

Grazing fees were designed to recover the cost of administering the resource, but were much despised by the permittees. A number of legal cases were brought against the Forest Service for charging or overcharging grazing fees, but the agency won most all of these (Rowley 1985). "On the Umatilla it was found that the grazing fees charged by the Forest Service amounted to about one-sixth of the fees charged by private owners for the same privileges. The charges for grazing lands bordering the Wenaha are between three and four times those charged by the government. The cost of running stock on the Wenaha is less than buying land and running stock on it, besides which a permittee is protected on the National Forest from fire and trespass." (Bright 1913: 33)

Sheep grazing increased in the Region during World War I as the demand for wool, meat, and hides grew dramatically. For a short period during the war, sheep grazing took precedence over any concerns about overgrazing of the rangelands. To this end the District Office on January 26, 1918, helped organize a new association called The Washington Forest Users Association at Yakima. Its purpose was to gain the cooperation of stock owners to "increase the capacity of the Forest ranges, and thereby increase the supply of wool and meat so essential to the welfare of our country at this time. The plan of the organization is the same as in the local associations, with the exception of the advisory board, which consists of one regular Forest user from each Forest in the State." (*Six Twenty-Six*, February 1918: 9-10) After the war, stocking levels were slowly decreased to minimize shock to the sheep owners.

Wildlife and Fish Management

During this era, most of the concern for wildlife revolved around huntable or trappable species, mostly deer and elk in the Pacific Northwest. There was a concerted effort to eradicate predator species such as wolves, coyotes, bobcats, bears, and cougars. Government hunters from the USDA Biological Survey were authorized to trap, poison, and kill untold numbers of these species. Sports hunters were also often given a free hand to shoot these predators, with states and counties offering bounties for the heads or ears. By the early 1940s, as a result, almost all the grizzly bears and wolves had been exterminated in the Pacific Northwest. Game wardens were employed by the state to enforce game laws, including overfishing. Often district rangers were responsible for enforcing game laws on the national forests.

The Wildlife Game Refuges Act of 1916 authorized the president to establish game refuges for the protection of game animals, birds, and fish on any lands purchased under the Weeks Act of 1911. It was also applied to other national forests. The earliest mention of state game refuges on the national forests in the District came from the Deschutes NF. The Paulina Refuge was established in 1913 to protect mule deer (it was opened for hunting in 1949). The Ferry County Game Preserve was established on the Colville NF between 1914 and 1916. "Intended as a refuge for deer, the 38,400 acre preserve became what Supervisor [C. C.] Reid called 'an especially favored hunting ground' due to lax law enforcement." (Holstine 1987: 55) It was abandoned by 1922. In the summer of 1920, a game refuge was set aside in the Foss River watershed on the Snoqualmie NF to protect mountain sheep. Then in 1922 the Whatcom County Game Commission set aside a game preserve on the Mt. Baker NF portion (Church Mountain area) of the county to protect deer, bear, and mountain goats.

For a short time, state game refuges were established on the Cascade NF. The Salt Creek Game Refuge (38,000 acres) was established in 1926 by the Oregon State Game Commission because of a request by a local sportsmen's club whose members were concerned about poaching by a large number of men employed to build the Southern Pacific Railroad over the Cascades. Another game refuge on the Cascade NF, the Three Sisters Game Refuge, was established in the late 1920s. The Canyon Creek Game Refuge on the Malheur NF was mentioned in *Six Twenty-Six*, as was the Dollar Mountain Game preserve in the Colville Forest, which "is to be abandoned, and open to hunting this fall. The preserve was established in 1929, soon after the Dollar Mountain fire from which the preserve was named." (*Six Twenty-Six*, October 1940: 8) Other game refuges, apparently all state refuges, were established in cooperation with the District and individual forests. By 1927, there were 2,319,921 acres of state game refuges located on the national forests in the District (USFS 1929).

The Olympic NF in cooperation with "sportsmen and nature lovers" and the Washington State Game Commission introduced mountain goats to the Mt. Olympus National Monument. In January 1925, nine or 10 goats were brought from Alaska in exchange for young Roosevelt elk. They were released at the foot of Mt. Storm King on Lake Crescent. The herd stayed small for several years, eventually becoming a management problem for the Olympic National Park.

The Forest Service in cooperation with the state fish and game departments stocked a great number of lakes in the high country from the 1910s to the 1930s. Special trucks were used when practical. Many of the high mountain lakes were stocked with fingerlings/fry by hand. Five-gallon milk cans were fitted with holes and screens on the lid to allow air to circulate over the water inside. When they were packed on horses and mules, the constant agitation of the cans allowed oxygen to enter the water. During hot weather, the fish cans were removed once or twice a day and laid into streams so the fish could have clean, cold, oxygenated water. Fish types planted included Chinook salmon, as well as rainbow, bull (Dolly Varden), cutthroat, eastern brook, and golden trout. On the Snoqualmie NF on October 22, 1916, Lake Dorothy was stocked with around 25,000 trout from Yellowstone Lake in Wyoming. On the Olympic NF, four rangers were reported to have carried "on the work of planting small, 2 ½ inch, Eastern Brook Trout in several of the streams and lakes in this region. So far, the total distribution has been around 18,000 in Bogachiel lake, Jackson, Turnwater, Snider and Hades creeks. These comprise a part of the 100,000 which are being held in storage at the Storm King Fish Hatchery." (*Six Twenty-Six* September 1935: 20) Over 100 lakes were stocked in the District.

Fire Management

Fire control was a problem that plagued large timberland owners, federal, state, and private wood lot owners. Between 1900 and 1933, the Pacific Northwest took the lead in this field. The story is a complex one, involving the aftermath of the 1910 fires in Region 1, the Weeks Act of 1911 and the Clarke-McNary Act of 1924, both aimed at federal and state cooperation for fire control.

Well documented by many writers and historians, the fires of 1910 burned some 2.6 million acres of national forests in northern Idaho and western Montana, and an additional 521,000 acres of private forest (Pyne 1982 and 2001). What is not well known is that there were thousands of acres of forest burning at the same time in the Colville region of Washington and the Blue Mountains and southern Cascades of Oregon. The fires burning in the Pacific Northwest in mid-August 1910 were serious:

> Acting District Forester George Cecil still considered the situation "so serious" that he elected to write "a special report rather than wait until the end of the month." For two months, no rain had fallen, save some "light" drizzles over the past week. Large fires burned on the Colville, the Wallowa, and especially the Crater national forests, near Butte Falls ... Cecil had a call for 500 more [Army] troops on the Crater and dreaded another "full month of the dangerous season" before him, while Washington fire wardens feared another Black Friday, like that of 1902 (Pyne 2001: 126-27).

In the 1910 fires, 160,000 acres burned in the Colville NF and 60,800 acres in the Crater NF. In all, about 700 million board feet of timber went up in smoke in the District. President Howard Taft authorized the use of Army troops to fight the massive fires within confines of national forest boundaries. District Forester C. S. Chapman commented on the use of Army troops: "the large number of serious fires made it practically impossible to secure sufficient local men to fight them. The federal troops were, therefore, secured ...These men performed invaluable services. Not only did they fight fire, but their presence in the country has a very desirable effect [in preventing arson]." (Cornwall 1911: 588) Over the years through to the present day, Army or National Guard troops would be called to firefighting duty and provide support to firefighting operations.

On the state level a series of acts were passed in both Washington and Oregon compelling private timberland owners to take action in the prevention and suppression of fire. Private industry cooperated with the state and federal agencies in fire suppression and detection, and a major educational movement alerted the general public to the dangers of forest fires. District Forester Cecil inaugurated Forest Protection Week in the Region during 1920. These legislative, administrative, and voluntary activities were paralleled by a revolution in firefighting equipment and new means of detecting fires. But this was not always possible. The 1902 Yacolt Fire area in Washington reburned and extended its reach to standing timber in 1918, 1922, and again in 1929. Little could be accomplished to fight the three massive fires—they were too large and there were not enough men to fight them. Big fires were eventually extinguished in many cases by the weather—rain or snow.

The first aerial drop of supplies to an ongoing fire was in the Olympic NF. Bill Ryan in the Olympic NF history book described this event:

> Lightning fires at the head of Tunnel Creek in 1929 was the first time that a fire crew was supplied by aerial drop. A crew of about forty men were working ... on two fires in rugged country without trails ... J. R. Bruckart, Assistant supervisor, arranged for two drops on consecutive days. Both canned and fresh supplies were packed in burlap bags together with blankets to offer some protection ... On the first [pass], the plane circled the camp to get a bearing. On the next circle, the plane door flew open and [Robert] McClay started kicking out the sacks with his feet while holding on to the inside of the plane with his hands ... These were free drops from an altitude of about 300 feet above camp ... Supplies came through in surprisingly good shape (Rooney 2007: 131).

The transition to modern firefighting took decades of learning by trial and error. Initially, rangers would climb mountains and, shading their eyes, look out across the forest to locate fires, and made do with available materials. In 1924, District Ranger Grover Blake on the Malheur NF observed a rather short lookout station on Myrtle Butte:

> An Osborne fire finder was set up on a wobbly table about 4 feet high, constructed of small, round stick wired together with emergency telephone wire. When a smoke was sighted, the lookout fireman would proceed to carry the table and fire finder to a spot from which the smoke would be visible through the trees. He would set the table down and orient the finder as best he could, as the table wobbled and shook, then take a [azimuth] reading and report. I sized up the situation and said to the Guard, "Don't you think we can rig us a better setup than this?" He thought it might be worth a try. So we felled two fir poles about 75 feet in height … [and dragged them] over to one of the tallest trees. I found some lumber and nails and we made a 50-foot ladder … At the top of the ladder we built a platform. Then we made a 30-foot ladder and pulled it up the side of the tree until it rested on the platform. Now we were up 80 feet and another platform was made.

Fire patrol with pilot, observer, and crew at Eugene air field, Oregon, 1925. USFS photograph.

> About 3 feet above the upper platform we cut the tree-top off and set up the firefinder on the stub (Blake c. 1960: 25).

Lookout structures were built beginning in the 1910s and 1920s. They ranged from "crow's nest" platforms built atop large trees, cabins built of the materials on hand, portable 12-by-12-foot prefabricated "Alladins," lookout houses with a cupola on top to house the firefinder, then tall wooden lookout towers with a cabin atop (later the towers would be often made of steel). The 14-by-14-foot D-6 cupola style, used for many years, was first designed by Lige Coalman, long-time lookout atop Mt. Hood, where the lookout house was built in the fall of 1915; it was abandoned in 1935 and fell off the mountain in 1941.

Many styles of lookout house and tower were later used throughout the Region. They were located on high peaks to cover all sections of the forests. These were supplemented by road patrols, primarily to catch fires caused by smokers or campers; by posted guards in major recreational camps; and by air patrols.

The Osborne firefinder, essentially a surveyor's transit sitting atop a plane table with a map, was initially developed in 1910-11 by Forester William "Bush" Osborne from the District Office. In the summer of 1911 testing began in the Mt. Hood NF on Battle Ax Mountain, Beatty Butte, Lookout Mountain, Mt. Wilson, and Olallie Butte. "To each of these, one of Osborne's [fire] finders was packed, by horse, by mule, and even on the backs of the forest guards themselves ... In 1915 a lookout and finder were established on the snowy crest of Mt. Hood itself." (Corning 1952: 19) The Forest Service in 1913 contracted Frederick Leupold, of the Portland firm Leupold, Volpol & Co., to have the firefinders manufactured at a cost of $25 to $36 each. After extensive field trials they were accepted by the Forest Service as standard equipment in 1918. A number of refinements were made to Osborne firefinders over the years, but the basic principles and design remain the same. (Osborne also developed the special panorama camera for fire detection used in the 1930s to take 360-degree panorama photos from atop each peak and lookout—see Chapter 6.)

For the ranger and forest guards on the fire line, the Pulaski tool and backpack pump cans supplemented the crosscut saw, the axe, and the shovel. A portable fire pump was first designed in 1911 or 1912 by Umpqua NF Supervisor Smith Bartrum and a representative from the Fairbanks-Morse Company. Although the first pumps were too heavy for forest guards or horses, various companies continued to improve on the pumps during the 1920s.

Communication from the lookouts to the ranger station was problematic in the early years. There were no radios or telephones in the earliest days. When a fire or smoke was located, attempts were made to signal the ranger station in the valley bottom where firefighting help was available by a variety of means, none very successful. Use of heliographs, which depended on the sun being out, was one way. Another was the use of wigwagging flags that the lookout positioned in certain ways to translate into Morse code. This method depended on clear weather. On the Cascade and Deschutes NFs, experiments beginning in 1919 used homing or carrier pigeons to send fire messages. The Deschutes NF had a pigeon loft built in Bend that housed 40 pigeons. "Messages have been carried to this loft from Crater Lake, Klamath Lake, Trail, and Portland, Oregon, as well as from all parts of the Deschutes Forest; birds have also flown to this loft from Prineville, Burns, Silver Lake and Vancouver, Washington." (*Six Twenty-Six* 1920: 2)

An interesting method tried in 1913 on the Crater NF was the use of dynamite blasts! Harold Foster from the District Office recounted the trials and tribulations of this system:

> In 1913 Trojan powder was provided for all the primary lookouts and a system of signals provided [seven sticks of powder to indicate a fire, then smaller blasts of five sticks to indicate the direction] … If it had been necessary to use it, I seriously doubt whether the listener could have distinguished the variations in time pauses on which the intelligibility of the message depends. … The observer would have no way of knowing whether the message were understood correctly, or even heard. … It was found that this system of dynamite blasts was not successful (Brown 1960: 171).

Before, and for a while after, World War I communications between the lookouts and the ranger station were limited to messages on foot, horseback, and carrier pigeon. Soon, however, an extensive (and expensive) system of field telephones, connected by miles and miles of No. 9 telephone wires, was used to communicate between lookouts atop the mountain peaks and ranger stations in the valleys below. Telephone wire had to be strung along forest trails and roads, among the trees and from poles, and across mountains and rivers. As trees often fell on the wire, breaking the connections, these phone systems needed continual maintenance and repair. Upkeep of the lines was sometimes a full-time job for the early rangers, especially after a big storm. Many of the phone lines allowed forest settlers to have phone access to the outside world. Rather like a party-line, everyone could overhear all the conversations. The

Ranger using tree phone, Santiam National Forest, Oregon, 1912. USFS photograph.

lookouts had both fire observation and telephone line repair duties. Two-way radios would come much later.

Just after World War I, the Forest Service contracted with the Army Air Service (Corps) to provide airplanes and pilots to spot fires from the air. In 1919, cooperative agreements led to experiments using airplanes to patrol for forest fires in California. This was quickly expanded to the mountainous areas of Oregon, Washington, Idaho, and Montana. Army pilots flew the open-cockpit DeHavilland bi-planes during the summer fire season while Forest Service spotters also aboard the two-seater plane looked for fires. The planes were initially stationed at Medford, to cover southern Oregon and northern California, and Eugene to cover the Cascades. In 1920, aerial patrols were also stationed briefly in Portland at the Sellwood landing field.

Beginning in 1920, each lookout house and tower was marked with large white triangles that enabled pilots to figure out where they were when flying over seemingly endless tracts of forest. These triangles laid out on the ground were 15 feet on each side with the apex pointing north and a number to indicate the map quadrangle and another number to identify the lookout site. During 1921, Portland-based planes began to look for fires in the "Olympic Blowdown Area" in Washington State. The system worked quite well for several decades until aerial patrols were replaced by lookout houses and towers atop the mountain peaks. (In the last few decades, most of the lookouts have been removed, replaced by electronic lightning detection systems, satellites, and aerial reconnaissance after major lightning storms.)

The fire season in Washington and Oregon usually lasts from late June to October. Standard fire plans were prepared for ranger districts, and approved by the forest supervisor. Presuppression involved a list of short-term personnel; public officials, such as sheriffs and state fire wardens; and individual local cooperators and their skills (such as cat operator, cook, packer, truck driver, clerk). It included also an inventory of fire equipment available. Each ranger district had a set of maps of the forest and district and other maps showing logical fire line locations, fire breaks, fire caches, camp sites, roads, and trails. Paralleling this was a fire organization in the District Office to help on large fires. This could and often did include hiring and transporting men to fight fires, and a listing of those in each national forest who were particularly skilled in fighting large fires and who could serve as supervisors.

The work of firefighting fell, to a large degree, on the temporary or short-term personnel. Increasingly, as forestry schools were established in the West, college undergraduates in forestry supplemented local men in this job, usually working on trail and road crews when there were no fires to fight. These were the people who occupied the lookouts, carried on patrols, or worked trail, and when fire broke out, "chased smokes" that is walking to investigate smoke columns and fought fires. Beginning in 1920, at Hemlock Ranger Station, Columbia NF, training schools were established for the short-term personnel. These schools lasted for a week to 10 days and were carried on in each ranger district. The temporary firefighters were given lectures on map reading, operation of the Osborne firefinder, fire detection, smoke chasing, firefighting techniques, photogrammetry, radio and telephone use, and maintenance of the phone lines. Then they practiced these skills in the field, under supervision. The general purpose and spirit of the schools were designed by F. E. Olmsted. The ranger, the fire control chief, and the supervisor generally ran the show, and often some of the staff from the District Office attended.

Private cooperation with the Forest Service began at an early period. Fires are not respecters of boundaries, and are a danger to both federal and private timber where tracts adjoin. The Western Forestry and Conservation Association promoted such cooperation, and it was financed by passage of the Weeks Act, which provided money to states for fire control on a dollar-matching basis. In the states as in the Forest Service, there was a shift to trucks and automobiles in 1918 and 1919. By 1924 portable water pumps were used on trucks and by a few field crews to put out small fires.

Chapter 5

Coping with the Great Depression and War 1930–1945

The beginning of the Great Depression was marked by the collapse of the stock market on October 29, 1929, a day now referred to as "Black Tuesday." Although the failure of the stock market was not the only economic reason for the decade-long depression, it was the most significant event in the minds of people in the Pacific Northwest and elsewhere. Federal employees took a 15 percent pay cut to stay employed. The economic meltdown was accompanied by one of the greatest droughts in American history, creating what was called the Dust Bowl in the central United States. Lumber sales plummeted and many timber companies went out of business.

But just at the worst of the depression, a new president was inaugurated. Franklin Roosevelt gave the country what it needed—hope for the future. Intending to get America going again, Roosevelt instituted massive federal spending on improvement projects, putting out-of-work men back to work, and starting new "alphabet" programs (CCC, WPA, ERA, NIRA, etc). Probably the most important thing that Roosevelt accomplished was lifting the confidence of the people. At the end of the 1930s, America was facing another crisis—World War II.

During this era, there were several changes to Forest Service symbols and badges. Badges were produced for fire guards that were made of nickel-plated bronze. In 1938 Chief F. A. Silcox approved a revision of the standard shield making the tree image in the middle longer/taller than in the original 1905 version. Since the late 1930s, there have been no additional changes to the image on the official badge and shield. All District Office names were changed to the present Regional Office designation on May 1, 1930.

Regional Foresters

District Forester Granger was succeeded by Clarence John Buck in 1930. C. J. Buck had earned a BA degree from Williams College and was a graduate of Yale Forest School in 1905 with a masters in forestry. After school, he served for three years in Region 5 and then in the Crater NF (now Rogue River-Siskiyou NF). In 1908, he was transferred to the new District Office in Portland and put in charge of lands. Buck served as regional forester from 1930 to 1939. His term was marked by the Great Depression and the work relief programs. His tenure also saw two major controversies, one over his authorization of selective cutting on dry-site Douglas-fir sites instead of clearcutting, and another about the Mt. Olympus National Monument and whether the area should become a national park. Buck was detailed for five months to the Washington Office in early 1938 to be part of a team working on a national recreation report. In 1939 he was transferred to Washington as inspector and assistant to the chief after a "run-in" with President Roosevelt over Mt. Olympus and the proposed Olympic National Park. Three years later Buck retired.

Buck was succeeded as regional forester by Lyle F. Watts. Watts had graduated from the Iowa State school of forestry in 1915 and earned a masters degree from Iowa State in 1928. He had worked, before coming to the Pacific Northwest, in Wyoming, Idaho, and Utah, and in 1928 organized and served for a year as dean of the school of forestry at Utah State Agricultural College. From 1931 to 1936 he worked at the Northern Rocky Mountain Forest Experiment Station, first as the senior silviculturist, then as the director from 1935. A year later, Watts was chosen as the regional forester for Region 9, headquartered in Milwaukee, Wisconsin. He became the R-6 regional forester in 1939, at a time when the Civilian Conservation Corps (CCC) program was being reduced and the war clouds were expanding across the horizon. Watts was deeply interested in development of the timber industry on a sustained-yield basis. He served as regional forester until 1943 and then moved to the Washington Office as chief. He retired from the Forest Service in 1952.

Watts was succeeded by Horace Justin Andrews. H. J. "Hoss" Andrews earned a masters degree in forestry from the University of Michigan in 1916. From 1914 to 1917 he worked on timber reconnaissance projects on the Santiam NF (now the Willamette NF), and various forests in California and Colorado, and he served in the Army Air Corps in World War I. He joined the Forest Service in 1930 as senior forest economist and director of the forest survey of Washington and Oregon conducted by the Pacific Northwest Forest and Range Experiment Station. In 1938 he returned to Michigan as a research professor in wild land utilization, but a year later came back to Portland as assistant regional forester in charge of the division of state and private forestry. He was appointed regional forester in 1943 and served until his death in 1951. He was a versatile, popular forester, well suited to serve in the shift from custodial management to intensive management.

New Deal Programs in the Region

CIVILIAN CONSERVATION CORPS (CCC)

President Herbert Hoover attempted in vain to remedy the crisis situation caused by the stock market crash in the fall of 1929. He felt that the government should intervene in the crisis only in a restricted way through the free enterprise system. Congress was not receptive to many of his proposed programs. The economy sank to the bottom of its greatest economic depression. During the closing days of the Hoover administration some limited funds were made available for putting men to work on conservation projects, largely related to road and trail work. In 1933, the Regional Office approved several forests to work closely with various counties to establish unemployment camps. Apparently, the workers were not provided with a salary; instead the counties and the Forest Service provided them with tents, board, clothes, tools, and tobacco. The Snoqulamie NF had a camp of 100 men who cut firewood for sale in Seattle. Other camps were set up on the Olympic, Umpqua, Siuslaw, and Willamette NFs. The camps were closed when the CCC camps started in the summer of 1933.

CCC Camp Glacier, Mt. Baker National Forest, Washington, 1933. USFS photograph.

President Roosevelt called for a full-scale use of work relief to conserve, protect, and renew natural resources. The Emergency Conservation Work Act was passed on March 31, 1933, only 10 days after Roosevelt took office. Roosevelt used the term "Civilian Conservation Corps" for this work. The act was extended by Congress and in 1937 was supplemented by formal establishment of the name.

Initially, the CCC program was administered by resource agencies in the Departments of Agriculture and Interior. The Department of Labor did the recruiting, while the War Department operated the camps and ran the educational program. The resource agencies carried on the field activities.

Regional Forester C. J. Buck was quite concerned by the mid-1930s about the pressure being applied by local and state officials and even the congressional delegations to place their chosen men in charge of CCC camp operations. During late 1934, Buck and John Bruckart from the Regional Office in the division of engineering worked together on the selection of supervisory personnel for the CCC camps. In the fall of the next year, the two men made a tour through the State of Washington to talk with several representatives and one senator. The politicians wanted more local men hired by the CCCs, but there were only a few additional jobs available. This did not make the delegation very happy. Buck issued a memorandum in May 1933 regarding the CCC chain of command:

> Forest Service personnel must refrain from criticism of camp management either directly to the Army personnel, or particularly to the enrolled men. In camp, outside of working hours, what the Army say, goes. If comment or suggestions are asked for by the Army, give them, but do not volunteer. … The enrolled men assigned to you are expected to deliver a fair day's work. Shirking or loafing should not be tolerated and should be reported with recommendations for separation to the Camp Commander. At the start, remember that most of your men are youngsters with little or no experience.

The CCC program had two main objectives. The first and most important was to find immediate and useful conservation work for hundreds of thousands of unemployed young men (there were no camps for women). The enrollment or enlistment period was for six months with the option of re-enrolling for additional six-month periods for a maximum of two years. The men were paid $1 a day with $25 per month of that sent home to their dependents, usually their parents—this money saved many farms and homes from foreclosure. The enrollees were allowed the remaining $5 per month, or slightly more if they

CCC Camp Oakridge, Willamette National Forest, Oregon, 1933. USFS photograph.

were leaders or assistant leaders, to spend at the CCC camp canteen or for other personal expenses. The government provided the enrollees with room, board, clothing, and transportation.

There were four distinct categories of enrollees. Most numerous were the young men or juniors between the ages of 18 and 25 (age limits changed to 17 and 28 in 1935), who comprised about 85 percent of the CCC enrollment. Another group was the local experienced men (LEM) who served as project leaders. These men were hired from the local communities and had often been previously employed in outdoor or woods work, and later worked as vocational teachers. The LEMs comprised about 5 percent of the total CCC enrollment. Veterans of World War I were another group of older men, a category that also comprised about 5 percent of enrollment. The last group was American Indians and residents from the U.S. territories, who generally had their own separate CCC companies and camps on their reservations or in their territories, where they could live at home and work on the nearby projects.

In the Pacific Northwest, the majority of the early CCC camps were located in or under the jurisdiction of the Forest Service (Williams 1983). Even the CCC camps on the O&C lands had Forest Service employees in charge of forest projects. For the first six-month enrollment period, 119 camps were established in the two states, of which 111 were under the technical supervision of the Forest Service. Eighty-four were on Forest Service land, six on O&C lands, five on state forests, seven in national parks, two in state parks, and 15 on private forest land. The two states had an average of less than 78 camps in the following three enrollment periods. The fifth enrollment period (1935) saw a doubling of the number of CCC camps in the Northwest, but this was followed by a reduction to previous levels.

The total number of CCC enrollees from Washington comprised 51,313 men, with 3,830 of the enrollees being American Indians and 7,174 camp officers and supervisory personnel. A total of 73,339 CCC men worked in the state during the program (1933-1942), representing almost 70 percent of the enrollees. The state had an average of 43 CCC camps per year. The estimated total cost for the CCC program in Washington was $76,655,597.

CCC enrollees from Oregon comprised 34,609 young men, including 2,767 American Indians and 6,820 camp officers and supervisory personnel. A grand total of 86,775 men were assigned to the camps in the state, thus 60 percent of the enrollees came from other regions. The state had an average of 51 CCC camps per year, and the estimated cost of the program in the state was $87,734,444.

Regional Forester C. J. Buck, in a letter to the forest supervisors, wrote that the CCC program would alleviate "the social devastation of unemployment." The increase in the number of workers would enable the Forest Service to tackle its long-talked-about forest development needs, and provide a large firefighting force and a tree-planting crew to help recover burned-over areas. The Washington Office issued improvement handbooks on the use of wood; the Regional Office set up site plans and landscaping standards; the national forests prepared site recommendations and selected places for recreation and other developments; and the ranger districts were the sites for projects. Plans for work were prepared in the Regional Office. Fred Cleator and Floyd V. Horton from the lands section worked closely with the national forests in the Region to implement the work plans.

The accomplishments of the CCC men included working on national forest projects such as reforestation, firefighting, insect and disease control, tree improvement, recreational improvements, truck trails (roads) construction, and the construction of needed administrative sites. As a Forest Service publication, *The Ranger,* in 1933 pointed out "there were no 'made jobs' just to give the boys work. And many a ranger, to his delighted surprise, found jobs being done this year that he had not hoped to get for many years, if ever."

The CCC men in the Pacific Northwest planted over 100,000 acres with trees, completed 37,000 acres of tree stand improvement, spent 1.1 million person-days fighting forest fires, and controlled tree pests on 1.2 million acres. They constructed 396 lookout houses, 281 lookout towers, and 147 trail and fire equipment shelters, and strung over 58,000 miles of telephone lines as part of an elaborate forest fire protection and communication system. They also built or improved 85,000 miles of truck trails (forest roads) as well as over 42,000 miles of foot trails. The construction of these lookouts, trail shelters, telephone lines, and trails provided greater opportunities for forest fire protection than ever before. The CCC could not build roads or highways because of a national agreement with the unions, but they could build a forest truck trail system. In doing so, they opened large areas of old-growth forest, making the national forests more accessible to recreationists and eventually timber management. The CCC camps were also responsible for greatly increasing the number of campgrounds and scenic vistas. The men also planted almost two million trees and shrubs in and around recreational and administrative sites.

Several instances of firefighting in the Pacific Northwest by the CCC stand out as major accomplishments. Soon after the CCC was organized in the summer of 1933, a huge fire broke out in the Coast Range of Oregon. What became known as the Tillamook Fire spread over thousands of acres of forest land. CCC workers from all over the Northwest were sent to the huge fire. In the fall of 1936 another huge fire, called the Bandon Fire, burned through acres of highly flammable gorse shrubland in the Coast Range of southwest Oregon, destroying thousands of acres of prime timber, and burned most of the town of Bandon along the coast highway (U.S. 101). Enrollees from CCC camps were called into action to fight the massive fires and protect life and property. In both cases, their efforts were successful in stopping the destructive fires, but massive destruction was left in fires wake.

The CCC enrollees in the Region were, for the most part, hard-working young men who wanted and needed work and money to help stave off the effects of the Great Depression. There were instances of CCC men showing daring and initiative. For example the R-6 newsletter for February 1938 recounted a story of a CCC hero from the Cascade Locks Camp on the Mt. Hood NF:

> The coveted CCC certificate of valor has been awarded by Robert Fechner, Director of the Corps, to Larry D. Smith … for saving the lives of persons from drowning … Assistant Leader Larry D. Smith while on leave May 22, 1938, in the company of Mr. and Mrs. Henry Phillips of Cascade Locks, Oregon, was crossing the Columbia River. In midstream the boat filled with water and capsized, throwing the trio into the icy water. When the boat came to the surface they clung to it. Smith realized that the weight of all three was too much for the boat, and concerned for the safety of his companions, especially Mrs.

> Phillips, an expectant mother, gave up his hold on the boat, saying he would swim to the Washington shore for aid; then began his swim through the cold, treacherous waters of the Columbia. A motor boat operator attracted by Smith's swimming, saw the plight of the three, and immediately put out from the shore. He rescued Mr. and Mrs. Phillips and then picked up Smith who by that time nearly reached the Washington shore.

The Regional Office, under the leadership of George Griffith in the office of public relations, organized a "showboat" crew for the two states—one truck for Washington, another for Oregon. The intent was to visit every main CCC camp to present a program of forestry-related motion pictures and lantern slides to as many CCC members as possible. The Washington truck was staffed by Albert Weisendanger and Norman Penick, the Oregon truck by Ole Fuller and Jack Jackson. They started their travels in the summer of 1933. The showboat program was designed to "present a vivid, instructive and worthwhile program. The material will bite right into the local situation, explain in interesting terms just why the particular types of jobs have been selected for the camps, and what the work means to forestry, both locally and nationally. This education program is sponsored jointly by the federal, state and private forest protective agencies." (*Six Twenty-Six*, May 1933: 29)

The CCC program and other work/relief programs were phased out after World War II began. Many of the facilities built by CCC workers still are standing today, a tribute to the hard work of the CCC men in the 1930s. The program was highly successful, and is generally regarded as the best liked of the New Deal programs.

OTHER RELIEF PROGRAMS IN THE REGION

Several other work relief programs also got their start during the Depression. The Works Progress Administration (WPA) worked with the CCC to build the stunning Timberline Lodge on the south slope of Mt. Hood. The CCC built the access roads and laid the foundation. The WPA built the structure and decorated the interior with wonderful carved wood and paintings, and provided many furnishings that are still in use today. The lodge was dedicated by President Franklin Roosevelt on September 28, 1937. The president was greeted by a Forest Service honor guard—four from the Regional Office, four forest supervisors, and four rangers. Other employees were stationed in strategic locations along the travel route, including Albert Weisendanger, who spent the afternoon by a culvert on the road to the lodge.

Another relief program that affected the Forest Service was the National Industrial Recovery Act (NIRA) that began in the fall of 1933, operating with special CCC funds. The laborers employed under the NIRA program were paid 60 cents an hour, while leaders were paid 80 cents an hour. On the Colville NF alone, the NIRA workers built some 140 miles of roads and 19 buildings including lookout houses and towers, and made many range and recreation improvements. NIRA workers on the Columbia NF were engaged in snag felling operations. The Forest Service furnished tools and camping equipment for the work. John Ray Bruckart (1980) wrote that the NIRA program "was a very successful program and it resulted in the elimination of many thousands snags in the old burns on the forest and made fire control on these areas much easier." The NIRA program reached its high point in 1935, but was terminated within two years.

A similar program called the Emergency Relief Administration (ERA) was created on May 6, 1935. Like the other New Deal programs, the ERA program was intended to get the unemployed back to work. All ERA workers were required to be hired from local county relief rolls. One ERA camp on the Olympic NF worked in cooperation with the Bureau of Fisheries to repair and improve the Quilcene fish hatchery. The crew worked two months during August to October of 1935 on the project. The work that ERA accomplished on the national forests was focused on presuppression fire activities and hazard reduction, as well as maintenance and related work. On the Umpqua NF, three ERA crews of nine men each were engaged in the summer of 1935 in making improvements to recreation sites, cleanup of roadsides, general maintenance, as well as making improvements to ranger stations. Even the Regional Office got into the game with an ERA crew working on a radio station project during the summer of 1935. The ERA program peaked in 1936, but gradually decreased until the U.S. entered the war in 1941.

SUBSISTENCE HOMESTEADS AND RESETTLEMENT CAMPS

The New Deal period was one of social and economic experimentation. One idea that was in vogue at that time was subsistence homesteads—government aid in placing rural dwellers on tracts of land with decent housing and sufficient land on which to make a living, as a remedy for rural poverty. An interesting development of this type was attempted in the Oakridge area between 1931 and 1938. Oakridge and Westfir were the two largest forest communities in the Willamette NF. As the Depression deepened, Ranger C. B. McFarland and Axel Lindh of the Regional Office in 1931 began to study ways that the distress might be alleviated. Two years later they issued a report titled "The Oakridge

Vicinity Recreational and Industrial Land Unit." The basic philosophy of the report was that loggers and mill workers should have sufficient land to have decent homes, raise gardens, keep chickens, cows and hogs, so that they could help them get through the hard times. It was believed that the national forest land near these depressed communities could provide that type of assistance to these desperate people. Meantime, a subsistence homestead division was set up in the Department of Agriculture. It evolved into the Resettlement Administration, and this in turn into the Farm Security Administration under Rexford Tugwell. In April 1935, Regional Forester C. J. Buck wrote to the Forest Service Chief Silcox recommending the Oakridge proposal and suggesting that an initial 143 acres be set aside for subsistence home sites. However, the plan fell victim to interagency rivalries in Washington. On August 26, 1935, Tugwell wrote to Roosevelt: "You spoke to me the other day about forest communities such as they have in the Black Forest [in Germany]. Since August 1, a plan for one of these has been at Works Progress [Administration]. It has been decided I understand to smother it there. I just want to be relieved of responsibility." In September, Earl S. Pierce of the USFS lands division in Washington D.C. wrote to Buck that the president felt the proposal "would not serve the purpose of the relief program as fully as desired."

At least five Resettlement Camps, staffed by WPA employees, were established in the Region during the 1936-37 period. The camps were located on the Siuslaw NF along the Oregon coast at Cape Creek, Cascade Head, Hebo, and Nestucca CCC camp locations, as well as the Tacoma Creek camp on the Colville NF. Ira Jones recalled his work at the Siuslaw NF camps:

Timberline Lodge, built by WPA and CCC, Mt. Hood National Forest, Oregon, c. 1940. USFS photograph.

> During the depression I was loaned to the Resettlement Administration and had charge of W.P.A. workers on the Siuslaw Forest. We had three main camps in former CCC Camp Buildings. They were located at Mapleton, Cape Creek, and Hebo. At one time we had 44 side camps with as high as 1100 men. All Foreman positions had to be approved by the Democratic organization. I remember one foreman sent out who was absolutely no good as a foreman. I fired him and got a call from the Portland Resettlement to come to Portland and get squared up with the State Democratic Chairman. I went in. The Chairman was at the Congress Hotel. I had known him many years at Pendleton, so I told him the man he had sent out as Foreman was no good as a foreman. He said "I know, but he is my father-in-law and I have to take care of him." However, he was not sent back to me (Jones n.d.).

Just as with the CCC projects on the national forests in the Region, men were assigned work on tree reseeding, removal of snags and windfalls, and cleanup of roadside vegetation. New construction of trails, telephone lines, and lookout sites, along with reconstruction and maintenance of forest trails and roads were also part of the projects. Resettlement crews worked on stream improvement projects, planting fish, and clearing summer home sites.

In one instance, the Oregon Coast Land Utilization Project, part of the Resettlement Administration, worked in the winter of 1938 to "rescue a river from virtual imprisonment" when they had to remove a sand bar across the outlet of the Siltcoos River next to the Siuslaw NF. A winter storm had built up a sand barrier some 300 feet wide and 10 feet across the mouth of the river that caused flooding of the nearby area and restricted salmon from accessing the river. Some 30 Resettlement workers were able to dig through a portion of the sand bar then, with two boxes of blasting powder, blast a hole that left the river free flowing once again (*The Resettler*, February 1937).

Another Resettlement Administration project was organized in the mid-1930s in northeastern Washington as a Soil Conservation Service project. Designated a "land use demonstration project" it involved 517,745 acres in Pend Oreille and Stevens counties adjacent to the Colville NF and Kaniksu NF (Region 1). The Colville NF added these resettlement acres in 1936 to the Colville and Newport ranger districts (Holstine 1987 and personal communication). (In the late 1960s, the Tacoma Creek Resettlement Camp became home to the U.S. Air Force's survival training program.)

Purchases of worn-out farms by the Resettlement Administration through the land utilization program would in the 1960s result in lands that were eligible for national grassland status. For example, by January of 1937, 91,384 acres had been purchased in Jefferson County in central Oregon. This area in the 1960s would become the Crooked River National Grasslands on the Ochoco NF.

Regional News

Tragically, Nelson F. MacDuff, supervisor of the Cascade NF, was killed on April 4, 1930. The *Forest Worker* reported the death: "[H]e was found dead from a pistol shot early next morning in a grove of young Douglas firs along a forest trail he was accustomed to travel. The manner of his death is unknown [although locals suspected it was suicide]" (*Forest Worker* 1930: 25). MacDuff Mountain was named in his honor the following year.

The first R-6 ranger training camp was established in the fall of 1931 at the Hemlock RS on the Columbia NF. The camp was described as:

> attended by 32 forest rangers from 19 different national forests of the two states, and will run from October 5 to November 3. The men selected are the younger, newer men in the ranger force, who are in the greatest need ... The purpose of the camp is to train rangers in the proper way to do their jobs, and the keynote of the camp will be "doing the job" rather than being told how to do it. How to estimate timber, scale logs, prepare government sale papers, how to lay out trails and build them, as well as telephone lines and lookout towers, how to prepare and organize for a large fire, and how to fight it, are some of the many things these young rangers will be taught. In fact, more than 120 subjects are to be covered during the camp (Guthrie 1931:3).

Among those attending were Ed Cliff, who later served as Chief of the Forest Service, and Axel Lindh, who went on to become the head of the division of lands in the Washington Office. John Bruckart recalled: "At the end of the training session, Allan Hodson requested the instructors to rate the first ten who were most likely to succeed during their Forest Service careers ... Ed Cliff was rated highest." (Bruckart 1980: 33)

In December of 1934, the Regional Office hosted the 2nd Pacific Northwest Regional Planning Conference in Seattle. Over 700 people registered from Washington, Oregon, Idaho, and Montana; many others attended but did not register. Pacific Northwest Research Director T. T. Munger presided over the land resources portion of the meeting, Regional Forester C. J. Buck spoke about "Sustained Forest Yield," H. J. Andrews talked about "Integrating All Land Uses," and S. A. Wilson discussed "Some Practical Aspects of Land

Planning." A similar planning meeting with the same subjects would be held 65 years later.

Gifford Pinchot revisited the Region in August 1937. He was greeted on the Siuslaw NF by Supervisor R. S. Shelley, who recounted the meeting with great affection and admiration for the old chief:

> I want you to know that he is the same old leader; his ideals undimmed by the years that have gone by. He still wields against wickedness and special interests and special privilege in high places, and for the common man in the lowliest place, a flaming sword of truth and righteousness untarnished by selfishness, corruption or greed … Gifford Pinchot himself claims no credit for the growth made by the Forest Service nor for the esprit de corps of its personnel. In the years that are ahead I ask for nothing better for the Forest Service than that we shall continue to have in full measure such inspirational leadership coupled with such breadth of vision and that fighting spirit which will never quit (*Six Twenty-Six*, November 1937: 5-6).

Chief Ferdinand Silcox visited the Region in the fall of 1938 and gave an address to the West Coast Lumbermen's Association in Eugene on September 10th. He talked about the need for more federal funds for fire control and warned that the timber industry needed to support long-term sustained-yield operations on private timber lands, otherwise the federal government would have to step in. Later, the chief met with employees from the Siuslaw and Willamette NFs.

On November 11, 1939, a news release noted that the City of Portland was presenting the Forest Service with the deed to Benson Park in the Columbia River Gorge. "The deed covers about 700 acres of land, all lying south of the Union Pacific Railroad Company's tracks. It includes Multnomah falls, with the park improvements, and Wahkeena falls and picnic grounds." The land had been purchased by Simon Benson on June 2, 1915, and then given to the City of Portland. Some improvements at the site had been constructed by the city, some by CCC crews.

The national forests in the Region changed slightly in 1943 when the Colville NF in northeastern Washington was moved from oversight by the Pacific Northwest Region to the Northern Region as a result of the supervisor's office having moved from Republic to Colville.

> Colville was the only logical place to put the new office, but the town was already the location of the Colville District Ranger Station of the

> Kaniksu National Forest (KNF), which was administered by Region One headquartered in Missoula. Having two regions administering two forests in one small town was ruled out of the question … Effective as of 1 July 1943, the U.S. Forest Service transferred 568,279 acres of the Colville National Forest … to Region One [transferred back in 1974]. Public Land Orders 162 and 163 assigned 241,952 acres of the Colville to the Chelan National Forest and 160,359 acres of the Kaniksu National Forest to the Colville (Holstine 1987).

Olympic National Monument and Park

President Franklin Roosevelt travelled by train in September 1937 to the Region to dedicate both Bonneville Dam on the Columbia River and Timberline Lodge. While Roosevelt was on his way west on the train, he visited Yellowstone National Park, where his daughter and son-in-law met him and traveled with him to Seattle. They spent some time with the President talking about the need for an Olympic National Park. From Seattle, they went by Navy destroyer to Port Angeles, where pro-park demonstration was staged for Roosevelt's benefit. The presidential party travelled by car around the peninsula between September 30 and October 1, looking at the potential for a national park, stopping overnight at Lake Crescent. A young boy in the area skipped school that day and when he returned to school his excuse was he "was catching fish for the President." The next day the party stopped for lunch at the Lake Quinault Lodge and, later, for a tree-topping demonstration at the Bloedell Logging Company operation.

C. J. Buck, regional forester for the Pacific Northwest Region, and John R. Bruckart, forest supervisor of the Olympic NF, were on hand to advise Roosevelt about the folly of creating a national park as they felt strongly that the standing timber in the area would be needed to produce lumber in the future. At the Lake Quinault Lodge, Buck had the Forest Service boundary sign moved two miles away to be on the edge of old-growth timber instead of overlooking a clearcut inside the forest boundary (Lien 1991; Richardson 1968). Their comments, however, instead of convincing FDR, antagonized him. In fact, Roosevelt believed that they were lying to him. The supervisor of the Mt. Rainier National Park, O. A. Tomlinson, who had not been invited by the Forest Service, was invited by Roosevelt to accompany the motorcade and provide advice. Both Buck, alternating with Bruckart, and Tomlinson rode with the President in his special car and talked with Roosevelt about the proposed park. Tomlinson's positive comments about the need for the park were nearer to the president's own feelings.

Not long after the presidential visit, Buck found himself under pressure from the White House. Irving Brant, a confidant of President Roosevelt, described the following scene regarding:

> FDR looked at me sharply and asked, "[I]s that fellow still there?" "Yes," I said. The president picked up the telephone and said, "Get me [Secretary of Agriculture] Henry Wallace." FDR held the phone. Then "Henry," he said, "what did you do about that fellow I told you to transfer out of Portland a year ago?" A long silence. Then, in a tone of incredulity: "You didn't?" Another long silence. And finally, speaking slowly and emphatically: "Well, I want it done." And he put down the receiver. It was done. On April 1, 1939 Regional Forester C. J. Buck was moved to the Washington office of the Forest Service as special inspector, which preserved his salary grade and took away his power (Brant 1988: 132).

While in Washington, Buck had another run-in with Roosevelt. He then resigned. Supervisor Bruckart was transferred to be Forest Supervisor of the Willamette NF in Oregon. It is unclear if he left the Olympic NF voluntarily or not.

A bill to establish the Olympic National Park was introduced in Congress during the following spring. Efforts at compromise were offered by the Forest Service, but the final decision was directly influenced by President Roosevelt. Acting Regional Forester Fred H. Brundage wrote to the staff of the Olympic NF on February 23, 1938, that "by direction of the President a new proposed national park boundary has been worked out by the Chief of the Park Service in cooperation with the Chief of the Forest Service." (Twight 1983) The amended bill was passed and signed into law on June 29, 1938. During the next two decades, several additions to the Olympic National Park were made from the Olympic NF and state lands; each was fought by the Forest Service and the State of Washington, but inevitably they lost each time.

World War II

The United States entered World War II on December 7, 1941. Within days the resources of the national forests, particularly the timber resources, were earmarked for the war effort. The number of timber sales went up; volume cut increased; the price per thousand board feet rose; and the total value of timber sales soared sevenfold. Forest Service employees could be hired at age 16, as the older teenagers joined the military. Recreation demand fell. Timberline Lodge at Mt. Hood was closed in 1942 for the duration of the war. Camping, hunting, fishing, and water activities were reduced. The war effort consumed the nation.

The Forest Service through state and private forestry was a cooperator with the War Production Board, which had been assigned the work of getting gasoline and other strategic wartime materials that were in short supply. The Timber Production War Project (TPWP)—often called the Teepee Weepee—was established under the leadership of Howard Hopkins to solve production problems for vital forest products. One of the first problems to be addressed was that sawmills across the nation were almost all closed, despite the growing demand for lumber. Basically, the problem boiled down to a severe shortage of labor. TPWP petitioned local draft boards to get the experienced workers needed to maintain sawmill production for the war effort. Nationwide, draft deferments were granted to some 5,000 men for the needed effort and more than 9,000 prisoners of war also helped increase lumber production. In Region 6, timber production increased from 862 million board feet in 1941 to a high of 1.488 billion board feet by 1944—almost a doubling in three years.

The Selective Training and Service Act of 1940 authorized the military draft, as well as providing that conscientious objection would be allowed. On February 6, 1941, President Franklin D. Roosevelt signed an executive order creating the Civilian Public Service (CPS) program as a way for conscientious objectors (COs) to serve their nation without directly contributing to war. Over 37,000 men entered the CPS program and the first CO camps were established at abandoned CCC camps in 1941. Three camps were established in the Region: No. 21 Cascade Locks (Mt. Hood NF), No. 56 Waldport (Siuslaw NF), and No. 59 Elkton (O&C Revested Lands Administration in Douglas County), along with several side or temporary work camps. The COs were principally members of five religious denominations—Amish, Brethren, Hutterites, Mennonites, and Quakers—as well as some from other denominations, and even a few who had no church affiliation (Cottrell 2006; Everson 1978-79; Justice 1991; Mills n.d). The CO men were not paid for their work and they had to serve until 1947, two years past the end of the war (Cottrell 2006); this was done in part to placate the public, who didn't want the COs to have an easier time of serving than did the soldiers, sailors, and marines fighting overseas.

In 1943, many COs volunteered for smokejumper duty. Sixty were chosen for this very dangerous work. The national forests in Regions 1 and 6 were able

Conscientious objector crew, Mt. Hood National Forest, Oregon, 1942. USFS photograph.

to carry on through the fire season with this help. But life was not easy for the CO smokejumpers. Communities, and even some Forest Service employees, did not like the fact that the COs were working in the national forests and frequenting nearby towns. Robert Painter, a Quaker CO, was sent to the Cave Junction Smokejumper Base:

> Painter was chosen by a group of smokejumpers who had been stationed at the CPS camp in Cave Junction, Oregon, to serve as director of their side camp. At the time, Painter was little aware of the ongoing problems that existed between some of the conscientious objectors and members of the Forest Service … The smokejumpers thought Painter could make peace between the groups … While at Cave Junction, Painter was kicked out of a church and beaten up by a gang. Other smokejumpers suffered similar indignities but such actions only strengthened his outlook. He was most upset by the fact that churches in the area refused to welcome CPS men (Cottrell 2006: 183-84).

In 1943, the Deschutes NF became host to a military post called Camp Abbot, located at the present-day site of the Sun River Resort in central Oregon. The camp included about 8,500 acres of the Deschutes NF. It was built during the winter of 1942 and completed in May 1943. Around 9,000 troops were stationed at Camp Abbot. Gail Baker, then staff officer on the Deschutes NF, reported that during the summer and fall of 1943 thousands of Fourth Corps troops engaged in maneuvers in central Oregon. "The maneuver area included portions of the Fremont, Ochoco and Malheur as well as the entire Deschutes. However, the Deschutes received the brunt of the war games activity." (Baker 1950: n.p.) The Army destroyed miles of forest roads and telephone lines in the war games, and started several forest fires that the soldiers had to put out. The camp was also used to house German and Italian prisoners of war, who were sometimes used in forest operations. Camp Abbot was discontinued in the summer of 1944, but it remained for several years as a military reservation. The Great Hall, built by the Army in 1944, still stands and is being used at the Sun River Resort.

As early as August of 1939, the Army tested an early aircraft warning service in California, western Oregon, and western Washington. During test runs on August 1 and 2, Army planes (representing potential enemy planes) flew over parts of the coastal areas. One of the "targets" was the Interstate Bridge between Portland and Vancouver. The system, using forest lookouts, telephone systems, and central information centers, worked.

Aircraft Warning Service (AWS) stations were established early in 1942 along the West Coast. The Forest Service mobilized in national defense, as did every federal and state agency in the two states. The Regional Office assigned Bill Parke to oversee the Oregon AWS sites and Lloyd Brown to oversee the sites in Washington. Selected lookout houses and towers were winterized for this use. An elaborate system of reporting every aircraft flying in the vicinity of each AWS site was initiated. Every AWS lookout reported to a central filter center, by phone or radio, all passing aircraft. At night, even planes that could not be seen, but heard, were reported. Observers—usually husband and wife teams in order to avoid cabin fever—kept 24-hour watch in shifts, 7 days a week. Winona Harner, an AWS lookout on Burnt Hill on the Olympic NF noted in her daily diary that on May 31, 1944, "Raining again for this last day of A.W.S. Mildred is glad it is ending but I'm not. Wes [her husband and AWS lookout] will go to fire guard school and then be fire lookout this summer here at Burnt Hill." (Rooney 2007: 108) The fear of air raids was realized in 1942 when a plane from a Japanese submarine dropped an incendiary bomb near Brookings on the Siskiyou NF, the only aerial bombing by an enemy airplane on the continental U.S. With the arrival of coastal radar detection, the AWS posts were phased out on June 30, 1943, for the inland forests and the following year for the coastal forests (Webber 1974 and 1992).

Just before the end of the war, a unique fire threat floated across the Pacific Ocean. The Japanese, in a desperate attempt to forestall the coming invasion of their homeland and to demoralize citizens of North America, deployed hydrogen-filled, paper balloons carrying bombs against the United States. The balloon bombs were, in retrospect, a slight diversion in the Pacific war front, but they were significant as they represented the first instance of intercontinental bombing in recorded history. The fire balloons, with suspended incendiary magnesium bombs, as well as regular bombs, were sent aloft from Japan during the fall and winter of 1944-45 to float on the newly discovered high altitude wind currents of the jet stream across some 6,000 miles of the North Pacific Ocean to the West Coast of North America.

Upon arrival over the western U.S. and Canada, the balloons were designed to drop their bombs, which the Japanese hoped would start huge forest fires. It was thought that soldiers would be diverted from military operations to fight the massive forest fires. It never happened. Of the estimated 6,000 fire balloons that were sent aloft, probably 2,000 actually arrived. They have been found in Alaska, Canada, U.S., and Mexico; most were found in British Columbia, Canada (58 landings discovered), followed by Oregon with 54 and Alaska with 38. A number of others were shot down by military aircraft over the Pacific before they made landfall. Two balloons floated as far east as Michigan, and one balloon bomb landed in the Cicero district of Chicago. Only one documented forest fire was started from the thousands of attempts.

There was loss of life associated directly with the balloon bombs. Near the town of Bly, Oregon, five school children and their teacher—Elsie Winters Mitchell, Sherman Shoemaker, Jay Gifford, Eddie Engen, Joan Patzke, and Dick Patzke—were killed by one of the bombs. On May 5, 1945, while on a church picnic trip, they found a downed balloon with unexploded bombs still attached on Gearhart Mountain on the Fremont NF. When they investigated the bomb exploded. These were the only known fatalities occurring within the continental U.S. during World War II as a direct result of enemy action. A *Seattle Times* report from June 1, 1945, explains the events surrounding the unfortunate deaths:

> The Rev. Archie Mitchell, minister of the Christian [and Missionary] Alliance church in Bly, Ore., was the only survivor from the church picnic. He and Mrs. Mitchell took five children in their car and picked out a shaded spot for lunch about 16 miles in the mountains. While Mitchell drove the car around by a road, the others hiked through the

> woods. "As I got out of my car to bring the lunch, the others were not far away and called to me they had found something that looked like a balloon," Mitchell related. "I had heard of Japanese balloons so I shouted a warning not to touch it.
>
> "But just then there was a big explosion. I ran up there—and they were all dead." The clergyman was so dazed from the blast and the shock of seeing everyone killed that he hardly realized two [three] forest service employees [Richard Barnhouse, George Donathan, and John Peterson] had heard the explosion and joined him.

A stone memorial with brass plaque at the Mitchell Recreation Area commemorates the unfortunate deaths. Ironically, on March 10, 1945, a paper balloon (No. 94) descended in the vicinity of the Manhattan Project's top-secret nuclear production facility at Hanford, Washington. The balloon landed on the main power line between Benton City and Grandview that fed electricity from Bonneville Dam to the building containing the reactor producing plutonium for the Nagasaki atomic bomb. The power line shorted out and shut the reactor down briefly until other power sources took over. It took three days for the balloon to be removed.

Gearing up to fight fires caused by the balloon bombing fell, in part, to the 555th Parachute Infantry Battalion. However, by the time they were deployed, the balloon bombing effort had ceased. The "triple-nickles," as they were called, was an all African-American parachute unit that was assigned in April 1945 to Army air bases in Pendleton, Oregon, and Chico, California. Their specific duties were part of a highly classified mission to fight fires caused by the incendiary balloons, as described in the 555th Web site:

> The first call for the Pendleton contingent came just a few days later on 20 July—to drop fifty-three men and two officers to fight a fire in the Meadow Lake National Forest in Idaho [probably refers to the Meadow Lake Fire, on the Salmon-Challis NF]. From 14 July to 6 October, the Chico and Pendleton units participated in thirty-six fire missions [apparently none caused by the balloon bombs] with individual jumps totaling twelve hundred. There were also casualties. In six months, more than thirty men suffered injuries from cuts and bruises to broken limbs and crushed chests. One typical report listed under "injuries": "1 EM (enlisted man) broken leg above knee, 1 EM knee out of place, 1 EM crushed chest." …

> On one jump in early August, one of the men suffered a spinal fracture. He remained on the scene throughout the fire fighting operation. Then, realizing his men were tired and short of food and water, he refused to burden them with the job of carrying him to the nearest airstrip. Somehow he managed to stand up and without help, walked straight-backed for eighteen miles to the strip where his units would be picked up by C-47 for the return trip to Pendleton. He spent weeks in the Walla Walla Hospital. Pure Guts! ...
>
> Tragically, one man lost his life. [His death occurred during a fire jump in the Umpqua NF near Roseburg, Oregon.] The ill-fated trooper [Private First Class Malvin L. Brown] had landed in the top of a tall tree. In attempting to climb out [of] his harness and lower himself with a rope that each man carried, he apparently slipped or lost his grip and crashed into a rock bed 150 feet below. It took three days for patrols to find his body. [He died on August 6, 1945, the same day the first atomic bomb was dropped on Hiroshima, Japan.] (www.triplenickle.com)

Timber Management

The timber industry underwent a series of major changes in the period between 1930 and 1945, changes that altered the nature of the industry and its relationship with the Forest Service in the Northwest and elsewhere. There was a major shift from private to public lands as a source of timber. In the first part of the period, many timber companies went under because of low prices and low demand for lumber caused by the Great Depression, though, as noted earlier, there was an increase in timber harvesting during World War II. Timber sales in the 1920s had reached 514 million board feet by 1929, just before the Great Depression. They fell drastically during the early years of the Depression to a low of 131 million board feet in 1933, rose slowly during the 1930s to 498 million in 1939, and rapidly during the war to a maximum of 1.488 billion board feet in 1944. This was the first time that timber harvests from the national forests in the Region exceeded one billion board feet.

In 1930, Congress passed the Knutson-Vandenberg Act, which put a portion of timber sale receipts into a trust fund to do work connected with reforestation work in the sale area. Because of limited timber harvest activities during the first part of this era, the act did not provide significant money for reforestation. That would change in the decades after World War II (see Chapters 6 and 8).

Sustained-yield, that is timber harvesting on a sustained basis, only cutting the amount that can be replaced by tree planting and new growth, became a management goal for federal, state, and industry forest lands (Mason and Bruce 1931). There were major changes in logging technology as railroad logging and river drives gave way to tractor logging and truck transportation. As a result, there was an increasing need for extensive road systems through the forests. The power chainsaw was beginning to come of age during this period, and by the end of the war falling axes and bucking saws were increasingly relegated to museums. Axes were and still are used in many forest operations and cross-cut saws are still used in wildernesses. Finally, there were major improvements in fire control, especially with the availability of CCC men to fight fires.

The National Industrial Recovery Act of 1933 called, in part, for expanding public funding for fire control, disease and insect protection, and better cooperation between federal and private forest land owners. The act set up codes of fair competition for industry; Article X, one of the codes, set up standards of forest practices. These practices included minimum standards of forest management and encouraged timber harvesting under the principles of sustained management. Many timber companies voluntarily complied with Article X to ensure better forest management, at least on paper. However, the lumber code was ruled unconstitutional by the Supreme Court less than a year later. Most of the states then turned to state regulation of timber harvesting practices, today called "best management practices," as a way to regulate harvesting, road construction, reforestation, and, in some cases, wildlife and fish protection.

The first federal law dealing specifically with sustained-yield operations came with the O&C Act of 1937. The O&C lands—railroad grants lands managed by the federal government—under the act were to be managed under the principle of sustained-yield, by not cutting beyond what can be regrown on the land. However, much of the discussion on sustained yield proved to be premature and overly optimistic. Sustained-yield for the Forest Service came with passage of the Sustained-Yield Forest Management Act of 1944, which authorized the Secretary of Agriculture to establish sustained-yield units (Loehr 1952). These units could be either cooperative units consisting of federal forest and private forest land or consist only of federal forest land. Cooperative sustained-yield units were to combine the management of federal timber land with private land to stabilize communities. Federal units, the other category, reserved national forest timber only for one geographic area, usually one community and one mill such as the Lakeview Federal Unit. The federal units had no joint management requirement (see Chapter 6).

Aerial photographs were used to locate and plot the best bodies of timber, estimate volume and species, and locate the best road sites, aided by the development of lightweight radio equipment that was used in both timber reconnaissance or survey and fire control.

Timber surveys were completed on many parts of the Region during the 1930s. Some experiments were tried with aerial photography in conjunction with the USDA Soil Conservation Service, but the most reliable was still the on-the-ground method. The Regional Office timber survey party during the summer of 1939, for example, travelled to the Wenatchee, Willamette, and Fremont NFs.

Power saws had been used before this period, but only as bucking saws, that is to cut a log that had been fallen into lengths. Experiments were made with both electric and gasoline saws in the felling of standing trees. Electric power saws were found to be unusable in the wet conditions that were characteristic of the westside forests. The gasoline power saw became the most practical new woods tool. In 1942 it was found that fellers with a power saw—they were all two-man saws in those early days—could fell a 7-foot Douglas-fir in 18 minutes, as against two hours for the crosscut. As a result, power saws gained acceptance after the war by the timber companies, but less so by the union workers.

Greater efficiency in the felling of trees and hauling of logs was also matched by greater utilization on the ground in logging operations. Lower stumps and

Felling ponderosa pine with experimental electric chainsaw on Hines timber sale, Malheur National Forest, Oregon, c. 1940. USFS photograph.

smaller tree tops became the rule. Because prices were rising, some timber operators began looking at older timber sales to salvage tall stumps or cull logs to make into plywood. The plywood industry in the region had been a very small one in the 1930s, but with the war there was a plywood boom as the product was increasingly used for airplanes, boats, boxes, and buildings. This enabled the use of lower grades of lumber and the better utilization of wood in the mills.

Cold weather was found to provide at least some control over mountain pine beetle epidemics. As early as 1924, extreme cold on the Deschutes NF was believed to have caused the death of at least half the pine beetles (*Forest Worker* 1933). In early December 1932, very low temperatures were experienced in the south-central Oregon forests and the northern California forests. The U.S. Bureau of Entomology, working with the USFS and National Park Service, found that prolonged exposure did kill the beetles, except where thick bark insulated them. In the summer of 1935, the Regional Office set out to determine the boundaries of the larger pine beetle infestations in the Northwest. A 15-man crew was sent out in August to start the survey work, and by the end of the season had covered some 3 million acres.

> The survey is a cooperative undertaking of Mr. Keen's office of the Bureau of Entomology and Plant Quarantine and the Regional Office. The work consists primarily of counts of beetle-killed trees on many miles of [forest] strip supplemented by careful analyses of insect conditions on numerous half-section sample plots … The beetle survey is not confined to National Forest timber but includes large bodies of privately owned ponderosa pine near the federally owned timber. The U.S. Indian Service is conducting a similar survey on the Indian Reservations in Oregon and Washington (*Six Twenty-Six* October 1935: 29).

The end result was to give location information for the CCC and ERA camps in their insect control work for 1936 and beyond.

In September of 1936 600 Douglas-fir seedlings were planted at Eagle Creek on the Mt. Hood NF by a troop of Boy Scouts "for the National Society, United States Daughters of 1812, State of Oregon." (*Six Twenty-Six* October 1937: 17). In May of 1937, the Olympic NF announced the first Christmas Tree Forest was being planted. The 40-acre tract was located in the northwestern corner of the forest. The experimental tree planting was expected to be cut after a five-year period. In 1940, the Mt. Ashland chapter of the Daughters of the American Revolution signed an agreement with the Rogue

River NF to have 1,200 ponderosa pine seedlings planted in the Ashland Canyon watershed. The work was completed by CCC boys. The DAR tree-planting efforts, known as the "Penny Pines" project, was dedicated on Flag Day (June 14) in 1940.

CLEARCUTTING VS. SELECTIVE LOGGING

In old-growth Douglas-fir stands the accepted silvicultural method of harvest at this time was clearcutting. This method of removing all standing trees from an area, used in Europe for many years, was desirable, since Douglas-fir reproduces best in the direct sunlight. In addition, large cleared areas simplify road construction, slash disposal, site preparation, and planting. Selective logging practices were tried for a number of years in the Region. Probably the first such sale on the westside Douglas-fir forests was on the Siuslaw NF in the early 1920s. As recounted by Supervisor Shelley:

> This sale was to H. H. Cook on the Waldport district in second growth Douglas fir. The trees to be taken were all marked before cutting and varying degrees of selection were employed on different plots. The logging was done with horses and on each plot all of the trees were marked, measured and tallied by diameters to the nearest tenth ... At the present time [1936], whenever Cook wants to buy more timber to supply local need, we make him a sale on this basis and go ahead with our selective logging. Since the date of this sale, May 25, 1923, it is believed that all of our second growth Douglas fir sales have been made on a selective logging basis (*Six Twenty-Six*, March 1936: 24).

In late 1934, after reviewing several research studies, Regional Forester C. J. Buck directed the national forests in Oregon and Washington to begin timber harvesting on dry-site Douglas-fir areas (essentially east of the Cascades) by selective logging, rather than by clearcutting. Soon a fundamental disagreement among Forest Service and academic researchers arose over the clearcutting issue. Two University of Washington forestry professors, Burt P. Kirkland and Axel J. F. Brandstrom, argued that "selective timber management" was economically advantageous, as loggers did not have to take every tree and that selective logging did not lay the landscape bare as did clearcutting. Forest Service researchers Leo Isaac and Thornton T. Munger, however, argued that selective logging was a short-term economic gimmick used during the Great Depression that would, in the long run, deplete the forests as only the prime trees would be taken from a stand, leaving the less-desirable species on site. Also, they argued that selective logging practices damaged the trees that remained on the site and that clearcutting was much better. Ivan Doig (1975-76) writing about "The Murky Annals of Clearcutting" said, "Professional foresters were honestly disagreeing about silvicultural alternatives, but mostly on economic grounds ... All in all, [it should] ... serve as a classic lesson that disputes over the use of our forests are not going to be decided on ecological merit alone. Nowhere near it."

These intellectual battles were played out on the ground in a number of experimental research studies. The Deschutes NF, for example, had a sustained-yield timber program that involved selective harvest. The Shevlin-Hixon timber company agreed in 1939 to experiment with selective cutting on both company and national forest timber operations. The experiment was scheduled to run for a year. After more than a decade of research it was found that selective harvesting was not as economical as clearcutting.

The selective logging method was used in various locations in the Pacific Northwest Region until the early 1940s, when C. J. Buck was given a directed transfer to the Washington Office and clearcutting resumed. The champions of clearcutting had their day, but within another decade, the issue would cause the Forest Service anguish during national debates over clearcutting on the Monongahela NF in West Virginia and the Bitterroot NF in Montana. As a result of the judicial and political battles that followed, clearcutting as a method was severely reduced.

Recreation Management

COPELAND REPORT

A national emphasis was placed on the importance of recreation in the national forests in the early 1930s. In 1933, pursuant to a Senate resolution by New York Senator Royal S. Copeland, the Forest Service issued an overall plan—*A National Plan for American Forestry*—generally known as the Copeland Report. The report was prepared under the direction of research chief Earle Clapp, and was critical of both federal and state commitments to forestry. The two-volume report set the stage for further actions by Congress and the president in changing forestry priorities. The report also recommended public regulation of private forestry, which set off another two decades of debate. Most importantly, the Copeland Report set the stage for putting people back to work through programs in the forests and grasslands that were designed for fire protection; insect and disease control; tree planting; and conduct research on forest management problems. Because of the Copeland Report, the Forest

Service was ahead of the game in coming up with projects when the Civilian Conservation Corps came into existence in the spring and summer of 1933.

The Copeland Report section on recreation, written by Robert Marshall, then with the Bureau of Indian Affairs' Indian Forest Service, briefly summarized the varied uses of the forests for recreation and urged a process of classification and analysis of the needs of the people. Marshall, co-founder of the Wilderness Society, had got his start as a forester on the Wind River Experiment Station in Washington in 1924 under the direction of Leo A. Isaac. His important report, which echoed ideas expressed by key Region 6 officers, formed the basic philosophy that the Forest Service used to manage the national forests for recreational purposes. Regional Forester C. J. Buck, writing for the Oregon State College Forestry Club paper and the *Journal of Forestry*, stressed the need for recreational planning, for building campgrounds, and the need for topographical and vegetation maps for planning purposes.

RECREATION AROUND THE REGION

Retaining trees along the increasingly popular highways that crossed the national forests had first become an issue in the 1920s. Sometimes called "leave strips," these areas were intended to screen management activities like logging from the eyes of forest visitors driving through the area. If these 300- to 500-foot corridors of old trees were left along the highways, it appeared that the forest was still untouched. Nationally, the Forest Service committed itself to preserving roadside strips, sometimes called beauty strips, along state, county, and forest roads. "While the proposal to acquire all timbered lands contiguous to highways … is regarded by the Forest Service as impracticable, the acquisition of privately owned forest within the boundaries of the national forests … will be accomplished … through [land] exchange." (Forest Worker 1931: 11) Still in practice today, these strips are now referred to as visual buffers. In 1931, Regional Forester C. J. Buck wrote a letter to *The Oregon Motorist*, outlining the regional policy of dealing with timber harvesting and scenery along the highways:

> For many years it has been a recognized principle in the handling of the National Forests that scenic values along important highways should be preserved and given first consideration in any cuttings which may be done in conjunction with sales of timber. When any tract of timber is reported upon and recommended for sale, the report and maps show the location of highways and the examiner makes recommendations

> as to the action which should be taken to make provision for the protection of scenic values and as to how the logging should be regulated to accomplish what is desired.

On the Deschutes NF, the Region in 1930 initiated a new program of purchasing roadside strips along the major highways. These roadside strips were successfully used to protect the forest scenery along portions of the Dalles-California Highway (U.S. 97), Willamette Pass Highway (OR 58), Santiam Highway (U.S. 20), McKenzie Highway (old OR 126), and along the Metolius River/Camp Sherman road. In the Regional Office, Fred Cleator continued his planning for highways and trails. The Civilian Conservation Corps provided the workers to put new recreation plans into action. This meant a vast increase in the number of campgrounds and facilities—campground shelters, garbage pits, toilets, tables, chairs, and the like. The CCC built many miles of trails, yet for the most part the new trails were not constructed for recreation but for administrative use, to get rangers and supplies to all parts of the forests. Yet they were increasingly used by recreationists. The CCC-made improvements enabled the Forest Service to keep up with demands for recreational facilities, and to plan for increased use in the future.

The early recreational uses of the forest—hunting, fishing, and camping, as well as berry picking, photography—continued. Automobile camping increased; use of horses, except in the back country, declined. Mountain climbing increased; the customary groups who engaged in climbing—Mazamas, Mountaineers, Obsidians, and Chemeketans—were increasingly joined by Californians "discovering" these new peaks. Near the close of the 1930s, a new recreation emphasis on wilderness and primitive area designation was proposed by the Forest Service and external groups.

On the Mt. Hood NF, probably the first banning of motorboats occurred in 1936 on Olallie and Breitenbush lakes. "The majority of people seem to be in sympathy with this restriction, although there are doubtless many owners of motorboats who believe that it is unjustified." (*Six Twenty-Six*, September 1936: 15). The forest guard at Breitenbush Lake received the following letter from a lake user that still resonates today:

> Another point upon which I might comment is the restriction of motorboats upon the lakes in your region … I should be very sorry to see the removal of the motor restriction. I know of no noise that is more disturbing than the din of an outboard motor and to have several such motors going at the same time on a small lake such as yours would completely spoil the quiet and restful atmosphere that now prevails.

Fishing at Packwood Lake, Columbia National Forest, Washington, 1937. USFS photograph.

> At best, with the opening of new roads, these mountain retreats will become spoiled to some extent by too many people, and I sincerely hope that the Forest Service continues to restrict the use of motors upon the high lakes of your region (*Six Twenty-Six*, August 1936: 15).

Secretary of Agriculture H. A. Wallace signed an order on January 13, 1938, that extended the boundaries of the Columbia Gorge Park Division on the Mt. Hood NF. The order added an additional 14,695 acres to the park area situated along the Columbia River Highway. The 13,873-acre park had been established by D. H. Houston, another Secretary of Agriculture, on December 25, 1915, for recreation purposes. A news release on January 28, 1938, described the intent of the expanded park area: "The area will be maintained in its natural state, subject only to such development as will make it of maximum free use and enjoyment to the general public; no use will be allowed or permitted which will interfere with these broad public purposes." Similar words and intent were incorporated into the Columbia River Gorge National Scenic Area and other national scenic areas in the Region.

With the advent of the CCC and WPA programs in the 1930s there were efforts to document archaeological and historical sites on the national forests in the Region. For example, on the Umpqua NF there was a WPA project to document old-timers' recollections by the use of trained interviewers. The WPA produced several score of these interviews. In most counties, there were WPA projects to document or inventory the historical records in the county courthouses. While many of these projects did not directly affect the national forests, the information provided background material for interpreting changes to the land and waters.

The Mt. Hood NF placed a number of interpretative signs to highlight the Barlow Trail along the south and west sides of Mt. Hood. An effort was made in 1936 by the Willamette NF to relocate and mark the pioneer trail of 1853 that crossed the summit of the Cascades near Willamette Pass. One old wagon was located and placed at the Lane County Museum in Eugene. Another wagon wheel, also believed to date from 1853, was found on the Snoqualmie NF near Naches Pass in the summer of 1937. A news release on August 4, 1937, described the wheel: "Its rusty, iron rim was hand-welded. Its huge oak hub was of ancient ox-cart type and its spokes long since loosened revealed the craftsmanship of an old time wheelwright." Archaeology sites were loosely documented on several forests, including the Siskiyou, Deschutes, and Fremont NFs. There were numerous instances of the forests working cooperatively with regional universities to record petroglyph and pictoglyph sites, excavate caves, and grave sites, and store found materials from the national forests in the Region.

In 1933, problems arose over the influx of large numbers of non-Indian berry pickers to the huckleberry patches on the Columbia NF. The Indians complained about the new competition, but the situation was taken care of by a direct meeting on August 24th between the Indians and the forest supervisor. Tensions were relieved for the moment. Supervisor Bruckart (1980) wrote that "later in the fall after the Indians had returned to their homes, many of the white people lingered on not realizing that after the rain started, the unsurfaced road to Trout Lake would become impassible for their vehicles. The Forest Service and Army trucks had to come to the rescue to bring their cars to the graveled road." Eventually gravel roads were built to the patches so that both Indians and whites relied on the automobile for transportation to and from the berry fields. On August 21, 1935, a tribal council was held at the Cold Springs Indian Camp in the Twin Buttes huckleberry fields.

> Chief Willis Yallup of the Rock Creek tribe and about 20 leading Indians from nearby reservations met with the forest supervisor and [Mt. Adams] district ranger. Expressions of good will and gratitude came from the Indians … About 300 Indians were present and these were chiefly from Rock Creek, Yakima, Umatilla, Warm Springs and Nez Perce Reservations. Mr. C. R. Whitlock, Superintendent of the Yakima Reservation was present and gave his support to the Forest Service … and show[ed] their appreciation for the large tract of land set aside for the Indians to camp on with exclusive use of the berries and pasture for their horses. [On the following day] A caravan of 26

> persons from mid-Columbia towns ... were taken on a tour about Mt. Adams by the Supervisor and District Ranger ... To add color to the trip, Chief Willis Yallup and son Thomas ... and Chief Joe Kup Kup of the Warm Springs tribe were taken along (*Six Twenty-Six*, September 1935: 9).

By the following season, the district reported that 4,290 campers had visited the area. Indians were outnumbered by white campers by more than two to one. Huckleberry production stood at 19,500 gallons during the picking season. During the 1938 season, almost 46,500 gallons of huckleberries were checked out of the area. By this time, Indians were outnumbered by white pickers on the berry grounds by almost five to one. At the time, the berries were sold at market for 50 cents a gallon, representing very good wages in the Great Depression. Huckleberry picking in the area continues to this day.

WINTER SPORTS DEVELOPMENTS

The 1930s also brought a new interest in winter sports, especially skiing, and the Forest Service began to plan for year-round recreation. Beginning in 1928, skiing enthusiasts began to urge the Forest Service to establish winter sports areas, and the Forest Service made a series of studies of the Region, working with outdoor clubs from the Hood River and Portland areas. During the middle 1930s, many small ski areas, both downhill and cross-country, were designated, developed, and strongly supported by local outdoor organizations and nearby communities. Such developments previously had been private, consisting of the use of lodges and cabins as headquarters for skiers—mostly cross-country skiers. In May 1930, Secretary Arthur M. Hyde instructed the Forest Service to issue a special use permit that authorized the construction of a cableway on Mt. Hood (*Forest Worker* 1931). But the ambitious project, which included a hotel, was never built as the Cascade Development company could not find the necessary financial backing. Several years later, Timberline Lodge and ski area would be constructed with federal money by the Civilian Conservation Corps and the Works Progress Administration. Ski lifts ascended Mt. Hood several thousand feet, but were never built to the summit. Mt. Hood rapidly became a national center for downhill skiing activities.

For the most part, these were cross-country areas, although several were equipped with crude rope tows. Many of these ski areas were developed initially with assistance of CCC and WPA workers, who often removed trees for short ski runs, built and maintained roads to the areas, and constructed warming huts and shelters.

On April 1-2, 1939, the Mt. Hood NF hosted national championships in downhill, slalom and combined, and the U.S. tryouts for the 1940 Olympic or international skiing federation world championships. The Regional Office and Mt. Hood NF personnel developed a winter sports uniform in early 1939, consisting of cap, jacket, trousers, and shoes. Included was a winter sports insignia—a white cloth patch with orange background and black lettering on each sleeve. This was the first winter sports design in the agency.

SKYLINE AND CASCADE CREST TRAILS

The Skyline Trail in Oregon and Cascade Crest Trail in Washington proved to be increasingly popular. In 1934 William L. Boyer made a reconnaissance of the trail and issued a detailed report. This included recommendations on relocation of hazardous or inconvenient stream fords; alternative routes over portions of the trail; evaluation of campsites and stock feed; recommendations for building of shelters, latrines, and garbage pits; and recommendations on trail signs and maps. Much of the work recommended was carried out with CCC help. The Oregon portion of the trail was essentially complete by the early 1930s, but there were still areas that were not connected.

Work on a continuation of the Cascade Crest Trail in Washington State was carried on during the mid-1930s. The location for the Washington portion of the trail began at Monument 78 on the Canadian border on July 15, 1935, and was completed on October 10 on the Columbia River at Sepsecan. Five field

Walking along the Skyline Trail near Breitenbush Campground, Mt. Hood National Forest, Oregon. Bob Unruh photograph.

parties undertook the task. They were instructed to "to lay out the trail so that it passed through scenic country, kept as close to the summit as practicable, and provided a wide, easy grade. The groups were also directed to note wildlife, check hunting grounds, and test the fishing in streams and lakes." (Holstine 1984: 12.5) Work was started soon afterwards by the CCC. The Forest Service wanted to continue the trail to the south in Region 5 (California), where the John Muir Trail had been built in the central Sierra Nevada. The three trail routes have changed slightly over the years, especially through purchase of lands for the trails in southern California to connect with Mexico. Today, the 2,650-mile three-state trail has been renamed as the Pacific Crest National Scenic Trail.

Fred Cleator in the Regional Office favored the building of shelters along the trails. The availability of CCC labor helped to implement this program, and scores of trail shelters were built along the Skyline and Cascade Crest trails. These were small, simple, open-front shelters in the Adirondack-style, built of whatever wood was available and roofed and sided with shakes. They were to be located about one day's hike apart. Other shelters of a similar type were built on the other forest trails for the convenience of Forest Service firefighters, packers, and others engaged in the business of managing the national forests.

WILDERNESS

In 1924, the first administratively designated wilderness for the Forest Service—the Gila Wilderness—was established on the Gila NF in New Mexico. In 1926, another special area was designated on the Superior NF: the Boundary Waters Canoe Area, which had been studied as a potential wilderness by Arthur Carhart in the early 1920s. An inventory of roadless lands larger than 230,400 acres (10 townships) was begun in 1926 under the direction of Leon F. Kneipp. There was, however, no general Forest Service policy on wilderness designation, and each potential area would have to be judged on its own merit. Generally, the Forest Service felt that wilderness designation would be unfair to dependent communities that relied more and more on the national forests for timber and range resources. In 1929, the Forest Service announced, under the "L-20" regulations, two new administrative designations: research reserves and primitive areas. Research reserves would be preserved for scientific and educational purposes, while the primitive areas would keep those lands intact for the "nature lover" and "student of history." However, these regulations did not prove to be very protective of the designated primitive areas. Rules and regulations governing use of the primitive areas were non-existent. By 1939, 67 areas—around 14 million acres—of national forests had been designated primitive. However, logging, grazing, and roads had been excluded from only four of these.

Prior to the 1930s, the management of the national forests basically amounted to custodial protection of the forests from fire. Very little timber harvest, road construction, or recreation development occurred. Then the Civilian Conservation Corps built thousands of recreation sites, administrative structures, and access roads on and across the national forests. Roadless areas were decreasing in size as developments impinged on the edges of the areas. But there were local forest and state concerns about the obvious changes to the forests. On August 31-September 1, 1935, the Siskiyou NF was host to Oregon's Governor, the State Board of Control, and a variety of other important state and federal people. They met at Gold Beach to

> discuss the advisability and feasibility of establishing a wilderness area. Interesting arguments were presented on each side both pro and con for establishing a wilderness area along Rogue River … Two major thoughts were agreed upon and these are that the Rogue River is a recreation area of more than state-wide importance [Wild and Scenic River designation would come decades later] and also that Curry County [residents along the Pacific Coast] … should have some highway connection with the interior part of the State (*Six Twenty-Six*, November 1935: 16).

The lack of protection for the designated primitive areas would have an impact on the perceived need for formal laws, instead of administrative regulations, that would fully lay out the rules for the protection of the areas. The U Regulations of 1939 allowed for three land designations: "U-1" Wildernesses (over 100,000 acres) that were designated by the Secretary of Agriculture; "U-2" Wild Areas (5,000 to 100,000 acres) that were designated by the Chief of the Forest Service; and "U-3(a)" Roadless Areas (that could be any size) that were to be designated by the Secretary if over 100,000 acres or by the Chief if under.

At this same time, national recreation director Robert Marshall, John H. Sieker from the Washington Office, and others were pushing for more primitive and wilderness areas. Marshall "insisted on throwing some kind of a boundary around every sizable piece of National Forest that didn't happen to have a road on it." (Cleator 1955: 2) Under Marshall's direction, the U Regulations granted much greater protection to wildernesses by prohibiting logging, road construction, and special-use permits. Also prohibited was the use of motorboats or the landing of aircraft except in emergencies or for administrative purposes. Marshall wanted the Three Sisters Primitive Area boundary expanded to the west to encompass the French Pete drainage. The boundary was changed, but the decision was revisited many times during the 1960s and 1970s as part of the French Pete controversy (see Chapter 8).

Marshall and Sieker visited the Region in September of 1939 looking for large areas of undeveloped forest land that could be designated as primitive or wilderness areas. The first area Marshall and Sieker inspected, along with Forest Supervisor Verne Harpham and District Ranger Fred Asam, was along the western slope of the Umpqua NF. Cleator played devil's advocate on this tour, arguing that the Umpqua NF should "leave commercial timber supplies for posterity ... Bob [Marshall] had great hopes but of course he was not expecting to succeed in setting aside all these vast reservoirs of timber as wilderness. His big argument was to stop road building in all these great contiguous areas until their Roadless and Wilderness values could be settled and agreed upon." (Cleator 1955: 4) Wildernesses on the Umpqua NF were not established until 1984, but several were within the old boundaries that had been mapped by Marshall and Cleator almost 50 years before.

> Marshall and Sieker then visited the Glacier Peak Primitive Area on the border between the Mt. Baker-Snoqualmie and Wenatchee NFs. Cleator, Wenatchee NF Forest Supervisor Gilbert Brown, ex-supervisor Hal Sylvester, Mt. Baker NF Supervisor Charles Flory, and five others joined the party. Cleator was interested in including Marshall into photographs of scenic vistas, but "each time he backed away remarking to the effect: 'We are privileged to look at a billion years of the Creator's work. Why should I, a mere human, have the temerity to compete with this masterpiece?' I had to respect such logic so gave up except for a few distant shots." (Cleator 1939: 9)

Marshall drew maps of the area and recommended that it be reclassified as a wilderness, but that decision would have to wait until 1960. The party arrived at Wenatchee on September 21, where at a local restaurant, "Bob contented himself with a double hamburger of raw, not rare beef. Next morning in the hotel Bob was too ill to get out of bed ... [later in the day] he was having convulsions ... [and was] put in the hospital at once." (Cleator 1939: 11) Six weeks later Bob Marshall died of a heart attack on November 11, 1939.

Wildlife and Fish Management

Game management in Region 6 continued to be most concerned with huntable and trappable species, especially deer and elk. Reports by the Region to the Biological Survey contained valuable information on the health of elk herds, status of predators, and winter kills. A cooperative study with the Washington State Game Commission and the Biological Survey on the Olympic Peninsula in 1935 looked at the Roosevelt elk and other big game species. Several observational shelters were built in three river valleys, which were probably the Dungeness, Hoh, and Quinault. Six salt licks and nine forage plots were established to study the feeding habits of elk. The first study report the following year indicated that virtually all elk killed by predatory animals were killed by cougars. Coming in second behind predators, diseases caused the majority of other elk deaths.

Emphasis on habitat management began to grow in the Region. In the 1930s, every forest in the Region was assigned the task of surveying for big game. Usually, this was completed in winter or early spring on cross-country skis. As usual, the Regional Office and the states were looking for unusual outbreaks of starvation, disease, and death in deer and elk species. There are many reports of ranger district personnel feeding big game, as well as birds such as quail, during heavy snow storms when the animals were unable to find food under the deep snow blanket. The Deschutes NF, in late 1936 or early 1937, conducted a survey of the Davis and Browns Creek area looking for a small population of white-tailed deer, and found 18 along Davis Creek and the Deschutes River. It was believed that as many as 50 were present on the forest. It remains unknown how the deer came to the area, as the only remaining populations of the Columbian white-tailed deer on the westside of the state are located along the Columbia River and in Douglas County, while the eastern-most forests in the Region contain a number of white-tailed deer, as well as mule deer.

Efforts to reintroduce beavers became a priority in 1935-37 for the Ochoco NF. Beaver had almost been exterminated by the fur trappers in the early 1800s, but by the early 1900s had regained some of their original territory. Yet legal trapping by the middle 1930s once again almost wiped out the species. A small number of beavers were trapped from areas where they were causing damage and relocated to other streams that needed restoration. Ranger Chris Criswell related a story about an accidental beaver transplanting on the Chelan NF in 1936:

> The State trapper came into the [Winthrop] ranger station late one night. He asked [District Ranger] Frank [Burge] if he could leave a beaver somewhere over night. He would pick it up and take it to a holding pen on the river down towards Twisp the next morning. Frank thought a moment and said, "Sure, let's use the office basement." Now a beaver can't 'go' unless they are in the water, so Frank sat a galvanized wash tub on the basement floor and filled it with water. He placed a piece of plywood so that the beaver could get up into the tub. The next morning Frank ran downstairs to see how the beaver was doing. We could hear him cussing and hollering. Frank had forgotten the outside door was wood. Mr. Beaver had gnawed a big hole through that door and headed for the Chewack River (Chriswell 1989: 15).

The Forest Service established a division of wildlife management in 1936. Several studies documented the role of predators in relationship to wildlife numbers. In part these research efforts were sparked by some disastrous population explosions in areas like the Kaibab Plateau in Arizona in the mid-1920s, due to overprotection of the deer and extermination of their predators.

The Region worked cooperatively with the USDA Biological Survey to eradicate predatory species that killed the deer and elk on the national forests by trapping and poisoning. The State of Oregon, at the time, paid a bounty of $30 for every wolf killed. The last of the wolves were exterminated from the national forests in the Region by the late 1930s. Coyotes were also killed in great numbers; cougars were targeted in all areas; and bears were shot and trapped across the Region. One account on the Olympic NF illustrates the war against predators: "predatory animal hunters on the Olympic Peninsula have waged a successful war against the cougar and bobcat this past winter [1935-36], a total of 38 cougars and 246 bobcats having been killed within and adjacent to the Forest since January 1. A bounty of $25 for each cougar and five dollars for each bobcat is paid by the Washington State Game Commission to hunters holding predatory animal hunters' licenses." (*Six Twenty-Six*, June 1936: 6)

The forests of the Northwest have been used by hunters and anglers alike since the late 1800s. Poaching of big game animals during the Great Depression was a constant problem—yet none of the killed animals were wasted. They filled the larder of families who were in need of food at a time when jobs were hard to find. Usually the district rangers were assigned the job of acting as game wardens on the remote national forests. One story about a poacher and a ranger occurred at Blue River near the Willamette NF. Manena Sparks Schwering told of an incident on the Blue River covered bridge:

> [D]uring the winter [probably in the mid-1930s] when meat was needed one of the local men [Dexter Sparks, Manena's father] crossed the McKenzie by boat. ... By late afternoon he had a nice small buck, but with the short winter day almost over, was unable to get home by dark … He decided to … simply to walk down the main street of Blue River and home with his deer on his back. ... He was just well entered into the covered wooden bridge at Blue River when he heard the unmistakable sound of a horse coming up the approach on the opposite end of the bridge. ... He was too heavily loaded to back out quickly ahead of the horse and rider, and he was certainly in no position to exchange a friendly greeting. So, doing the only thing that occurred to him ... he backed quietly against the side of the bridge and leaned back ... the weight of the deer pulled him off balance and down into the space between the guard rail and the bridge wall. He was thus jack-knifed [and] ... could not free himself. ... It was subsequently learned that it had been the Ranger [Smith Taylor] returning home and the large smear of blood on the bridge was a reminder that brought chuckles for weeks afterward when noticed by family or friends (Schwering n.d.).

Generally, the largest deer and elk are found in the high mountains. While individuals often drove their vehicles into the mountains to find a camp spot, others travelled on horseback, usually with guides, to find the best hunting locations. In the early winter of 1945, a sudden snowstorm in the Blue Mountains of Washington changed the hunting situation for scores of elk hunters. Gerald J. Tucker, district ranger at Clearwater RS on the Umatilla NF, wrote about the terrible snow storm and rescue of the hunters:

> The elk season that opened in the Blue Mountains of Washington, November 4, 1945, was destined to be one that nobody connected with it would ever forget ... And even the weatherman had no real inkling of what the storm would be like or how long it would last … On Friday, when it had raged for more than 48 hours … I knew then we were in for trouble … [Packers tried to get their clients out, but the snow drifts prevented them from leaving the mountains.] At this point, we encountered an obstacle of a kind that often hampers such rescue operations and raises touchy questions. How much aid

> are lost or stranded hunters entitled to expect from public agencies? [This is still a concern sixty years later.] How far can county highway departments, state game agencies, or the U.S. Forest Service go in spending money on search-and-rescue work? ... [The county and Army got involved with snow plows and supplies.] At 7 o'clock that night, Monday, November 19, when the storm had raged for 13 days, the trapped hunters drove the muddy road into Pomeroy, where almost no snow had fallen. There were no serious casualties ... It was 2 p.m. on Tuesday November 20, when we finally reached Pomeroy with the last of the hunters and vehicles (Tucker 1964: 20, 22-23, 107).

Fish management continued much as it had during the previous period. Fish stocking in mountain lakes was still a priority. Fingerlings were obtained from state fish hatcheries, while stocking the young fish in the more remote areas had to rely on Forest Service rangers and CCC men, as well as packers and sheepmen, to transport the special fish cans on mules and horses. From 1933 to 1939, CCC enrollees planted 4,145,650 fish fry on the national forests in Region 6. The forest with the largest plantings was the Fremont, with some 1,380,000 fry planted, followed by the Chelan (570,000), Whitman (490,000), Columbia (372,000), and Olympic (322,250). The other forests made up the remainder. One interesting story about transporting the fish on pack animals came from the Siskiyou NF:

> When the Agness district began its program of fish planting ... the boys found that fish in transit via water can, must be kept splashing about in order to preserve their supply of breathable oxygen. Therefore while loading the cans of fry, those pack animals already loaded could not be allowed to stand still because of possible suffocation of the fish, and Art [Peterson] was going to be more than helpful—he grabbed a handful of rope and started to lead out. The water splashed on her [the pack animal's] neck plus the rattle of the cans in their frames was too much for ole "Blue" and she took off—dragging the disgruntled Peterson behind—across the cattle guard, into the driveways and part way back before coming to a snorting stop (*Six Twenty-Six*, August 1939: 14).

In early 1938, the new salmon and trout hatchery at Tyee Springs was opened seven miles above the Hemlock Ranger Station on the Wind River Highway. It was a cooperative effort by the CCC to build the hatchery, the Bureau of Fisheries to provide money and expertise, as well as run the operation, and the

Columbia NF to host the site. In the first year saw 3 million Chinook salmon and 600,000 trout raised at the hatchery. The adult salmon, used for their eggs as well as food for the young fishes, were caught in the Little White Salmon River. This hatchery is still in operation today.

The Mt. Baker NF in the late summer of 1941 experimented with planting fish by airplane. On August 24, 16,000 cutthroat trout fry were dropped by airplane into Crater Lake, a small mountain lake on a branch of the Suiattle River. The operation was carried out through cooperation with the Wildcat Steelhead Club of Sedro-Wolley. The fish were dropped in gallon cans that were equipped with a block of cedar attached to the bottoms of each can. The wood served as a float to invert the cans, thus "liberating" the fish fry.

Range Management

Grazing continued in the time-honored manner, but gradually became less important as an economic activity in the Region. The major reason was a decline in the amount of grazing land. Grazing camps that had formerly supplied 30 days grazing now could supply 16 days at most; meadows formerly covered with grass were now full of lodgepole pine. Studies made in 1933 and 1934 indicated that tree reproduction was encroaching on range land at the rate of 2.5 percent per year.

On July 18-20, 1936, a "show-me trip" on the Colville NF highlighted controversies between domestic livestock versus wildlife. The group of about 30 prominent sportsmen, stockmen, newspapermen, and conservationists traveled some 250 miles by car and another 21 miles on horseback inspecting the ranges. "After this trip there were none present who could offer any suggestions for the betterment of our range conditions." (*Six Twenty-Six*, August 1936: 9)

In May of 1936, the Region scheduled a range-management training camp at the Ochoco RS east of Prineville, Oregon. A news release on May 5, 1936, that could have been written 70 years later, noted that:

> the training camp will be under the general direction of A. H. Hodgson, assistant regional forester in charge of personnel management, assisted by R. C. Lindberg, training supervisor. Subjects scheduled for discussion will include the stockman's problems; social and economic problems involved in the use of range lands; plant identification; soil erosion; range survey; range inspection; management plans; wild life management; beaver transplanting; and stream and lake management.

There was still a need for forest managers to become familiar with the grazing allotments on each forest. Chris Chriswell, on the Umatilla NF in 1938, related a story about the Tamarack cattle and horse allotment:

> I hadn't realized the darn cow range was so big. We traveled about 25 miles that day to cow camps, salt grounds, important forage areas and water improvements. All the time Henry kept up a running commentary on everything. He gave personality sketches of all the permittees. The man was a walking encyclopedia. All the time he was effortlessly slouched in the saddle. I walked the last 5 miles back. It was several days before I could sit comfortably. Henry hadn't been trying to show off. He hadn't tried to wear me down. He had just simply and seriously showed me over the allotment. Wasn't that what I had asked him to do? (Chriswell 1989: 34)

By 1945 the grazing activity that had once been the dominant economic activity in the Region was much reduced. Relations between the Forest Service and grazers, herders, and camp tenders remained cordial, even friendly. Camp tenders aided the Forest Service in stocking mountain lakes with fish, and on occasion herders and camp tenders reported on and assisted in putting out fires. The Forest Service, with CCC help, continued range improvements. Range research continued; the early studies made by John Kuhns in 1916 were used as a basis for study of changes in carrying capacity. Most of the allotments showed a drastic reduction of available feed. After more research some areas were opened to cattle and horses and closed to sheep.

Engineering

In 1932, a track-mounted tractor equipped with a blade received Forest Service approval, mostly for trail construction. Soon there were many manufacturers of such equipment and the "bulldozers" or "cats" (short for Caterpillar, one of the companies building the bulldozers), as they became known, were readily used to build roads into logging areas. Gravel roads and bulldozers caused truck hauling to supplant railroads and river log drives for log transportation.

The Regional engineering division in 1930 began a project to develop a transportation plan for fire control purposes. The plan was completed within three years. It was more than adequate for fire planning purposes during the CCC era, but proved inadequate after World War II when roads were needed to access the timber in the Region. Ray Grefe wrote that the transportation plan "paid dividends in systemizing road programming" during the 1930s. He also noted that engineering "worked out a system of visibility mapping from lookouts, using panoramic photos. This proved valuable in assaying the effectiveness of existing lookout points and selecting additional points needed to fill the gaps." (Grefe 1990: 276)

During the Great Depression, engineering in the Regional Office grew in terms of workload and work force. Ward Gano recalled that prior to the mid-1930s "Division of Engineering had been organized primarily to handle mapping, surveillance of Roads and Trails activities on the Forests, and liaison with the Federal Power Commission (FPC) on water power projects." (Gano 1990: 499) That situation began to change in 1934 when Regional Engineer Jim Frankland "set up an architectural section headed by W. I. 'Tim' Turner ... This unit developed standard plans for offices, warehouses, guard stations, shops, residences, and others, in a distinctive 'Cascadian' architectural style, for construction by the CCC forces. The same group ... later developed the design and plans for the Timberline Lodge WPA project." (Gano 1990: 500)

Ward Gano from the Regional Office was the leading figure in the design and construction of Timberline Lodge. The initial survey was made on the site in 14 feet of snow during the spring of 1936 by Gano, who became the resident engineer, assistant architect Linn A. Forrest, and landscape architect Emmett Blanchfield. Forrest worked on the design drawings for the lodge while Gano participated in the structural design. Fortunately, the summer was a very long one, enabling the workers to get the building enclosed before winter and the worst of the cold weather. The Regional engineering division also developed a bridge and structural design section, that designed and oversaw construction of the Magic Mile Ski Lift at Mt. Hood in 1938-1939.

Regional Office's division of engineering was placed in charge of establishing the CCC camps and projects. Ray Grefe documented the work:

> [S]election of campsites, construction of camps, procurement of tools and heavy equipment, approval of work programs, allocation of funds, and approval of expenditures. The Division of Engineering provided overall supervision, guidance, and inspection. Elliott Roberts was detailed to the State of Oregon to head up its program and Lloyd Brown to the State of Washington. Three men in Engineering were authorized to approve CCC vouchers. A stock of vouchers 12 to 24 inches high was par for the course day after day ... The architectural section under W. I. (Tim) Turner did an excellent job in designing a new style of forest architecture to replace the conventional gray buildings up to this time.

In the fall of 1934, the Regional Office established special recreation schools that focused on recreation structures. These three-day schools were oriented to the CCC camp foremen and superintendents, and Forest Service supervisors and rangers, as well as Regional Office representatives. The school covered such topics as "Camp ground design, site planning in detail, and the great variety of improvements made up the heaviest course of study. Discussions on … how to keep toilets clean without nailing shut, cautious clearing, reducing sign nuisance, etc." (*Six Twenty-Six*, December 1934: 29)

One unique structure built in the high country was the Dee Wright Observatory on the Willamette NF, a tribute to one of the unsung workers of the Forest Service—the packer. Horse and mule transport were universal in the early days, as the packers and pack animals carried supplies to camps, carried in firefighting equipment, and stocked mountain lakes with fish fry. Some of the packers were legendary story tellers. Dee Wright was one of these men. A short, stocky man, he regaled the CCC boys with his stories. In 1917, while on the Mt. Hood NF, he had been kicked in the chest by his favorite mule, "Dynamite." A heart attack took his life in 1934 while he was rowing a group of CCC men across the McKenzie River. The lava rock observatory at McKenzie Pass, then under construction, was named for him.

Several of the virtually unknown functions of the division of engineering were the Regional Radio Laboratory, the Sellwood Shop, and the Regional Equipment Laboratory. The Radio Lab was established in 1931 at Tacoma, Washington. Its function was to develop portable radios for Forest Service field work. By the mid-1930s the Radio Lab was moved to Vancouver, then to the Sellwood area of Portland, and contracted work with Radio Specialty, an electronics company in Portland. Howard Vollum, a 1936 graduate of Reed College, worked for Radio Specialty. He and several Radio Lab employees established a gigantic electronics firm—Tektronics (Gray 1982 and Mayer 2005). Lawrence Waggener related how the Sellwood Shop came into existence:

> A lengthy search produced [located] the idle streetcar barns and two-story office waiting room building owned by the Rose City Traction Company … In the fall of 1939, the Forest Service purchased these properties … for $37,500. Regional Equipment Pool and repair facilities were carried on at these premises from 1939 until early 1953. The work grew … to 101 shop and 15 equipment pool personnel during the 1951-1952 [1953] Korean War. In addition to usual Forest Service equipment needs, the organization was repairing and

> refurbishing all kinds and types of military equipment (amphibious vehicles, tankers, 10-ton wreckers, trailers, and tank retrievers, etc.) (Waggener 1990: 491).

The Equipment Lab located at the Sellwood Shop was under the supervision of Ted P. Flynn. Ray Grefe described the operation that was designed to develop very specialized pieces of equipment that were unavailable through commercial manufacturers:

> This work ... resulted in production of a variety of items to serve road and trail construction, fire control, timber management, and other activities. Among them were the trail compressor, trail tractor, trail beetle [a midget trail tractor that was later adopted by the Army for airborne troops], trail mule, trail grader, and trail patrol grader (the latter developed by L. A. Waggener). Road equipment included a portable rock crusher, rock ripper, front end and motor patrol blade, and a portable chipper. The original heavy logging tractor [called the Tomcat] and the sno-motor [also adopted by the Army] were early development items. Most of this equipment served its purpose well and was later taken over in whole or in part by commercial companies. No attempt was made to patent any of these items, the purpose being to try them out and get them into commercial production if possible (Grefe 1990: 277).

Fire Management

Comprehensive fire protection had developed in the 1920s, aided by a series of federal acts that laid the basis for federal-state cooperation in forestry, especially fire protection. In the 1930s, as we have seen, the CCC built miles and miles of fire breaks and fire lines and truck trails, and provided thousands of men to fight fires.

Twenty-, thirty-, forty-, or fifty-men crews were organized and were given special training in firefighting. They could be rushed to any fire by truck. The "one-lick" method of firefighting was developed on the east side of the mountains, and then applied to the west side. It is essentially an application of the assembly-line technique to building a fire line. Instead of one firefighter being responsible for a segment of the line, a crew working at intervals of 8 to 20 feet worked as a team to build the line. Moving like a long snake through the forest, these specialized crews were able to build a fire trail rapidly and effectively. Many miles of access roads, partly to further timber sales but also for fire control, were built in this period, as were hundreds of miles of new trails. The crawler tractor and bulldozer were increasingly used to build fire lines where terrain permitted, and planes were used for fire spotting and for dropping tools or food.

On August 14, 1933, a fire was accidently ignited during a logging operation in the Coast Range of northwest Oregon during a time of low humidity and strong easterly winds. It quickly spread to 40,000 acres; 11 days later the fire exploded or "blew up" to almost 250,000 acres in less than 24 hours. A column of thick smoke from this Tillamook Fire deposited ash on ships as much as 500 miles away. Heavy ash and cinders fell on Oregon beaches. Rain halted the spread of the devastating fires and they were finally extinguished with the help of the CCC on September 5 after burning around 240,000 acres. Simultaneously, the Wolf Creek Fire started on August 24. Located just north of the Tillamook Fire, it burned a little over 43,000 acres. Portions of the Coast Range forest in the Tillamook area burned in 1939 and again in 1945. Three years later Oregon voters passed a constitutional amendment authorizing $12 million in bonds to rehabilitate the Tillamook Burn. A huge reforestation project began in 1949, using helicopters to distribute seeds. More than 72 million seedlings were planted by ground crews including, state forestry workers, prison inmates, and many volunteer school groups. Much of the Tillamook Burn became the Tillamook State Forest on July 18, 1973.

In 1935, the Forest Service adopted the "10 a.m." policy: as soon as a fire was spotted, it was to be controlled by 10 a.m. the following morning, or if that failed, then 10 a.m. the next morning, until the fire was out. Every fire, large

Francis Lufkin preparing for experimental jump at Winthrop, Chelan National Forest, Washington, 1939. USFS photograph.

or small, was treated the same way (Pyne 1982). This policy persisted in the Forest Service for decades. "The 10 A.M. policy guided federal fire protection until the National Park Service broke ranks in 1968. Ten years later the Forest Service altered the old policy to allow for natural fires to burn in wilderness sites and for controlled burning to reduce fuels and improve habitat." (Pyne 2001: 268)

In the mid-1930s, the Regional Office funded a region-wide project to photograph every forest for fire purposes. The project, began in 1933, used special panorama cameras that could slowly swivel 360 degrees. The camera was invented by William Osborne, the same person who invented the Osborne fire finder. Albert Arnst from the Regional Office was project director. He and six two-man crews traveled extensively through the Region taking photos. Before the project ended in 1938, panorama photos (three from each point) had been taken from every high point and peak—14,000 photos from lookout houses, towers, and high places where lookout houses were never built, like Mt. Thielsen, and 200 photos from emergency lookout sites. Many times the photographer climbed existing lookout houses and towers—sometimes even on the roof—to get the view as seen from the lookout. These 160-degree photos (with a small overlap with adjacent photos) were then inscribed with horizontal azimuth and vertical lines. The 5-inch tall by 39-inch wide photos were mounted in special three-fold, heavy paper folders. Each photo folder offered a complete panorama of what was seen from that location. When the photo folders were returned to the forests, duplicate copies were sent to every lookout. With these panorama photos, a lookout could exactly describe where a smoke or fire was located, thus more quickly enabling firefighters to reach the scene.

Every lookout ground house and tower seems to have its share of stories. Early in the period there were still a few lookout trees, but by the mid-1930s most had been replaced by buildings with many modern conveniences including radios, stoves, beds, stools, and limited storage. Outhouses were typically some distance from the lookout and fire tools were usually kept in a separate building or shelter.

The Monument 83 lookout house in the Pasayten Wilderness, Wenatchee NF, located almost on the U.S.-Canadian border, is actually on Canadian soil. A news release dated October 24, 1937, reported that "last year representatives of the Canadian government requested that the American flag not be flown from Monument 83 lookout house. Needless to say, the request was complied with … so a flagpole was erected some 125 feet … on the American side, which made it eligible for the stars and stripes." An unusual incident, as recounted by Ray Kresek, involves the lookout at Monument 83, who was not named:

> During his final season, he was standing on the tower catwalk "watering the flowers" at the very moment that Thor decided to eject a lightning bolt his direction. It struck the roof five feet away and burned a hole in it … When the time came to board up the windows and head south for the winter, the old-timer refused to … He said he would walk the 54 miles to town … After a week [he was found] … curled up inside the dark, boarded-up cabin at Monument 83. He refused to budge, and the government couldn't evict him … for he was on international soil. (Kresek 1988: 187)

In the 1930s and 1940s many telephone lines used by the lookouts on mountain peaks were still widely used in the national forests in the Region. These telephone systems were used by many homesteaders and even remote resorts as the only way to talk with people outside the national forests. Radio sets were experimented with in the 1930s, but for the most part they were unreliable, especially in steep, mountainous terrain. Wendall Jones (personal communication) reported that "old Fire Control Officers had no faith in radios and liked the privacy of talking direct on the phone to his lookouts and guards, even though the phone line was a party line at times."

Because many of the forest fires in the West were started by lightning in inaccessible locations, the Forest Service experimented with firefighters parachuting to fires before they became large and out of control. The first experimental jumps began in 1939 at Winthrop, Washington, on the Chelan

Telephone line repair, Snoqualmie National Forest, Washington, 1921. USFS photograph.

NF. "A total of 58 jumps were made over terrain including airport, meadows, open hillsides, subalpine forests, lodgepole (tall and thickets), tall timber on flats and hillsides, and open and timbered ridges. Accuracy ranged from a 'bullseye' to a distance of a quarter of a mile from the target." (*Six Twenty-Six*, December 1939: 18) By the summer of 1940, the smokejumpers, as they became known, were operating out of Winthrop and the Moose Creek RS on Montana's Bitterroot NF and made their first jump on a fire on the Nezperce NF in Idaho (Moody 2007). The successful operation proved that smokejumping into remote, rugged areas was feasible. During World War II, the lessons learned from smokejumper training methods of jumping into heavily forested areas would prove useful to the new military paratrooper units like the 101st Airborne.

The Regional Office newsletter had a short article about how law enforcement was becoming a part of fire control operations. The report came from the Siskiyou NF:

> The past two years … we have had a special man assigned to law enforcement work, in the person of Lee Blevins … Lee probably is especially gifted in this work. … This past year he has been successful in digging up evidence on last year's [1933] fires, which in one case brought an indictment, and in another case was sufficient for the penitentiary board to withdraw their parole on a man who was let out on [after] serving a 4-year sentence … Mr. Blevins has been successful in convicting three of the worst offenders on the Siskiyou Forest (*Six Twenty-Six*, February 1934: 15-16).

As in World War I, many Forest Service men were called to the armed services in World War II. The shortage of field men was met by volunteers. The call went out in state and federal governmental publications and in the newspapers for volunteers. For example, *The Oregonian* for March 22, 1942, carried headlines reading "On the Vast Wilderness Tract Foresters Are on the Lookout for Bombs and Enemy Fire Bugs." Leon F. Kneipp of the Washington Office told of the need for lookouts to work with state and federal foresters and for 40-person fire crews. Forest Service Reserves, as they were called, were wanted for temporary and even a few permanent positions. Women, older men (non-draft age), and even high school boys were actively recruited. A news release on April 14, 1943, indicated that 246 women had been hired,

> [T]o fill forest fire protection positions next summer on the national forests in Oregon and Washington … School teachers and wives of

> men in the military services have been employed in large numbers. The women will be stationed on lookouts to detect fires, or will patrol campgrounds, register forest visitors, drive fire trucks, serve as alternate fire dispatchers, cook for fire crews, attend the telephone switchboard and do clerical work in rangers' offices. At lookout stations where the fire guard's duties are limited to fire detection, and do not involve searching for and fighting forest fires, women lookouts will be employed ... One fortunate forest ranger has reported that he has employed a versatile young woman who can handle his pack string of mules, drive the fire truck, operate a fire pump, write shorthand and type his official letters.

Men between the ages of 18 and 60—the lower age later dropped to 16—who were not eligible for the draft were also called upon to apply for work. Outdoor clubs like the Mazamas and Mountaineers called on their members to use vacation time to work for the Forest Service or state forestry departments. On the Deschutes NF, 365 boys registered for a weekly one-hour Fire Emergency Corps course that emphasized forest fire safety and suppression of small fires. Most of the forests in the region had similar programs. Volunteers staffed 32 fire stations in the Mt. Hood NF and one in the Siuslaw NF in 1943.

SMOKEY BEAR PROGRAM

The cooperative forest-fire prevention campaign—a joint venture of the Forest Service and state forestry officials—was organized during the war, when it became vitally important to protect the nation's timber supply. In 1942, forest managers' concern about forest fires was reinforced by a Japanese submarine's shelling of an oil field near Santa Barbara, very close to the Los Padres NF. Ongoing war efforts drained firefighters and heavy equipment. Protection of the nation's forest resources became a matter of national importance. The Forest Service wanted to encourage the general public to participate in forest-fire prevention. The first step was taken and the Cooperative Forest Fire Prevention Campaign was formed.

During 1942 and 1943, wartime slogans were used on forest-fire prevention posters. Walt Disney's Bambi was used for the poster distributed in early 1944 and was a great success. Because of the growing popularity of the program, the Forest Service and Wartime Advertising Council decided to choose an animal to represent forest-fire prevention. Smokey Bear was described by Richard Hammett, director of the Wartime Forest Fires Prevention Program, for the first time on August 9, 1944, as having a "nose short (Panda type), color black or brown; expression appealing, knowledgeable, quizzical; perhaps wearing a campaign (or Boy Scout) hat that typifies the outdoors and the woods." Blue jeans were added later. The bear was named "Smokey" after "Smokey" Joe Martin, who was the assistant chief of the New York City Fire Department from 1919 to 1930.

Albert Staehle was asked to paint the first bear, which was completed in 1944 and distributed the following year. This first poster of Smokey showed him pouring water on a campfire. In 1946, Rudolph "Rudy" Wendelin left the Navy and returned to the Forest Service as an artist, working closely with the volunteer advertising agency on Smokey Bear posters. Rudy become one of the best-known Smokey Bear artists and soon became known as the "caretaker of the Smokey Bear image." After his retirement in 1973, Rudy continued to paint Smokey and act as a program consultant.

On June 27, 1950, a badly burned young bear cub—the most famous survivor from a massive fire on the Lincoln NF—was moved to the National Zoo in Washington, D.C., where he became the living symbol of Smokey Bear. In May 1975, the original Smokey Bear was retired from public duties. After he died quietly the following January, Smokey II took his place. In the summer of 1990, Smokey II died. There are no more living Smokey Bears at the National Zoo.

Research Work on the National Forests

The Willamette NF became a major center for research to study logging methods in old-growth Douglas-fir ranging from pure stands 170 years old to almost 500-year-old trees. Experiments showed the need for flexibility; no one system could be applied overall to a large area (Geirer 2007, Skovlin 1991).

The 7,520-acre Pringle Falls Experimental Forest was established in 1931 on the Deschutes NF. It was enlarged in 1936 with the addition of the Lookout Mountain unit, making a total area of 11,058 acres. The experimental forest was established to provide demonstrations of silviculture as well as insect and disease control on the ponderosa and lodgepole pine forests of central Oregon (Youngblood 1995).

The Wind River Experimental Forest was formally established in 1932, although studies had been conducted for more than 20 years at the site. Thornton T. Munger had come to the Wind River area in 1909. He established a tree nursery, an arboretum, growth and yield plots in Douglas-fir sites, and a tree heredity study. Research work at the 10,000-acre experimental forest has concentrated on studying the old-growth Douglas-fir forests, as well as

Wind River nursery Douglas-fir research study plot, Columbia National Forest, Washington, c. 1930. USFS photograph.

silvicultural methods to manage them (*Forest Worker* 1932). A research natural area was established at the site in 1926, and was renamed for Munger in 1977.

The Cascade Head Experimental Forest was established in 1934 on the Siuslaw NF (Adams et al. 2004). The land around Cascade Head along the Oregon coast was the historic home of the Neschensney or Neschesney tribe. Under the Indian Reorganization Act of 1934, about 700,000 acres on Cascade Head were sold in the mid-1930s to the Portland YWCA, and some came eventually into the Siuslaw NF (see Chapter 7). The Siuslaw NF Web site described the unique area:

> When the Cascade Head [Experimental Forest] was established, the area was primarily covered with a forest that grew up after the huge Nestucca Fire of the late 1840s. Stands of spruce and western hemlock that survived the fire are found in the Neskowin Crest Research Natural Area. The Nechesney Indians burned some of the forest close to the ocean in the early 1900s. Some of the more gentle country to the east, homesteaded by European settlers and abandoned in the early 1920s, now supports evenaged, single-canopy forests with dense shrub understories. The forested ecosystems include productive young, mature, and old-growth stands of Sitka sprucewestern hemlock and Douglas-fir forests with riparian areas and streambanks dominated by red alder. Western redcedar is found occasionally. Some of the highest growth rates and greatest volumes per hectare for any temperate forest

in the world are reported for this area. Experimental clearcutting, shelterwood cutting, thinning, and salvage from large windstorms have affected about 25 percent of the forested area.

The 28,000-acre Starkey Experimental Forest and Range was established by the acting Chief on July 11, 1940, on the Umatilla NF. The Starkey Experimental Range, as it was originally called, was the first such experimental range in the Region, and worked cooperatively with the Oregon Department of Fish and Wildlife. The area within the fenced 40 square miles was considered typical of rangeland in the interior West. Research at first emphasized range management, including livestock grazing methods and rangeland improvement. There have been major studies of the interactions of grazing animals, natural vegetation, elk and deer, hunting, and riparian management (Adams et al. 2004, Rowland et al. 1997, Skovlin 1991).

RESEARCH NATURAL AREAS

Research Natural Areas (RNAs) were newly established on several national forests in this period. RNAs, as opposed to experimental forests and ranges, are designed to preserve rather small, intact areas on the national forests as examples of different ecological niches for both scientific research work and for general educational purposes.

During this era, the following RNAs were established:

- Metolius RNA on the Deschutes NF was established in 1931 for the study of ponderosa pine forest types on flat and steep slopes;
- Quinault RNA on the Olympic NF was established in 1932 to preserve tracts of western hemlock, Sitka spruce, western red cedar, and Douglas-fir;
- North Fork Nooksack RNA was established in 1934 on the Mt. Baker NF as an example of Douglas-fir and western hemlock at mid-elevations;
- Wind River RNA was established in 1934 on the Columbia NF to study the old-growth Douglas-fir and western hemlock forests;
- Ochoco Divide RNA was established in 1935 on the Ochoco NF for the study of mixed conifers;
- Pringle Falls RNA located on the Deschutes NF was established in 1936 to preserve tracts of ponderosa pine and lodgepole pine with bitterbrush communities;
- Port Orford Cedar RNA on the Siskiyou NF was established in 1937 as a sample of old-growth Port Orford cedar;

- Neskowin Crest RNA was established in 1941 on the Siuslaw NF for the Sitka spruce and western hemlock stands growing adjacent to the ocean;
- Goodlow Mountain RNA on the Fremont NF was established in 1942 to exemplify the transition from sagebrush to ponderosa pine;
- Coquille RNA located on the Siskiyou NF was to provide examples of old-growth Port Orford cedar stands.

On June 10, 1938, a 3,000-acre special botanical area within the boundaries of the Agness Ranger District of the Siskiyou NF was set aside by the chief of the Forest Service. The Big Craggies Botanical Area was designed to protect the plant *Kalmiopsis leachiens*, a dwarf rhododendron found only in this location. "Plant collectors and propagators have threatened to exterminate this rare species by their activities in gathering the plant during the last year." (*Six Twenty-Six*, August 1938: 14)

Chapter 6

Ramping Up Production 1946–1959

When World War II ended in August 1945, the nation was ready to move forward with a renewed energy. The Great Depression was over, the United States (with allies) the victor of a two-fronted war, prosperity was "just around the corner." There was nothing that could not be conquered, no matter how small or large. Forestry schools around the nation were training thousands of new foresters and other technically trained professionals under the auspices of the GI Bill. As Forest Service budgets increased, the forests in the Region began hiring these new graduates in great numbers. Then there was a lull in hiring as the new positions were all filled, but during the mid and late 1950s hiring again increased. For the Forest Service, and the Bureau of Land Management (BLM), this meant opening some of the federal forests to higher and higher timber harvest levels, intensive timber management, miles and miles of new forest road construction, and an upward spiral in outdoor recreation use.

Until the end of the war the timber industry had viewed the national forests as huge sources of cheap timber that needed to be kept off the market so that they could keep private timber prices high. But after the war the industry wanted more national forest timber to supplement heavily cutover private forest lands. The opening of the national forests to timber harvesting and increased road development after World War II to meet the needs of the growing housing market would have consequences for the future.

This was an era of inventories and planning, particularly in timber and recreation. There were analyses of workloads and administrative capabilities to utilize available workers to do their jobs better and be more cost effective. This era was the real beginning of intensive forest management—multi-year timber action plans, insect and disease control efforts, controlling competing vegetation in replacement tree stands, and establishing seed orchards to grow better trees. The Region had practiced extensive forestry from Pinchot's era, but now intensive forestry was being applied at a large scale on all ranger districts on every forest in the Region. This new commodity-focused period in Forest Service history is sometimes referred to as the "hard hat" era. Tasks were accepted and taken on by the rangers and field workers with gusto. In the aftermath of World War II, there was nothing that could not be accomplished with enough money and manpower.

This important era set the stage for the coming of environmental laws and regulations of the 1960s and 1970s. The Forest Service during this postwar era was internally focused. The agency did not hear what the public wanted or could not respond quickly to change, and this would repeat itself in the 1970s as the seemingly endless wilderness debates hit the agency head-on (Chapter 8).

Regional Foresters

Horace J. Andrews became regional forester in the Pacific Northwest Region in 1943 (see Chapter 5). His tenure was marked by the successful war effort and the postwar boom, the rush to open the national forests for rapid development through increased timber sales and accompanying road construction for forest access and truck hauling of logs. In addition, the Sustained Yield Management Act of 1944 led to the legal agreements after the war for the Shelton Sustained Yield Unit in Washington and several federal units in Oregon and Washington. Andrews served until his untimely death in a car accident on March 24, 1951, and was succeeded by J. Herbert Stone.

Stone had broad experience in the Forest Service from 1926 to 1967, including as regional forester in the Southern Region (R-8) in 1946-1951. In April 1951, Herb Stone became the regional forester for the Pacific Northwest Region. His time in this role was marked by increasing timber sales, the exchange of some O&C lands in the western Oregon forests with the BLM in 1956, passage of the Multiple-Use Sustained-Yield Act of 1960, the Wilderness Act of 1964, and the National Historic Preservation Act of 1966. Stone retired on June 2, 1967, and was succeeded by Charles Connaughton (see Chapter 7).

Regional News

A Regional Office news release in early 1946 noted that the Region's 19 national forest supervisors held their "first post-war meeting in Portland, Oregon, March 19-27 … Timber management takes the spotlight position on the agenda." Regional Forester Andrews, it was reported, said, "Fullest possible use of national forest timber to meet the needs of community stability and welfare is a primary consideration of the forest service ... At the same time, we must harvest this timber crop under sustained yield principles." The news release added that "grazing, wildlife, and postwar recreational use of the forests will be given careful consideration," but by implication not on the same level as timber. Another meeting was held the following year where "Major attention will be given to technical and administrative costs of timber sales, made acute by the current pressure for public timber to meet postwar logging demands … Of special importance to supervisors from east of the Cascade range is the problem of livestock range management."

On August 13, 1948, the name of the North Pacific Region was changed to the Pacific Northwest Region by Chief Lyle F. Watts. The R-6 Administrative Digest of August 25, 1948, noted that the "new name was suggested by Dorothy Keith of the Division of Wildlife and Range Management."

Chief Lyle Watts visited the Region from July 29 through August 6, 1948. Regional Forester Andrews and Station Director J. Alfred Hall of the Pacific Northwest Experiment Station accompanied the chief to portions of the Rogue River, Umpqua, Siskiyou, Willamette, and Siuslaw NFs. A Regional Office news release noted that Chief Watts "complimented the Region on the good job … in meeting the problems of a greatly enlarged timber management program, and pointed out the possibilities of improving silvicultural practices in the timber stands of southern and southwestern Oregon." He also "pointed out the big job ahead of the region and the Service in developing its recreational facilities. Mr. Watts was impressed by the degree of deterioration in campground facilities because of lack of funds for development and maintenance."

In 1954, the Regional Office moved out of the Post Office Building to 729 N.E. Oregon Street, across the Willamette River from the Portland downtown area. In 1959, because of increasing space needs, the division of engineering was moved out of the Regional Office into the neighboring Omark Building.

The Regional Office provided several workshops in 1946 covering "conservation and outdoor education" for teachers at the Rustic Inn near Cle Elum (this was the fourth consecutive year) and at Hoodoo Ski Bowl at Santiam Pass (second consecutive year). Total attendance at both camps was 174 teachers. In addition, the Regional Office provided conservation education assistance to several colleges and universities in the two states. During 1959, the Regional Office Information and Education section (I&E) provided two exhibits to the Oregon Centennial Exposition in Salem, and took part in six conservation education workshops where over 200 teachers from both states participated.

EMPLOYMENT AND HOUSING

During the heyday of getting out the cut, with new employees hired, it soon became clear that there was not adequate Forest Service housing for this rapid growth in employees. A Regional Office survey of housing needs for each ranger district was begun in 1955. The survey, which studied office space, residences, storage facilities, and utilities, found problems at almost every ranger station.

"It was found we need at least $7,000,000 for this project. It is particularly important that adequate housing be provided for timber sale personnel if we are to reach our allowable annual [timber] cut in a planwise manner. This critical situation was alleviated somewhat during the year. Eighty-eight housetrailers were purchased from other government agencies. They are being used by families in rapidly expanding work areas. This provides temporary relief in the most critical places until more suitable, permanent housing can be obtained." (*Pacific Northwest Report* 1955)

Following World War II, the Forest Service was viewed by most employees and those who wanted employment in the agency as a great place to work. As the needs of the agency grew, especially in Region 6, because of increasing timber harvesting and road construction needs, many new forestry graduates saw the Forest Service as their first choice for a long career. Plus, once hired, they were often assigned to small, remote ranger stations where there was a comfortable feeling of closeness and camaraderie among the rangers, employees, wives, children, and the communities. Regional Forester Ted Schlapfer described the scene:

> One of the things that inspired me to stay dedicated to the Forest Service was its own dedication. Not just willingness to … put in long hours, but to look for new ideas and be involved in developing new technologies … and the kind of motivation that they developed in the individuals to recognize the responsibility of your service to the general public … was outstanding (Schlapfer 1984: 15).

The Region began a district ranger workload analysis in the mid-1950s. The analysis said that "a ranger's work load should be such that he can adequately plan, direct, supervise and participate in managing the resources. Work-load studies revealed an overload [of work] on a number of districts in the region." (*Pacific Northwest Report* 1956: 16) As a result of this study, four new ranger districts were created—Hoodsport on the Olympic NF; Willard on the Gifford Pinchot; Blue River on the Willamette; and Metolius on the Deschutes. This was the era of "splitting" districts into more manageable units. Later would come the "lumping" era, to reduce costs.

For example, the Willamette NF increased the ranger districts from six in the late 1940s and early 1950s to nine by 1959, and is now back down to four (Rakestraw and Rakestraw 1991). Another example from the Mt. Hood NF was recounted by Dick Worthington, a transfer to the forest in 1956: "When I went to … the Mt. Hood Forest [it] had been singled out for attention on reorganization. It was felt that the ranger districts were too big, and that they should be make them smaller so that the ranger could be more knowledgeable about all of the things that were going on. … The Mt. Hood was destined to have eleven ranger districts … then there was a plan to make thirteen districts. Thank goodness that was never implemented." (Worthington 1984: 20)

By 1958, workload analysis was expanded to fire control, timber management, wildlife management, watershed management, land adjustments, information services, and personnel management. Engineering and clerical workloads would be studied later.

WOMEN EMPLOYEES

The Forest Service employed few women. Some Forest Service wives found creative ways to find work in the agency—and get paid—as well as to be closer to their husbands while they worked in the field. These Forest Service wives found themselves at a slightly different place from women in previous decades, becoming an important part of management even though they were usually unpaid workers.

Gertrude Conroy became the first woman to hold the position of regional auditor. The news release on January 16, 1951, noted that "a Portland woman has taken over a job formerly thought as being pretty much reserved for men … This is the first time that a woman has held this important post in the Pacific Northwest region." Conroy started her career in the Forest Service in 1918 as a stenographer-clerk in the District Office and worked her way "up the ladder" in the fiscal section.

Another successful woman was Nella Hulet. She was promoted to the Regional Office from the Snoqualmie NF to work in the Information and Education section in 1956 under the direction of Jack Wood. A news release dated May 5, 1956, described her work: "In her new assignment Mrs. Hulet will provide information service covering national forest work and natural resource conservation to various women's organizations … She will also assist public schools in carrying out conservation programs." This appointment, apparently, was the first position the Regional Office dedicated to what today is called environmental education.

Land Status and Ownership

The Siskiyou NF was split on June 3, 1947, cleaving off much of the southern end of the forest in California and combining it with parts from the Klamath and Trinity NFs to form the new Six Rivers NF. However, the current Rogue River-Siskiyou NF still manages over 87,000 acres in northern California.

On June 15, 1949, the name of the Columbia NF was changed to the Gifford Pinchot NF to honor the late Forest Service chief. Gifford Pinchot III was present at the signing of the name change document by President Harry Truman in the White House. The chief's widow, Cornelia Pinchot, spoke at the dedication of the new forest name at La Wis Wis Campground on October 15, 1949. Forest Supervisor Chris Chriswell commented: "It turned out to be quite a project for us. Not only did we have to change all the signs but Lee Prater and Dick Moser came out from Washington [office] to photograph much of the forest and make a one reel movie." (Chriswell 1989: 72)

As part of a national effort to improve efficiency, the Wallowa and Whitman NFs were administratively combined into the new Wallowa-Whitman NF. The effort was concluded by July 1, 1954. As part of the combination, the Blue Mountain Ranger District (RD) was transferred to the Malheur NF. Another forest name was changed in 1955—the Chelan NF was changed to the Okanogan NF on April 4, 1955; as part of the change, the Chelan RD was assigned to the Wenatchee NF.

Congressional hearings were held regarding the future status of the Oregon coast sand dune area as early as 1957, but there was little progress on legislation. By 1959, Oregon Senators Richard Neuberger and Wayne Morse proposed that a 35,000-acre national seashore be created. This national seashore proposal, to be under the administration of the National Park Service, included 13,000 acres of land administered by the Siuslaw NF, 500 acres of Oregon state park land, 15,000 acres of Crown Zellerbach's Tahkenitch Tree Farm, and 4,000 acres of freshwater lakes. (The final resolution of the issue would be in 1968; see Chapter 7.)

The 1954 Act that set in motion the termination of the Klamath Indian Reservation was amended in 1958. (Portions of the Klamath Reservation would three years later become the new Winema NF and the Klamath Marsh National Wildlife Refuge; see Chapters 7 and 9.) The amendment authorized the sale of

> 617,000 acres of the Klamath Indian Forest to private purchasers. This acreage consists of eleven blocks of sufficient size to practice sustained yield forestry, and a number of small, scattered fringe tracts that are being sold independently. ... The amended law also required that all the tribal lands be reappraised ... If all these units are not sold before April 1, 1961, the United States will purchase the remainder (including the Klamath Marsh) at its appraised value, up to a total of $90 million. Such federally purchased forest land would be added to the national forest system (Pomeroy 1959: 37).

In 1954, Congress settled a longstanding dispute over the management of O&C lands within the national forest boundaries by arranging a partial land exchange between the Forest Service and the BLM. In the exchange, which occurred in June 1956, the national forests in the Region transferred 241,000 acres to the BLM, while the BLM transferred 243,000 acres to the Forest Service. The land exchanges affected the western edges of the Mt. Hood, Willamette, and Umpqua NFs, northern edge of the Rogue River NF, eastern edge of the Siuslaw NF, and the northeastern edge of the Siskiyou NF. In effect, the land transfer had the effect of "blocking up" national forest land and leaving many of the intermingled lands to the BLM.

In 1954, the Forest Service became responsible for managing approximately 10 million acres of "land utilization projects" (referred to as LU lands). These LU lands, acquired by the federal government during the Depression years of the 1930s, were in many cases relinquished or abandoned farms, with most located in the Great Plains. The original intent was that the Forest Service act as interim manager pending final disposal of these lands. Discussion over the future of these lands continued until 1958, when a revised national policy transferred around six million acres to states and colleges. In 1960, many of the earlier LU projects became the first national grasslands in the Forest Service (see Chapter 7).

There was some discussion by the Sierra Club in the late 1950s about the creation of a national park along the crest of the Cascades in Oregon. In the October 1959 issue of the *Sierra Club Bulletin,* author David Simons suggested that national parks status for the high Cascades was better than the current array of wildernesses that were not linked together. His proposal was to have a 970,000-acre Cascade Volcanic National Park extending from Mt. Jefferson on the north to Diamond Peak on the south, covering the headwaters for the Santiam, McKenzie, and Willamette river drainages. "The Cascade Volcanic National Park would protect a fine balance of land types, including the vital semiprimeval gateways and wilderness forests … [it] would greatly increase public appreciation of the region … A long period of public concern could end. The national park concept is a simple, direct one with a vigorous tradition of public backing." (Simons 1959: 13) The national park idea was not well received, but it would be revived in the 1960s by Senator Wayne Morse and others (see Chapter 7).

Timber Management

In the 15 years of rapid development of the national forests since the end of World War II the Forest Service was leaning heavily toward management of timber as the primary goal of national forest management (Kaufman 1960). Since most of the foresters of that era had similar backgrounds in forestry education, it was not surprising that timber harvesting would become the predominant view of the agency as the rank-and-file foresters knew they were "doing the right thing" and they were happy doing it:

> The patterns of [timber harvest] increases each year was firmly established. Timber had always been a large and very important subject in the minds of foresters, particularly those in the Forest Service. But never before had the actual management of timber required so much time and energy and attention, to the point where the timber program itself (rather than its theoretical objectives) seemed to dominate the agency's consciousness and workday alike. Timber was such an active program by 1952 that any ambitious young forester could see that in the Forest Service, timber was where careers were to be made (Clary 1986: 125).

On the Colville NF, as one example, timber harvesting was growing considerably after the war. The Colville had large areas of burned-over forest from the 1920s and 1930s fires; by the 1940s the burns were in young standage conditions. Harvesting was ramping up on remaining old-growth trees. The annual harvest in 1940 was just over 2,670,000 board feet, increasing to 12,530,000 board feet in 1945 as part of the World War II effort. That increase was about to grow even more dramatically in the 1950s:

> By 1953 the maximum allowable cut of about 27 million board feet per year was being taken from the [Colville National] Forest. Receipts, or profits, from timber sales increased fantastically. As with most national forests in the early 1950s, the CNF for the first time showed a profit in receipts over expenditures. W. H. Ibenthal, who became supervisor in 1957, thought the maximum allowable cut for the Forest was set too low … Ibenthal launched an in-depth study of existing timber stands on the CNF. Foresters and statisticians working on the project produced a new timber management plan which was adopted in the early 1960s. The plan boosted the maximum allowable cut on the Forest to over 90 million board feet per year [by 1965 the cut timber cut was a record 100,297,000 board feet, a growth of almost 50 times in 20 years!] (Holstine 1987: 93).

Log train crossing the Cowlitz River, Gifford Pinchot National Forest, Washington, 1949. USFS photograph.

The technology of cutting and moving timber from the woods to the sawmills had changed dramatically. Highly efficient chainsaws came into increasing use. There were still a few logging railroads still in operation through the 1950s in Oregon, but they eventually were replaced by roads. With the increased emphasis on timber production, the number of timber sales jumped dramatically. Extensive and intensive forest management was beginning in earnest. Near the end of the war the Forest Service had told Congress that the national forests, especially in the Pacific Northwest, could take more of the national timber burden; if Congress would give the agency more money (basically for timber planning and roads), the national forests could greatly increase the amounts of timber. Congress responded and timber sales increased as the demand for lumber continued (Clary 1986).

Use of funding under the Knutson-Vandenberg (K-V) Act of 1930 became more important in reforestation work by the end of the 1950s. Significant also was use of funds authorized under the Brush Disposal Act of 1916, which were increasingly used for reforestation, erosion control, and road maintenance work. Former Regional timber director Wendall Jones reported that "there were constant efforts by GAO and some legislators to greatly reduce the entitlement funds to the Forest Service, but none had success." (Jones personal communication)

A little-understood administrative definition—allowable cut—would come to haunt the Forest Service for decades. The agency defined and used the allowable-cut concept from the early 1950s as a measure of what was needed to ensure that sustained-yield forestry was being practiced on the national forests. The allowable-cut idea was conceived as the "ceiling" or maximum against which timber harvesting on the forests would be measured. The timber industry, on the other hand, wanted the agency to use the allowable cut as the "floor" or minimum of timber harvesting volume. "Turning the traditional definition of the policy on its head, industry lobbyists insisted that the failure to achieve maximum allowable harvests of timber meant a failure to achieve sustained yield." (Hirt 1994: 132)

Timber harvest on the westside forests in the Region increased steadily during the late 1940s and 1950s. In 1945, the timber harvest level was 1.284 billion board feet, rising to 1.495 billion by 1950. In 1951, timber cut from the national forests in the Region for the first time exceeded 2 billion board feet. By 1955, timber sales in the Region amounted to 2.633 billion board feet. Purchasers paid over $37 million for the timber, or an average of $14.16 per million board feet. This amounted to 94.5 percent of the annual allowable cut for that year. The first time the Regional timber harvest exceeded three billion board feet was in 1959 (see Appendix E).

During 1955, some 10,276,000 seedlings and transplants were produced at the Wind River and Bend nurseries. During the same year, nearly 21,500 acres were reforested, with around 20,200 acres of the total financed by K-V funds, sometimes referred to as sale area betterment funds. In addition, 5,500 acres of stand release and thinning and nearly 34,999 acres of pruning were accomplished by use of K-V funds. The total area successfully reforested in the Region by planting and seeding by 1955 amounted to 175,141 acres. By 1959, the Regional planting program had jumped to 18 million trees, most of which were grown in nurseries in the Region. During the same period, the acres reforested almost doubled to 40,660 acres.

The Forest Service began using a completely revised timber sale contract, effective January 1, 1956. A national news release on March 22, 1955, indicated that "in drawing up the new form the Forest Service timber management officers were assisted by representatives of the timber associations and of forest industries which buy national forest timber. Timber buyers and the Forest Service have long recognized the inadequacies of the old timber sale contract which has been repeatedly amended since it was first written." Such standardization of contract language across the nation allowed the Region and timber purchasers the ability to legalize timber sale contracts to eliminate conflicting language, better define terms, and it gave the agency the ability to enforce the signed contracts with purchasers.

Logging road erosion, Umpqua National Forest, Oregon, 1957. USFS photograph.

In December of 1958, the Small Business Administration (SBA), established in 1953, entered into a memorandum of understanding with the Forest Service to give preferential treatment to small business concerns in the sale of timber sales. The act provided that certain sales be set aside for small operators as defined by SBA so that small operators would get a fair share of government contracts. The Forest Service manual noted that small businesses could better compete "if Government loans were available to finance the construction of certain roads … Accordingly, the Forest Service and SBA agree … to make available road construction loans to successful bidders for National Forest timber sales who meet SBA size, eligibility, and credit requirements."

Timber production in the Region was unexpectedly boosted during the 1950s when the Corps of Engineers began building dams on several forests. Each reservoir area had to be cleared and cleaned of all trees and brush to "full pool level." Since many of the dam locations were on national forest lands, the job of removing the vegetation fell on the Forest Service, working cooperatively with the Corps.

CHANGING TECHNOLOGIES

Changing technologies changed how the forests would be used, as well as how they looked. Power saws, eyed with suspicion in the late 1930s and early 1940s, were increasingly used after the war. High lead and full suspension logging systems were tested for yarding of logs as they caused much less damage to the forest soils than crawler tractors or simply dragging the logs across the landscape by donkey engine. In the central and eastern parts of the Region, big wheels or arches pulled by tractors with one end of the logs suspended were used to move logs along the forest floor.

There was an increase in the number of small sales. Small operators—often called "gypos"—with a tractor and bulldozer, a portable loading rig, and power saws could set up a profitable business. Many of these operations were owned by World War II veterans who saw great opportunities in the national forests in the Pacific Northwest. Numerous small to medium sawmills opened across the region. For example, in Jackson County, Oregon, near the Rogue River NF, 39 new sawmills were opened in 1946. Such sawmills quickly became the mainstays of hundreds of small towns and medium-sized cities. The national forests in the Region were becoming integral parts of stable communities, as they often provided the logs that were needed to supply the sawmills for long-term employment and profit. For example, the newsletter *Hines Log* from Hines, Oregon, in the August-September 1947 issue noted the "arrival of the first truck load of logs from Ochoco national forest Tuesday, May 13, opened a new chapter in the operations of the Edward Hines Lumber Co. with additional guarantee of adequate [timber] supply." Sawmills and plywood mills also began using new technologies to increase productivity and keep costs down. Compared to the number of men it would have taken to produce the same amount of lumber before the Great Depression and World War II, employment in the forests and mills was down, but production was up, and so

High-lead yarder, Mt. Hood National Forest, Oregon, 1956. Photograph by Bob Unruh.

were the profits. For the first time, national forest timber was quickly becoming an integral part of mill capacity and employment.

Fully managed forests, using intensive forestry techniques, were to become the standard model of growing second-growth forests. Intensive forestry practices in this era included site preparation, hand or machine planting, spraying, fertilizing, thinning and pruning, pre-commercial thinning, commercial thinning, then final harvest, after which the cycle would restart. There was a huge need in the Region to sustain the increased timber harvesting effort on the national forests. As discussed earlier in the chapter, tree nurseries and seed orchards produced millions of seedlings for replanting in both fire-burned and timber sale areas. For example, the Denny Ahl Seed Orchard on the Olympic NF was designed to "produce frequent, abundant, and easily harvested crops of genetically improved tree seed from known superior parent trees." (Rooney 2007: 141)

Aerial spraying to kill competing vegetation, usually brush species in tree plantations that would retard tree growth and thus reduce the long-term

Salvage logging by the Fall Creek Lumber Company, Willamette National Forest, Oregon, 1947. USFS photograph.

production of timber, began in the mid to late 1950s. For example, on the Rogue River NF in the summer of 1958, an aerial application of 2,4-D and 2,4,5-T, in a solution of diesel oil and water, was used to treat four areas totaling 120 acres in the Butte Falls area (Brown 1969). Other forests were also beginning to use these aerial herbicide sprays to "release" young trees where their growth was felt to be inhibited by competing brush or grass species. Soon this method became established in standard timber management practices for a "managed forest."

Tied to these efforts was the need to have complete access to the timber stands through an extensive network of roads of various standards, new types of low-impact logging equipment, and prompt salvage of insect- or fire-damaged trees. The objective of the increased timber harvesting was to produce the greatest volume of board feet of timber on a sustained-yield basis. Much of America in this era tended to agree with this industrial forestry management philosophy.

Unmanaged, wild, or old-growth forests could not be expected to yield the same amount of timber at harvest time. The Forest Service, as well as the timber industry, began referring to the old-growth forests as overmature, unproductive, decadent, disease-and-insect-ridden, fire hazards. Little thought at the time was given to the ecological and cultural values of retaining old-growth trees and forests—that would come several decades later (see Chapter 9). On the other hand, managed forests were viewed as young, dynamic, healthy, and fast growing. Converting the old-growth forests to second-growth forests, based on research, could boost the annual allowable cut. Generally, the managed forest trees in the Douglas-fir forests, as well as the pine forests east of the Cascades, were thought to be ready for harvest at 80 to 120 years.

INSECTS AND DISEASE

Two major types of insect and disease outbreaks occurred in the Region in the 1950s—spruce budworm and blister rust. Treatment for spruce budworm, which affected Douglas-fir and white fir trees, was generally confined to the Ochoco, Malheur, and Wallowa-Whitman NFs, although as early as 1950 spraying was carried out on the Willamette NF. Aerial spraying, using DDT mixed with fuel oil, was carried on with the cooperation of the State of Oregon and private forest owners. Wendall Jones recounted that during the Willamette NF spraying in 1950, "I was stationed at Fish Lake Remount Station the summers of 1950 and 51. The mosquitos [sic] were real bad at Fish Lake, so the Ranger had a plane make a pass over the lake to give them a shot of DDT. It worked." (Jones personal communication) About 621,000 acres were treated in 1955, with an increase to more than 818,000 acres in 1958. The total area treated for budworm infestations from 1949 through 1955 was 3,840,000 acres.

Blister rust is a Eurasian fungus that attacks sugar pines and white pines and was found in various locations in the Region for decades. Many blister rust control crews were employed to fight off the fungus. The treatment work was difficult and backbreaking and usually involved grubbing or digging out the wild currants and gooseberry plants that served as alternate hosts to the blister rust and by burning infected and dead trees. An infected area of 6,162 acres was treated in 1955 on the Rogue River and Umpqua NFs. The BLM also treated a larger area adjacent to the Siskiyou NF, while the National Park Service treated two smaller areas on the Mt. Rainier and Crater Lake NPs. In 1959, also on the Rogue River NF, a white pine seed orchard was established at Jim Creek on the Prospect RD to assist in the development of blister rust resistant pines.

By 1956, the balsam woolly aphid, also called by its Latin name *Chermes*, was becoming a significant problem, attacking over 355,000 acres of mostly true fir species in the two states, increasing to 599,000 acres in 1957. The most severe problem was in the Mount St. Helens area on the Gifford Pinchot NF, where 47,680 acres were affected. Salvage of the trees to remove the eggs and larva was the only method of control.

Bark beetles were, in 1956, at a very low level. By 1957, the black-headed budworm made a devastating appearance on over 253,000 acres on the Mt. Baker and Snoqualmie NFs.

TIMBER SALE PLANNING

Planning for road construction and timber harvesting got a real start when the Forest Service was revising existing national forest "working circle" timber plans to account for the increasing demand for national forest timber. Working circles were principal watersheds, both on and off the national forests, at around 100,000 to 500,000 acres that were used for calculating the timber production capacity of each area. Historian David Clary described the scene in 1946:

> Planning was also important. The Washington office told the regions in April that "the objective at this time is to have approved management plans within the next 10 years for all working circles in which timber can be marketed" ... [David T.] Mason assigned a new staff member to the "development of some new general instructions on management plan preparation."... A lot of timber-management plans were revised

> during the first years after the war, all of which went to Washington for review. A large number increased the allowable-cut levels, often by considerable degrees. The reasons were manifold, including updated information on timber inventory or regeneration, improved harvest methods and equipment, increasing access to timber over greater areas, and, probably, the general administrative sentiment favoring the maximum potential contribution to the national timber economy. Not untypical was the new plan for the Coeur d'Alene National Forest, which proposed an allowable cut in white pine that was 300 to 400 percent higher than the levels set in 1939. Plans that did not calculate timber so as to permit the greatest annual allowable cut were returned to the Regions for revision (Clary 1986: 122).

During 1955 seven revised timber management plans in the Region were approved by the Washington Office during the year. The revised plans, reflecting reinventories of standing timber, in seven working circles, resulted in an increase of 100 million board feet of annual allowable cut. More reinventories were begun on an additional 1.2 million acres, with all forests in the Regions scheduled for completion in 1960. The increases in allowable cut were calculated because of "new logging equipment and methods, the utilization of tree species heretofore not harvested, and the use of a greater proportion of individual trees." (*Pacific Northwest Report* 1955: 3)

On the Mt. Baker NF, the timber reinventory suggested a cut raise from 145 million board feet to 222 million board feet. Chris Chriswell, supervisor of the forest at the time, commented in his memoirs:

> What a mistake this was going to be! … Much of the inventoried timber not only formed the foreground of great scenic beauty but ran right up into it in stringers between the avalanche paths … We were clear cutting on the slopes of Mt. Shuksan, long a famous mountain that graced many a calendar and illustrated the covers of national magazines. Cutting units were opening up ever higher on all sides of Mt. Baker itself. Some of the higher cutting units exposed solid rock walls with little soil on them. There were two basic problems fast becoming evident. How could we manage such steep rock areas for production of timber and should we be messing up scenery that was so superb as to have national importance? ... In order to reduce the forest's allowable cut we had to reduce the area classed as commercial timber in our plans.

> Raising the cut of a National Forest based on reinventory had been easy to do. It increased the work load, increased the budget and accelerated the road and bridge program. The organization expanded with new employees. Business management expanded. To lower the cut was a very different thing … The worst part of it was the accompanying reduction of employees causing a lowering of morale on the forest. These reductions in cut took about 12 years and we lived in an atmosphere of constant controversy, not only with our forest users and in the political arena but within the policy making of the Forest Service itself … I found out we were not alone. All Regions were facing similar problems (Chriswell 1989: 100, 121).

During the 1950s, planning became a standard of the ranger districts and national forests. Administrative historian Herb Kaufman described the plans as he found them during his study (and that generally apply today):

> [T]he Washington office requires every Ranger district to have a plan; there is none without a fire plan and a timber plan...Occasionally, a supervisor may insist that his Rangers prepare plans for functions not covered by [the two plans]. ... Sometimes only one or two districts in a region or a forest may be deemed by regional or forest officials to have a volume of business of a given type to warrant the formulation of an explicit plan ... Some districts have as many as half a dozen or more (including, in addition to the mandatory timber and fire plans, land use, watershed, wildlife, recreation, grazing, transportation, planting, and others).
>
> Plans, at least as they are treated in the Forest Service, are preformed decisions. They set long-range (eighty to a hundred or more years for a function like timber management; five, ten, or twenty years for others) quantitative and qualitative goals, break these down into shorter-range objectives, and sometimes reduce these to annual targets. They spell out the steps and stages by which the goals are to be achieved, including the methods of operation, and priorities by geographical area, in each district ... the format of the plans is normally suggested by guidebooks and handbooks, if not by the [Forest Service] Manual and its supplements and by individual instructions from higher headquarters. And staff assistants from forest headquarters commonly participate extensively in every phase of plan formulation—checking field work, reviewing calculations and estimates, offering and vetoing ideas (Kaufman 1960: 99-100).

Planning efforts were underway to address increasing demands on forest resources in the Region. The effort, which presaged the Multiple Use Act of 1960, was described as follows:

> Ever increasing demands upon the national forests to furnish timber, forage, recreation, water, and other products has brought into focus conflicts among the various uses. A larger program of land management planning was started this past year. It consists of gathering pertinent resource data, patterns of use, and other important related information. Following an analysis of this factual data, conflicts in use are carefully weighed. The preference use or uses to which the areas will be dedicated will be those best designed to satisfy long-term public needs for all of the resources of the national forests. This type of land management planning was started on at least one ranger district on each forest (*Pacific Northwest Report* 1956: 7).

One such ranger district was the Fort Rock district on the Deschutes NF. It was one of four districts in the U.S. to be designated as an intensive management district. A special land management plan for the district was approved on April 22, 1958. The long-term planning efforts to meet the public needs, what became known as the highest and best use, were overridden by timber objectives. When figuring the economic benefits of all the resources available, timber harvesting almost always won out. Wendall Jones noted that "this was due to the knowledge base and capabilities for evaluating the timber resources were well in place while no one had come up with programs to evaluate the other resources at that time." (Jones personal communication) It was only decades later that the other resources could play on a more even field through the land management planning process during the 1980s.

In 1958, the Forest Service issued the results of the nationwide timber resources review. This extensive national study, begun six years earlier, and prepared with the assistance of other federal, state, and private organizations, found that the nation needed to produce more timber to meet expected demands. The Region and state forestry departments in both states co-sponsored a series of small ownership meetings to learn more about the problems woodland owners had in producing timber. More than 1,000 persons attended 11 area meetings.

Timber sales were changing as well. Gone were the days of selling the stumpage of entire drainages, with timber harvesting designed to continue over decades. From the 1920s to the 1940s, most timber sales and cutting boundaries, along with the roads or railroads necessary to remove the logs, had

Pattern of clearcuts, Willamette National Forest, Oregon, 1963. USFS photograph.

been made by the purchasers. The district ranger's work was in reviewing the purchaser's plan. By the early to mid-1950s, these layout designs were made by the Forest Service experts.

Ira Mason, in charge of timber management in the national headquarters, "was persuaded by 1948 that clearcutting should be the rule for Douglas-fir, if the goal of reproduction for sustained yield was to be achieved" (Clary 1986: 182). Clearcutting became the preferred method of timber harvesting in the Region by the mid-1950s for most, but not all, species. The decision about the wide-spread use of clearcutting was based on a number of factors including better growth of Douglas-fir seedlings in open areas; in addition, it was cheaper to cut all the trees in a sale area, conduct site preparation for seedlings, and then plant the seedlings in cleared areas. Additionally, there was less need to reenter the areas once the plantations were started. Much of the clearcutting was conducted on timber sales less than 80 acres in size, but some larger units were authorized. Partial cutting and shelterwood harvests were becoming common on the drier eastside forests, but such methods were also tried on the westside.

At the same time, there was a growing concern that the Forest Service was clearcutting too many areas and building roads in areas that were also used for recreation. This issue and others about resource priorities would involve many outdoor groups, timber industry organizations, the Forest Service, and Congress, and would result in the Multiple-Use Sustained-Yield Act of 1960.

SUSTAINED-YIELD UNITS

The Shelton (Washington) Cooperative Sustained-Yield Unit 100-year agreement was signed in 1946, effective in 1947. The Simpson Timber Company on the Olympic Peninsula and the Olympic NF was the only cooperative unit ever established. Basically the Simpson Timber Company had an estimated eight more years of old-growth timber that could be used by sawmills in Shelton and McCleary before the company would exhaust its timber supplies. The cooperative agreement essentially saved those two logging communities and allowed the building of Camp Grisdale, a logging community that was to last 40 years.

Around the Region there were numerous attempts to establish sustained-yield units under the act of 1944. When the law was passed it was assumed that communities would flock to join the Forest Service to help stabilize their communities. On the westside forests, one attempt involved the private Hill timber property in the Sweet Home area that was checkerboarded with timber land on the Willamette NF. This was land originally in the land grant from the Willamette Valley and Cascade Mountain Military Wagon Road (see Chapter 1). Another attempt involved the Pope & Talbot holdings near Oakridge on the Willamette NF. Pope & Talbot in 1946 purchased "30,000 acres of virgin timber lying along the middle fork of the Willamette River … This holding is checkerboarded with some 160,000 acres of Willamette National Forest timber … It is this timber that Pope & Talbot proposes to pool with Uncle Sam's, under Public Law 273." (Say and Baser 1952: 29) The Pope & Talbot land purchase was from remaining Oregon Central Military Wagon Road grant lands (see Chapter 1). Both attempts failed.

A public hearing on a proposed Hood River federal sustained-yield unit was held in the community of Hood River on August 11, 1952. A press release at the time noted that the "Hood River and Cascade Locks Chambers of Commerce, who are the applicants for the unit, believe that an assured supply of timber for the maintenance of local plants will lend considerable support and stability to the economy of Hood River County." Another proposal, this one from the southwest Washington communities of Stevenson, Carson, and North Bonneville, as well as Home Valley, Skamania, and adjacent areas, was received in the Regional Office in November 1953. Both were turned down.

In 1955 an earlier application for a Wind River Federal Unit was turned down after a hearing that determined that the proponents failed to qualify on the basis of need. Apparently, several informal requests for federal units were received, but they also did not qualify. In 1958, "the secretary announced that it was departmental policy to establish no new sustained-yield units of

Loading Christmas trees at Shelton, Olympic National Forest, Washington, 1950. USFS photograph.

either type, though established units would be 'continued for the present'." (Clary 1986: 145) Historian Clary went on to comment that the "unexpected opposition from unexpected sources" (timber industry and local communities) to these sustained-yield units, as they were intended to promote community stability, "should have given the agency pause, causing it to reexamine its objectives and to redefine its vision of the public interest; but it did not … the agency, sure of the rightness of its cause, brushed aside any objections from its opponents … It failed to consider whether the objections were valid; and it did not really even listen to what its opponents had to say." (Clary 1986: 146)

The failure in these cases was fear of monopoly, as small timber owners disliked the long-term sale of timber to a large owner without competitive bidding. It did not take long for the timber industry and the affected communities to realize that what was good for one company sawmill and town was bad for another. The 1944 act became associated with un-American, anti-free enterprise monopolies, and even "communist-inspired" social engineering.

In a press release by the Regional Office on March 3, 1950, it was announced that Homer J. Hixon had been given responsibility for "management planning, including such work in connection with cooperative sustained and federal units under the sustained yield unit act." Chief Lyle Watts, one of the main proponents of sustained-yield units, "quit urging that more units be established. He could not bring himself to admit that the program was dead ... In 1953, after Watts had retired, the Department of Agriculture effectively buried the program by requiring communities, rather than the Forest Service, to initiate proposals, reserving approval to the Secretary of Agriculture." (Clary 1986: 145) Only five federal sustained-yield units were established across the national forest system: Vallecitos, New Mexico (Carson NF); Grays Harbor, Washington (Olympic NF); Flagstaff, Arizona (Coconino NF); Lakeview, Oregon (Fremont NF); and Big Valley, California (Modoc NF). Only the Lakeview unit is actively operating today.

Recreation Management

The period after 1945 was marked by major changes in American recreational pursuits. The development of light-weight equipment by outdoor equipment companies like REI in Seattle and new outdoor sports meant that many more people could enjoy the forest offerings across the vast Pacific Northwest. Increasingly, there was a growing need to set aside areas for such activities as downhill skiing, cross-country skiing, wilderness hiking, travel by trail bikes and off-road vehicles, pack trains, backpacking, water skiing, whitewater boating, stream and lake fishing, driving for pleasure, picnicking, camping, and the like.

In order to meet this rapidly accelerating and expanding demand the Pacific Northwest forests had to construct more developed facilities for those enjoying the forests. "[S]ome additional money was appropriated by Congress for sanitation and care of public campgrounds. The increase in funds to our region enabled us to maintain about one-third of the campgrounds satisfactorily and to replace or add 195 pit toilet buildings, 743 tables, and 484 fireplaces. In addition, a number of campground water systems were improved." (*Pacific Northwest Report* 1955:3) New dams and reservoirs also caused an increase in recreational visitors. Rivers that were once free-flowing were now blocked. Recreation use changed from fishing along the bank to special river boats to larger, powered lake boats. Swimming, almost unheard of in the cold mountain streams, became popular in the reservoirs that often had sun-warmed surface water. More campgrounds and boat-launching ramps were constructed. On several reservoirs, such as Detroit Lake on the Willamette NF, special use permits for construction and operation of docks and resort facilities were authorized.

Fishing at Buck Lake Organizational Camp, Umatilla National Forest, Oregon, 1956. USFS photograph.

Anglers at Lost Lake, Okanogan National Forest, Washington, 1952. USFS photograph.

Recreational demands on the national forests were increasing; millions of new visitors used the national forests and parks. In 1952, recreation use in the Region amounted to 3,810,000 visits; three years later the recreation visits had increased to 5,192,496. Winter skiing in both states was fast increasing. Ski area developments across the Region in the mid-1950s included two new rope tows at the Ski Bowl on Mt. Hood; new Poma lift and lodge at Snoqualmie Pass; new Poma lift, rope tows, and double chair lift at White Pass; a new chairlift at Stevens Pass; ski lodge at Panorama Dome at Mt. Baker; ski lodge, rope tows, and Poma lift at Tomahawk near Klamath Falls; and a new ski jump at Spout Springs. Crystal Mountain, Inc. applied in 1958 for a permit, which was granted, to develop a 1,400-acre winter sports area near the Chinook Highway (U.S. 410) in Washington. Also in 1958, the Bachelor Butte ski area was dedicated on January 3. A national championship ski jumping contest was held at the Leavenworth ski area in 1955 and 1959, with the last ski jumps made in the late 1970s.

One of the premier ski areas in the Region was and remains Timberline on Mt. Hood NF. "Oregon's famous Timberline Lodge on Mt. Hood is now under the operation of Richard L. Kohnstamm. The lodge was closed a portion of last year through cancellation of the previous operator's permit because of unsatisfactory operation. Mr. Kohnstamm has done an exceptionally fine job of rehabilitating the Lodge and in providing services of which the public may be proud. One of these is the construction of a new chair lift, the upper terminal of which is located adjacent to the west wing of the Lodge." (*Pacific Northwest Report* 1955: 3)

In 1956, the National Park Service launched its "Mission 66" program, designed to restore and improve the national parks over a 10-year period. The Forest Service, feeling a sense of competition with the Park Service over recreational opportunities, began "Operation Outdoors" the next year. It was a five-year program with increased funding that was designed to improve and replace many of the older CCC-built structures and to expand recreation facilities and opportunities to meet the increasing recreational demand on the national forests. The program was successful at increasing appropriations at the congressional level for recreation on the national forests.

There were changes also in the region's conservation and environmental groups. The Pacific Northwest Region had worked well for decades with such groups as the Mazamas, Mountaineers, Obsidians, Chemeketans, and local outdoor or sportsmen's clubs. After 1950, some clubs expanded from regional to national status, with the Sierra Club the most notable of these. The club had been, in the past, mainly a regional organization, with headquarters in California, and had had little interest in the Pacific Northwest. Under the leadership of David Brower, the Sierra Club began a program of national expansion with the aim of becoming the dominant force in wilderness preservation. In 1954, a Pacific Northwest chapter was established in Seattle (McCloskey 2005).

Modern-day efforts to preserve the Columbia Gorge area began to make headlines in the early 1950s. The Governor of Oregon was involved in creating a commission of citizens to study the issue. Chris Chriswell related a story about one of the citizen members:

> This lady, a past manager of the Battle Ax Inn at Government Camp had been appointed by the Governor of Oregon [in 1953] to chair a

Getting a lift at Willamette Pass Ski Area, Willamette National Forest, Oregon, 1952. USFS photograph.

> commission. Her name was Gertrude Jensen … She was a large woman who approached everything head-on and the Lord take the hindmost … A minister was going to sell the timber on his 40 acres [in the Gorge] and she gave him such a raunchy tongue lashing that he was in shock for some time and forgot his timber sale. We were waiting in the wings and were able to effect land exchanges on some of these parcels of land. Fortunately for us we were on Gertrude's side. I should not ever want to cross her (Chriswell 1989: 92-93).

WILDERNESS

The Glacier Peak area within the Mt. Baker and Wenatchee NFs was proposed by the Region in 1959 to become a wilderness. Regional Forester Herb Stone commented, "[E]stablishment of the Glacier Peak wilderness area is in harmony with the multiple-use concept of national forest management." This was still the era of administrative wildernesses established by the Forest Service and the Secretary of Agriculture, not congressionally designated wilderness as established by the Wilderness Act of 1964. Chris Chriswell in his memoirs relates a story about Regional Forester Herb Stone visiting the forest in 1957 to see the Glacier Peak area. On horseback, Chriswell, Stone, and others had plenty of opportunities to personally see what the proposed wilderness was really like:

> Ranger Ross [Roscoe] Files organized a pack trip out of Glacier that would take us over three high passes in the roughest country to be found ... The trail started on Ruth Creek at about 1,500 ft. elevation and we climbed into Hannegan Pass at over 5,800 ft. From there we dropped down to a low elevation again on the Chillawack River, camping at the old U.S. Cabin ... The next day we forded the Chillawack River and headed up Brush Creek to Whatcom Pass where we were again over 5,000 ft. ... The next morning we rode down 14 switchbacks to Little Beaver Creek and the Twin Rocks Shelter at 2,900 ft. We again climbed to the third pass at Beaver Pass and camped at the shelter there in a dense mountain hemlock forest ... The fifth day we rode along Ross Lake, across the dam, over a swinging bridge ... over Diablo Dam to the Diablo G.S. on the Skagit River. Herb Stone dearly loved to ride in the backcountry (Chriswell 1989: 102-103).

The proposal for Glacier Peak wilderness, covering areas on the Mt. Baker, Wenatchee, and Okanogan NFs, was roughly the same area that Bob Marshall thought should become wilderness in 1938 (see Chapter 5), including the entire roadless area between Stevens Pass and the North Cascade Primitive Area at Ross and Diablo lakes. On September 6, 1960, the Secretary of Agriculture dedicated the Glacier Peak Wilderness. However, the issue lingered and would finally culminate in 1968 when the North Cascades National Park was established.

The French Pete controversy was a wilderness issue that had lasting effects on the Region. In 1950 the question of creating a wilderness in the Three Sisters area came up. Bill Parke, in the Regional Office, helped to organize field trips by the Forest Service and various hiking clubs to examine the area. The proposal by the Willamette NF was to eliminate the west side of the Three Sisters from any wilderness consideration. In addition, they would add several wild areas (roadless areas between 50,000 and 100,000 acres—at the time wildernesses were areas over 100,000 acres) near Diamond Peak, Waldo Lake, and Mt. Washington. The Wilderness Society and most other organizations accepted the Forest Service multiple-use plan.

In 1954, public notice was given of this proposal: to classify 196,709 acres as the Three Sisters Wilderness, and to establish the Diamond Peak and Mt. Washington Wild Areas on the Willamette and Deschutes NFs. The question of the acreage to be included in the Three Sisters Wilderness resulted in a spirited controversy and a public hearing was held in Eugene on February 16 and 17, 1955. The Secretary decided on February 3, 1957, to establish the Three Sisters Wilderness and to exclude 53,000 acres in the French Pete area. Wilderness status was not to be considered for this area, but scientific, ecological, and scenic values would be protected until a management plan had been finished.

The Willamette NF began developing multiple-use plans for the area. These included some timber sales, particularly in the areas where there was insect infestation; plans for stream rehabilitation; and setting aside a number of natural and special interest tracts.

In 1959 the Sierra Club issued a major blast against the Forest Service, accusing lumber interests of wasteful practices, and suggesting that the national demand for lumber was supposed to be on the decline. The Sierra Club advocated the creation of a Volcanic National Park, extending from Olallie Lake in the north to Diamond Peak in the south, totaling 972,000 acres. But the park idea died.

Watershed Management

In the mid-1950s, a number of water-power and flood-control dams were constructed. Hills Creek Dam on the Willamette NF was a Corps of Engineers project, while the Skagit River and Ross Lake projects developed by Seattle City Light were continuing on the Mt. Baker NF. Seattle City Light was apparently required to provide money to hire a customs officer where the dam reservoir would get close to the border. When the gate was closed at the border the "Customs officer was glad to open it when handed a pint of good U.S. bourbon. He called it, 'Canadian – U.S. Goodwill.' After several years he died and a City Light supervisor at Hozomeen reported that 'Goodwill' had finally killed him." (Chriswell 1989: 114)

The Pacific Northwest Power Company applied for licenses to build two dams on the Snake River below Hells Canyon. The dams were to be called the High Mountain Sheep (see Chapter 7) and Pleasant Valley dams. In 1955, preliminary road, trail, and recreation plans for these proposed dams and reservoirs were begun by the Region through the Wallowa-Whitman NF.

The Corps of Engineers in various reports from the 1930s and 1940s recommended building dams across almost every large river in the Region for flood control, as well as hydroelectric and irrigation purposes. A series of other dam projects was proposed in the two states during this time, with dams authorized in many river locations during the 1950s. Most of these dams were moved to a different location or never constructed, often as a result of local opposition.

An example from the Willamette NF involves the Corps' proposal for at least two dams across the main McKenzie River in the late 1940s. One was in an off-forest location near Gate Creek (never built), while the other would have flooded many small communities in the upper McKenzie Valley, and damaged prime trout habitat. After considerable opposition from the affected people, the Corps presented an alternative: instead of one large dam on the upper main stem of the McKenzie it would build several smaller dams on the tributaries, Blue River and the South Fork McKenzie River. These were built: Cougar Dam on the South Fork was authorized in 1950, construction began in 1956, and it was completed in 1964. Blue River Dam, also authorized in 1950, was completed in 1969. Later, another dam and diversion tunnel were built on the upper McKenzie, this one for power generation—the Carmen-Smith project, which was approved by Eugene voters in a special election on April 28, 1959, and completed four years later.

> There was discussion about management of the surface waters of these new reservoirs—Forest Service or Corps of Engineers? For the most part, the Corps manages the water, while the Forest Service manages the adjacent land. It is still unclear which agency manages the land that is exposed during the winter months during the drawdown of the reservoirs every fall. An unusual event on the Wallowa-Whitman NF occurred on May 18, 1956—a dam collapsed. Baker Valley has just experienced its most hectic week in history. At 3 p.m., May 18, the dam at Goodrich Reservoir went out, dumping 55 million gallons of water down the mountain side to the valley below. In four hours it fanned out through the valley to the Union Pacific Railroad where it stood four feet deep on the hay meadows. Goodrich Reservoir is situated on the forest, ten miles west of Baker at the far end of the city water system … Fortunately no residents were injured …We have not yet been advised as to the cause of the failure of the dam (*N. W. Forest Service News*, May 30: 1).

A division of watershed management was established in the Regional Office during 1957. The new division was described as having "the objective … to bring about the best possible management of soil and water. The division has three sections: Soil and Watershed Management, Hydrology, and Small Watershed Projects. Work in watershed management should be further strengthened by the addition of a soil surveyor and a hydrologic engineer in the near future." (*Pacific Northwest Report* 1957: 7)

Mineral Management

Mining in the Region underwent some major changes during the postwar period. National need for minerals led to a revival of mining activity, but postwar legislation created a new era of controversies. In Washington and Oregon, of 5,988 claims recorded, covering 184,673 acres, only three were actually producing minerals. The claims had 1,945,842 board feet of timber on them, worth, by 1949 figures, $13,246,060; under the law prior to 1955, the mine claimant had the right to also cut the timber. Pumice claims in the two states covered large timber holdings, including wilderness, and blocked access roads through the refusal of claimants to give rights-of-way. A flurry of mining claims was experienced on the Fremont NF, caused by the discovery of uranium, a strategic material used in the production of atomic bombs. Around 1,000 uranium claims on the Fremont were made in the summer of 1955 alone, and other forests experienced a smaller number of uranium claims. Oil and gas lease permits were received during 1955 with 80 percent of the total on the Siuslaw NF, 15 percent on the Ochoco NF, and 5 percent on the Fremont and Snoqualmie NFs.

Cleveland Van Dresser wrote a story for *American Forests* concerning the mining claim situation facing the Forest Service in the early 1950s. He wrote that "the Forest Service has no quarrel with legitimate miners. In fact the Service encourages in every way possible the discovery and development of the nation's mineral resources. But when a horde of 'phonies' descends upon public domain [national forests] and by means of legal chicanery assumes title to it—that is definitely something else again." (Van Dresser 1952: 9) The author went on to describe a situation on the Gifford Pinchot NF:

> 43 mining claims [for pumice] … were voided in the Gifford Pinchot National Forest of Washington … the claims had been declared [by the BLM, then in control of mining rights] null and void for failing to meet even the minimum requirements of the mining law. The 43 claims of 160 acres each covered 7160 acres of timberland in the Randle, Washington area. The BLM administrator also announced that legal steps had been started against 72 other placer claims in the same area covering 7600 additional acres (Van Dresser 1952: 9).

On the Siskiyou NF in the spring of 1952 a shooting incident by a disgruntled miner led to the death of a state policeman. Dick Worthington recounted the tragic story:

> There had been a major confrontation between a miner and a trail crew ... The miner was a recluse-type fellow and he shot the heel off one of the trail crew's boots. ... He probably shot in the ground to scare him ... At any rate, they left immediately. The fire control officer got the state policemen and they went back to confront the miner ... as they walked in [to his cabin], he stepped out from behind the tree and shot the cop and killed him. Then he ran back to his cabin and packed a bunch of things and disappeared. ... That winter ... the miner gave himself up ... Incidentally, he went to the insane asylum in Salem (Worthington 1984: 16-17).

Milvoy Suchy was appointed as the regional mineral examiner in late 1952. A news release on December 12, 1952, described the newly created position: "Suchy will work on a backlog of patent applications for well over a hundred mining claims ... In addition, he will examine and report on scores of active claims, many of which are reported to be seriously retarding orderly development of national forest resources. In cases where mineral values are not found, claims will be contested."

A scandal occurred in southwest Oregon on the Rogue River NF with the Al Serena or Buzzard Mine, where patents for quartz rock with gold veins were given to the company over Forest Service protest in January 1954. However, the scandal spread and Congress held investigations of the claims in 1956. In part, the Al Serena was important in the rewrite of the mining laws to prevent abuses through passage of the Multiple Use Mining Act of 1955. Other mining claims that were legal, but suspicious, involved pumice claims on the Willamette NF that it was thought were used to steal timber within the Three Sisters Wilderness.

In 1955, the Multiple Surface Use Mining Act, sometimes called the Surface Rights Determination Act, helped prevent such abuses of mining laws and curtail abuses that had interfered with managing national forest lands. This was the first major change in the 1872 Mining Act. Timber cutting on a claim within the national forest was to be carried on under Forest Service regulations. The Forest Service was given the authority to manage and dispose of the timber and other surface vegetation on the claims. In other words, the title of the claimant was only to the mineral except for trees on the claim to be used only for work on that same claim for mining purposes.

The law also removed common varieties of sand, gravel, stone, pumice, and cinders from materials that could be claimed and placed them under the Materials Act of 1947. It also provided that all titles be registered with the

BLM; previously, records of claims had been scattered in state and numerous county files. An important feature of this law was that, after proper notice, mining claimants could be requested to prove the validity of their claims. This procedure quickly eliminated several thousand abandoned mining claims on the national forests. The law also gave authority for access across claims to adjacent lands.

In addition, the Region initiated "test areas" to investigate unpatented claims to determine the surface rights under this act. The test areas included the North Fork of the Snoqualmie River in the Snoqualmie NF, in the Skykomish RD also on the Snoqualmie, the Little North Fork of the Santiam River and Quartzville Creek areas on the Willamette NF, and the Lawson Creek claims on the Siskiyou NF.

Wildlife and Fish Management

Major shifts also occurred in regard to wildlife management in the period after the war. The rush of students taking advantage of the G.I. Bill helped to produce many students in the relatively new field of wildlife management (using the direction established by Aldo Leopold), and schools of forestry increasingly added optional courses in wildlife management. By 1955, 10 of the 18 NFs in the Region had written wildlife management plans for their forests.

Cooperation between the states of Washington and Oregon and the Forest Service was necessary since fish and big-game resources were under jurisdiction of the state fish and game commissions. The Forest Service managed the habitat, the states the fish and wildlife species. The relationship needed to be formalized, and by the mid-1950s, cooperative agreements had been made. The Forest Service agreed to aid the states in enforcing fish and game laws, giving them facilities for work, making wildlife inventories, and helping in management. For their part, the states agreed to share facilities with the Forest Service, to ask Forest Service permission to introduce exotic species and to use poison against predators, to advise about any changes in policy, and to give annual reports. In 1959, the Newberry Crater on the Deschutes NF was declared a game refuge by the state. The Forest Service also worked closely with other federal agencies concerned with wildlife such as the U.S. Fish & Wildlife Service.

The accepted methods of timber harvest tended to favor increases in deer and elk populations. Areas opened by clearcut logging led to grasses and shrub browse, which the big game animals thrive on. Lightning and Indian-set fires had formerly made the openings, but now clearcutting and site-preparation fires took their place. In areas where deer injured seedling conifers, managers experimented with chemicals to render the browse unpalatable and placed plastic protectors on vulnerable seedling tops. Older, unthinned second-growth Douglas-fir forests, with their dense canopies, provide only limited forage but do provide cover from heat and cold, as well as protection from predators. Stands that have been properly thinned are much better at providing important shade in the hot summer months and protection during the critical winter months. The old-growth habitat tends to provide a mixture of shade and protection, as well as spots of grasses and browse.

In 1958, personnel from the Oregon Game Commission, Pacific Northwest Forest and Range Experiment Station, and the Fremont NF cooperated to study the Silver Lake deer herd. The "study should furnish basic information on deer-livestock relationships, forage use, game numbers, and hunter success." (*Pacific Northwest Report* 1958: 9) This was also the year that the Oregon Game Commission instituted the unit management system to better distribute big-game hunters across the state. Similar studies, using fencing, were underway on the Olympic NF to measure the damage caused by elk grazing on natural and artificial tree regeneration.

The stocking of fish in streams and alpine lakes continued. Timbered buffer strips were to be left to protect stream temperatures. Streams were protected by prohibiting the felling of trees into creeks and rivers, as well as yarding logs across streams; removing debris and old log jams; prohibiting crossing streams for skid and spur roads unless the roads were on bridges or major culverts; and replanting of exposed soil areas. But probably the first recorded study of fish kill by spraying was observed on the Malheur NF in 1958. "A loss of fish life occurred in two streams as a result of the spruce budworm spray project … according to studies conducted by the Oregon Game Commission. After a thorough investigation by both agencies, it was determined that the loss was caused by double spraying the drainages, inadvertently. A plan is underway to prevent this on future insect spraying projects." (*Pacific Northwest Report* 1958: 13)

A growing problem in some locations on the national forests was (and remains) the illegal introduction of unwanted or invasive fish species. Diamond Lake is a 3,000-acre prehistorically fishless lake on the Umpqua NF. The Oregon Department of Fish and Wildlife first planted rainbow trout in the lake in 1910. Production took off and anglers flocked to fish for the fast-growing rainbows. But the trout fishery began to decline in the 1940s and 1950s when another introduced fish, the tui chub, multiplied. Tui chub minnows are native to Klamath Lake and they were used by some anglers as bait fish. Apparently, these small fish were accidentally or purposefully spilled into the lake. In 1954, Diamond Lake was drawn down about eight feet then poisoned with the chemical rotenone, a pesticide derived from plant roots, killing every aquatic species. After a major fish restocking, Diamond Lake enjoyed a boom in recreation, several times exceeding 100,000 anglers per year.

Special care was taken in the Region to protect nest trees of bald eagles under the direction of the Bald Eagle Act of 1940 (golden eagles were not protected under law until 1962). This same type of special management was also to be carried out around reservoirs in conjunction with the Corps of Engineers and the states.

Some animals, however, were not protected by either federal or state laws. For example, a news release from October of 1950 said that the "porcupine has been declared to be a renegade of the forest and local U.S. forest service officials ask hunters and other forest travelers to shoot them on sight … porcupines have murdered or maimed many young pine trees, in one of the worst rampages in recent years … [Fred] Kennedy has these tips for hunters who want to help get rid of 'Porky': Use a .22-calber long-rifle hollow-nosed bullets." Other techniques were also employed, including the use of strychnine in baited traps that could only be entered by hungry porcupines. This epidemic of porcupines, and their killing, persisted for years in the Region.

Range Management

By 1955, the Region provided forage for almost 90,000 cattle and 165,000 sheep. Seeding of range lands was accomplished on some 13,000 acres, with another 2,200 acres on private lands under a cooperative agreement. A range analysis program was initiated: "To put measurement of range conditions on a scientific basis, the region has started installing lines of permanent study plots, called range transects. Recurrent observations at the same point will give reliable data on the trend of range conditions. Progress in this work has been made, but only about 15 percent of the needed transects have been installed to date." (*Pacific Northwest Report 1955*: 5) The analysis program was 29 percent complete by the end of 1957 and expected to be completed within three years. These analyses by forest considered big game and livestock range needs, as well as mapping.

A large number of range improvements were completed in the Region in this period. For example, in 1957 construction was completed on 178 miles of new fences; 114 water developments; 106 stock ponds; 8 corrals, 11 miles of ditches and pipelines, and 18 fenced observation plots. This work was just a small portion of the overall maintenance of similar projects.

Engineering

Engineering in the Region changed quickly after World War II. Timber management was becoming the dominant use and engineering was needed to help ramp up production. "[B]etween 1945 and 1980, most Forests in Region 6 went from literally primitive transportation systems to full development to meet Forest Service management objectives. … Housing and facility needs grew in proportion. More road access encouraged more public use of the Forests and, thus, expanded recreational facilities needed to be engineered." (Unruh 1990: 556) But not all could be fixed by just hiring more people in engineering. C. E. Remington recounted what the division of engineering was like in 1954:

> I reported … March 1, 1954, at Portland, Oregon, as the Assistant Regional Engineer for Region 6. I arrived in Region 6 in time to be whisked off to the Wind River Experiment Station where a Forest Engineers' meeting was in progress … Region 6 had made it a practice for years to assign any forestry graduate—and many who were not graduates of any school—to the position … of Forest Engineer. Many carried the title whether they wanted it or not, and whether or not they might have graduated from the only two schools that claimed to graduate Forest Engineers—Oregon State and the University of Washington. As a result, even after I arrived in the Region, you could count the actual graduate Engineers on the fingers of one hand and still have enough left over to scratch with! (Remington 1990: 473)

Roads continued to be constructed to open the forests for timber management, as well as to provide access across the forests for transportation of forest products and recreationists. Most of the forest road work was completed through timber sale contracts. Yet this caused problems as the timber sale operators had different standards as to what they wanted for roads. "One of the biggest pains Forest Engineers had to deal with during the rapid development of the road system was the prudent timber sale operator road concept … A purchaser could only be required to build a standard of road that a prudent operator would build to remove the timber for a given sale. The Forest Engineer was expected to plan and get built the ultimate standard road needed for National Forest management purposes. Needless to say, the two standards were rarely the same." (Unruh 1990: 557)

In 1955, according to the Regional Office report, 550 miles of new timber access roads and 205 miles of reconstructed roads were completed, with an additional 80 miles of new, reconstructed, and other roads completed using federal road and highway funds. Other accomplishments in 1955 included 128 miles of trails, and maintenance on 12,000 miles of roads and 14,000 miles of trails.

The Olympic NF was getting heat from the Washington Fish and Game Department in 1956. The state agency, as well as fishing organizations, were increasingly concerned about the number of landslides in the Olympics caused by road construction on the west side of the Olympic NF. Forest Supervisor Dick Worthington described the problem:

> The Olympics is very unstable country … So when you go and build roads in this stuff, you often create problems. If you don't follow all of the prerequisites of road building, for example, don't sidecast dirt on real steep land, don't put wood material in your fills because it rots and then you will have slides, get the water drained off the road—rules of thumb like this—then you are going to have problems. … [some] loggers … were more interested in getting logs out then [than] they are building roads, so if you don't have somebody that is knowledgeable and is a capable administrator, you … wind up with a poor job, and that was what was going on on [in] the Olympics … It was no wonder that the fish people were upset with us … We had a bunch of meetings, and I wound up having a confrontation with the timber industry over roads (Worthington 1984: 29-30).

During the 1958 field season, electronic computers were used for the first time for road design work. This electronic system was used that year to design

President Harry Truman and Gifford Pinchot III at the renaming of the Columbia National Forest as the Gifford Pinchot National Forest, Washington DC, 1949. USFS photograph.

155 miles of roads which "has resulted in the release of a considerable number of men from office work for field work." (*Pacific Northwest Report* 1958: 20) The computer road-design programs "started quite modestly with transverse notes, expanded as new programs were worked out, and finally included all elements of road design, including cuts and fills, quantities, and plotting of notes. Machine work from 1958 to 1966 was done on the BPR [Bureau of Public Roads] computer, and beginning in 1966 on the Forest Service machine." (Grefe 1990: 279)." Computer design in 1958 amounted to 50 miles of road, increasing to 1,100 miles by 1960, and 2,100 miles by 1966.

Engineering also worked with the Gifford Pinchot NF to redesign and reconstruct the water system for the Hemlock Ranger Station, adjacent Experiment Station, and the regional training school buildings.

The Regional Engineer Division in early 1954 completed a study of how the equipment management shops on the forest around the Region were operating. The overall direction of the forest shops came from the Regional Sellwood Shop that "had an organization of some 60 to 70 mechanics, and each shop had a smaller contingent. They had complete control of all equipment and, in addition, maintained an equipment development center that was second to none in the Service." The study recommended that downsizing was needed to get the forest shop groups under control. "All of this had come to a screeching halt in the spring of 1954, and the RIF [reductions in force] actions were very difficult to accomplish. … the small remaining shop was left in Portland with repair mechanics left on each Forest." (Remington 1990: 473)

New recreation maps were needed, especially to show the miles of new roads used to access the timber sale areas. Between 1947 and 1955, some 44,000 square miles of planimetric mapping on a scale of 2 inches to the mile was accomplished. Additional mapping continued in subsequent years; by 1962 some 100,000 square miles had been completed, covering all the forests in the Region. The USGS continued to improve their mapping of the national forests during this era, but had great difficulty in keeping up with all the new roads being built.

Fire Management

Fires continued to plague the Region during the late 1940s and 1950s. Lookout houses and towers, the mainstay of fire detection for decades, were beginning to be phased out, partly because other means of detection were more efficient. Increasingly, patrol planes and helicopters were used to detect fires. Road travel now was possible to or near many fires. In more remote areas, smokejumpers still dropped to the fire or helicopters brought in firefighting crews. The R-6 *Administrative Digest* on July 14, 1948, noted that "Jack E. Handy ... and Francis Lufkin ... tested a Bell 47B3 Helicopter on July 3 to determine to what extent it could be used in fire control work ... Tests were made from the Pasayten Airport." The helicopter made a successful landing and take-off at an elevation of 7,400 feet, and they determined that the helicopter could prove useful for setting a smokejumper and his equipment into a high-elevation forest. Wendall Jones recounted his experience in moving firefighters by helicopter, "I participated in an 'experiment' of moving a crew by helicopter in 1950—one crewman at a time. And it took longer to get the job done than if the crew had started hiking up the hill at the same time." By the early 1950s, helicopter work in the Region was restricted to fire reconnaissance on big fires. But, Jones continued, "use of helicopters for fire detection was very limited because the job could be done very well using small fixed wing planes at a much lower cost." (Jones personal communication)

The R-6 *Administrative Digest* on October 19, 1949, noted that "the Forest Service airplane maintenance and repair shop was moved from [the] Portland-Columbia airport to [the] Troutdale airport last week. A hanger located at the west end of the Troutdale airport, leased from the Port of Portland, will accommodate the shop and provide winter storage for our six airplanes. The new location will continue to be known as the Air Base Shop."

One story about firefighting in 1952 comes from the memoirs of Chris Chriswell, who was stationed on the Mt. Hood NF. A slash fire on Second Creek escaped from the location it was supposed to be burning. "Again, we hired Burnsiders [i.e., mainly alcoholics from the Burnside area of Portland]

Flat Top lookout with Billee Web, Gifford Pinchot National Forest, Washington, 1949. USFS photograph.

and set up a fire camp. … Burnsiders were always interesting characters. A bus load of them stopped at the office in Estacada and the driver went to get directions to the fire. He made the mistake of leaving the bus door open. The men saw a tavern across the street and soon the bus was empty. The driver had quite a time rounding up that bunch." (Chriswell 1989: 91)

By 1955, the Regional fire aircraft fleet consisted of six planes—three were used to transport smokejumpers and the other three for passenger travel and aerial reconnaissance work. A smokejumper satellite base was established in the spring of 1959 at Redmond to serve forests in the central Oregon area, supplementing the smokejumper bases at Illinois Valley on the Siskiyou NF and Intercity on the Okanogan NF The Forest Service airplane fleet provided planes for scouting and reconnaissance work, transport planes to fly in men and supplies, and helicopters. The R-6 forests constructed hundreds of helispots on the forests where helicopters could be used to transport firefighters and equipment. During the 1958 fire season, the Region had two helicopters under contract for fire work.

The first use of aerial tankers in the Region was during the summer of 1957. Many of the airplanes were converted crop dusters and World War II bombers, now their bomb bays full of a mixture of sodium-calcium-borate and water, rather than bombs. An experimental air tanker base was established at Wenatchee in 1957 with the borate mixture first being dropped that summer on a fire in the Wenatchee. The following year, 14 air tankers were stationed at Wenatchee, La Grande, Medford, and John Day. They dropped over 122,000 gallons of the sodium-calcium-borate mixture and another 77,000 gallons of just water in 1958. During 1959, additional air-tanker bases were established at Redmond, Lakeview, Eugene, and Troutdale.

Research Work on the National Forests

During the same period, national forest research came of age as the research stations and new experimental forests conducted studies to find better ways to harvest trees, construct new roads, and measure the effects of logging and roads on streams and watersheds. Research during this period (only partially covered in this history) underwent some fundamental changes. One major shift in emphasis was the study of the relationship of people to the out-of-doors. This included anti-litter campaigns, studies of patterns of outdoor recreation, and an examination of the social aspects of camping. Research centers were established at Corvallis, Olympia, La Grande, and Roseburg.

New research natural areas (RNAs) were established in the Region by the chief of the Forest Service during this postwar era. Several were established in the 1940s: the Abbott Creek RNA, located on the Rogue River NF, was established in 1946 to study mixed conifers, especially sugar pine forests; Cedar Flats RNA on the Gifford Pinchot NF was established in 1946 to study stands of western red cedar in valley bottom habitats; Lake Twentytwo RNA, established in 1947 on the Mt. Baker NF, was estavlished as an example of old-growth western red cedar and western hemlock; Long Creek RNA, also on the Mt. Baker NF, was established in 1947 as a compliment to the Lake Twentytwo RNA, as they both were set aside for research on western red cedar and western hemlock stands; Meeks Table RNA, located on the Snoqualmie NF, was established in 1948 as an example of ponderosa pine and pinegrass communities in eastern Washington; and the Persia M. Robinson RNA on the Mt. Hood NF was established in 1948 for the study of ponderosa pine and mixed pine-Douglas-fir forests.

In 1948 the Blue River Experimental Forest (renamed the H. J. Andrews Experimental Forest on July 26, 1953, to honor the memory of deceased Regional Forester Horace J. Andrews) was established. The "Andrews" was used initially to study methods of harvesting Douglas-fir. This involved experimenting with clearcuts to determine the optimum size of clearcut areas, partial cutting, and studying the respective advantages of artificial versus natural regeneration. In 1949, a harvest of 10 million board feet was made, with studies of the optimum size of clearcuts, road building, and siltation. (Geirer 2007)

In other areas, the research sites studied the use of various chemicals in brush control; the use of fire retardants; how best to handle flood damage after the severe flood of 1948; and utilization projects to determine methods for using waste for fiberboard and pulp. The Entiat Experimental Forest was established in 1957 on the Wenatchee NF to study the effects of road building and timber harvesting on small streams. In 1970, a fire roared through the area and since then studies have focused on wildfire recovery. In 1951, the 4,647-acre South Umpqua Experimental Forest was established on the Umpqua NF. It was intended to provide research on mixed evergreen forest types. The size was reduced in 1967 to 650 acres. (Adams et al. 2004)

The Pacific Northwest station conducted the first commercial thinning on a Douglas-fir plantation in the Region during the spring of 1951 on the Olympic NF. As described by Elmer Show, "the experimental thinning was confined to a five-acre area within the 1,500 acre Snow Creek plantations, established in 1927 adjacent to the old Snow Creek Guard Station." (Rooney 2007: 141) Six years later, the Olympic NF was the site of the first clonal seed orchard in the entire US Forest Service.

Chapter 7

"Full Speed Ahead" or "Get Out the Cut" 1960–1976

When much of the country was looking at the sky as the new era of space exploration was getting underway, the Forest Service was looking at the ground and how best to manage the forests for the present and future. The 1960s through the middle of the 1970s was an era marked by the "can-do" attitude of the Forest Service. This was a time of moving the agency into intensive timber management that adhered to the industrial forestry model of forest management.

Building forest access roads and "getting out the cut" were the primary focuses of the agency in the Pacific Northwest. Foresters and professional engineers got out massive amounts of timber to achieve increasing timber volume goals. New professionals began to come into the work force at the national forest level, including forest engineers, landscape architects, foresters trained in silviculture, and wildlife biologists. It was an era of high morale, big budgets, lots of new employees, and major accomplishments.

Regional Foresters

Charles Connaughton began his Forest Service career doing silvicultural and watershed research. Later, he was the director of the Southern Forest Experiment Station. He was appointed as the regional forester in the Southern Region in 1951 and then in the Pacific Southwest (California) Region in 1955 before being selected as regional forester for the Pacific Northwest Region in 1967. He was a strong advocate of multiple use and had great skill in reconciling competing uses of forest land and in working with competing groups. Near the start of his tenure, the first Roadless Area Review and Evaluation (RARE) was begun under the direction of the Wilderness Act of 1964. Connaughton served as regional forester until his retirement in 1971. He was succeeded by his deputy, Rex Resler.

Rexford Resler worked on the Siuslaw and Rogue River NFs, and on the Willamette NF, where he became deputy supervisor in 1962; he was appointed supervisor of the Malheur NF in 1965. In 1968 he went to the Washington Office, where he worked in the recreation division, then was reassigned as the assistant director of timber management. In 1970 he became deputy regional forester in the Pacific Northwest Region and the following year was appointed regional forester. He was faced with implementing the new National Environmental Policy Act of 1969 and the Clean Air Act of 1970. Resler was transferred to Washington in 1972 as deputy chief, retiring in 1978. He was replaced as Pacific Northwest regional forester by Theodore Schlapfer.

Ted Schlapfer started working in the Pacific Southwest (California) Region. As R-6 regional forester, he became involved with the end of the RARE analysis and recommendations for wilderness. In addition, Schlapfer contended with implementing the new Resources Planning Act (RPA) in 1974, the Monongahela decision against clearcutting in West Virginia, the National Forest Management Act (NFMA) in 1976, and a new roadless area review (RARE II) that was begun in 1977. Schlapfer retired in 1977 and was succeeded by Dick Worthington (see Chapter 8).

National Changes in the Forest Service

Lyndon B. Johnson's administration looked at ways to improve management of the federal government. An Organizational Review Committee was organized in the spring of 1964 to review the management and policies of the Forest Service. The team, headed by Edwin Deckard, spent one year visiting the Portland Regional Office and the other Regions except Alaska, looking into

NEW PATCHES AND UNIFORMS

In the 1960s, a cloth shoulder patch was authorized for wear on the left shoulder of official uniform shirts and jackets. The first authorized patch, issued in 1962, was flat on the bottom and sides, but rounded on the top. A curved overhead bar was added to designate which national forest or other office the wearer was from. The current Forest Service shield patch was introduced on March 20, 1975. Special law enforcement patches have been used since the 1970s, although law enforcement badges resemble those of other federal agencies.

Dress uniforms for women were officially adopted in the late 1960s. All of this was part of an effort to present a new image of the employees to the public.

> In 1969, I remember a Forest Service uniform policy review. Uniform components and designs were being changed. Of course, we didn't have to worry about uniforms for women [there were no official women's uniforms at the time]. Too bad we didn't have a woman on the review team—we men would have looked better. In 1970 while working on the Snoqualmie National Forest, I had an assistant, Mabel Gillard … She was so ecstatic when they announced that women no longer had to wear dresses or skirts to work. From that day on, she always wore pants or slacks! (Cooper personal communication)

the appropriate size of ranger districts, national forests, and regional offices, and recommended that these be standardized throughout the national forest system. The report noted a great range in the number of ranger districts per forest, ranging from three to 13 and in size from 211,200 to 3,118,900 acres. It recommended that an optimum ranger district would have between 10 and 90 employees, manage 300,000 to 600,000 acres, and operate on a budget of $150,000 to $400,000 per year.

The report also found that the number of employees in the various regional offices varied considerably. For example, employment in Region 7 was 1,421 employees, while Region 6 had 7,414 employees. Region 7 stood out in the analysis as a place for reorganization or elimination. The Chief and staff decided to split Region 7 between Regions 8 and 9, which took place in late 1965 and early 1966. The report also proposed top-level changes for the agency, including establishment of an associate chief, a deputy chief for the national forest system, and directors for the northeastern and southeastern state and private forestry areas. These changes were made and still survive to this day.

The Public Land Law Review Commission (PLLRC) was established in 1964 by Congress to study and recommend changes or additions to the nation's land laws. The PLLRC report (*One Third of the Nation's Land: A Report to the President and to the Congress*), printed in June 1970, made 137 recommendations. The commission was critical of the Forest Service for spending too much time and money on administering natural resources on the national forests. The commission, with a strong commodity bias, recommended that "dominant use" should replace multiple use as the "highest and best use of particular areas." An internal government plan was proposed during the Nixon administration to reorganize all the federal agencies into standard regions based on military regions.

> [E]very agency was told to wind up conforming to this structure. When it was announced the Forest Service, of course, was immediately told that they were going to have to conform to this regional structure, which meant in their case, since they had a regional office in Missoula, and the region that would cover Montana was going to be in Ogden, Utah, that the Missoula regional office would be closed. ... Oddly enough, there would be no office in Albuquerque, New Mexico, which was a regional office in the Forest Service for Arizona and New Mexico, that would have been transferred to Dallas, which would be a new federal region. And there would have been relatively few forests in that

region, too. Anyway, the idea no sooner got out, was announced, than, I don't think the ink was dry [and the idea was dead] (Wolf 1990: 2).

The PLLRC task force report to standardize the Regions was rejected in 1972 by chief and staff, and instead they directed that the Regional Offices had to be reorganized. Former Regional Forester Jeff Sirmon noted that "the sudden move to establish three Deputy Regional Foresters at each Regional Office—National Forest System, State and Private Forestry, and Administration—was to help stave off efforts to combine the Forest Service into Interior or a new Department of Natural Resources." (Sirmon 1990: 680) Apparently by adding new deputy regional foresters they could take more of the management burden from the regional foresters and thus manage the Regions more effectively with smaller staffs and divisions under each deputy.

Public Law 92-82, signed into law in August 1971, authorized cooperative law enforcement between the Forest Service and local police authorities. A Regional Office news release on September 1, 1971, noted that "such cooperation could include reimbursement to a state or its subdivisions for expenses incurred in connection with law enforcement on National Forest System lands."

Regional News

In the summer of 1960, the Regional Office was reorganized. A Regional news release on May 13, 1960, described the changes: "[T]he Division of Recreation and Lands will be split into two separate divisions. The Division of Recreation with Frank B. Folsom as Division Chief ... will retain the responsibility for national forest recreation management, minerals management, special lands uses, and special land classification studies. The Division of Lands with Russell P. McRorey as Division Chief will be responsible for procurement of rights-of-way, land adjustments and classifications, boundary survey and posting, and land status records management." The rights-of-way section had previously been part of Timber Management.

In the summer of 1961, a new position of deputy regional forester was established. Alfred Spaulding was promoted to the new position on July 9. A news release about the new position noted that "the workload has been expanding and this will continue as pressures for greater use of the National Forest resources increase." The news release continued "the existing organizational pattern which provides for an Assistant Regional Forester in charge of each of the major fields of activity will be continued." The Regional

Governor Mark Hatfield and astronaut Walter Cunningham in the lava beds, Deschutes National Forest, Oregon, c. 1965. USFS photograph.

Office moved again, in December 1965, to a new headquarters site in the old Multnomah Hotel in downtown Portland at 319 S.W. Pine Street, where it remained until 1992.

Lassie and "Ranger" Cory Stuart (played by Robert Bray) from the TV show came to the Region in 1964. Filming for the series was, in part, conducted at Blue River on the Willamette NF. Lassie returned for a visit in 1967 as the production company filmed scenes at Timberline Lodge, Wind River Nursery, Broughton Lumber Company flume in the Columbia Gorge, and during log rafting on the Columbia River.

Walter Cunningham and four other astronauts journeyed to the lava fields near Bend as early as August 1964 as a test for what they might find on the moon. Others arrived on October 8, and six more in November. Cunningham and a team of experts from the Manned Spacecraft Center in Houston visited the Deschutes and Willamette NFs in the summer of 1967 to walk among the lava beds at McKenzie Pass, pumice areas near Crescent, and the obsidian flow at Newberry Crater. Although the first moon landing was five years away, the astronauts wanted to be prepared for any eventuality. Twenty-two astronauts, including Stuart Roosa, a former Siuslaw NF employee, visited the Paulina

Peak area on the Deschutes NF in July 1966. Roosa, who had done survey work on the road up Mary's Peak and was a former smokejumper at the Siskiyou Smokejumper Base at Cave Junction in 1953, went to the moon in the Apollo 14 mission in 1971. As pilot of the command module he orbited the moon for 33 hours while others walked on the surface. He carried hundreds of tree seeds supplied by the Forest Service; those that have since been planted have been called Moon Trees—one was planted at the Siskiyou Smokejumper Base on June 12, 1976.

The Region became an interagency cooperator dealing with electrical transmission lines. One such project occurred in 1965: the Northwest-Southwest Intertie transmission line developed by the Bonneville Power Administration and Portland General Electric Co. Authorized by legislation in 1964, the line from the Columbia River to the California border crossed the Deschutes and Ochoco NFs and Crooked River National Grassland. Smaller electrical transmission lines were constructed between the John Day Dam and the lower Willamette Valley, and others were built along transmission corridors in other locations in the Region.

"Managerial Grid," an employee-effectiveness training program, was introduced to the Region in 1967. Designed for agency employees at all levels in the organization, it used a team approach to make better decisions. The "grid work" included identifying personal management styles, working as a team, and a brief personal evaluation by peers and a psychologist. The personal evaluation was dreaded by some employees. The week-long course was described as follows: "The Grid enables a manager to better understand himself, his organization, and points the way to developing more skillful managers and a more productive organization. Some 700 Forest Service employees from the Pacific Northwest Region, from assistant regional foresters to ranger district assistants, have participated in the program's initial seminars." (*Pacific Northwest Report* 1967: 22)

The Managerial Grid training emphasized working together to solve problems, talking issues through, coming to joint decisions. Everyone who passed through the course talked "grid talk"—especially notable was an exercise that evaluated everyone's personal leadership styles, such as "9-1" or "9-9" or other combinations, indicating their primary methods of interaction with peers and employees. Soon other training courses were available, especially those that were useful in giving foresters, engineers, and wildlife employees hands-on experience with social interactions. For many this was quite a change from dealing with just natural resource and engineering matters.

One highly sought-after course was the "intensive semester" where a small number of employees were brought together, in Portland and sometimes in Washington DC, to interact with academic and community leaders, as well as officials from the Washington Office and Congress. The intensive semester (referred to by some as the "expensive semester") was organized by the Regional training-development expert Chuck DeRidder. In this month-long training, after being taught about political and interpersonal aspects of their everyday work, trainees were sometimes left in unfamiliar social settings in different parts of the country and forced to innovate and in some cases survive. Refresher sessions were used to reinforce the ideas and experiences learned. After 1995, the Region did not renew the intensive semester idea in large part because of the high costs involved.

Other management training programs were established to better professionalize the forest and ranger district employees. Training of employees, used since the 1920s to teach fire control techniques, had expanded by the 1960s to include training for new professionals and administrative management, and training in road and bridge design, business management, safety school, first aid, personnel management for federal executives (a Civil Service School course), and civil rights, as well as numerous functional staff training sessions.

The certified sale administrator program was also offered. The course covered both engineering and timber aspects of the timber sale contracts. Those completing the program became certified sale administrators. Another program was the siviculture certification program that was offered by joint agreements with Oregon State University and the University of Washington, where enrollees spent a term or semester attending. To attain certification, enrollees were required to pass a test that represented a somewhat complex, real problem and offer solutions before a board of examiners at the regional level. Logging engineers were trained at Oregon State University beginning in the early 1960s.

Women and Minorities

Women employees in non-traditional positions were becoming an increasing presence on the national forests in the Region. Anne Webb Heisler graduated with a degree in forestry from Washington State University. She worked one summer as a lookout on the Rogue River NF then went on as a junior forester on the Mt. Hood NF. She became the public information officer on the Mt. Hood NF, then moved to the Regional Office as the environmental education coordinator. One story that Heisler told about the difficulties of field work was recounted by William Morse:

> For one thing, the men I work with are in superb physical condition. They bound up the side of the mountain like marathon runners, while I huff and puff a hundred yards behind ... Last fall, I was coming down a steep slope. ... It was a cold day, and I was bundled up in warm clothes and rain gear. The hill was covered with down logs and bracken fern about three feet high. I dropped behind, and I hurried to catch up, my foot slipped on a down log—and bang! There I was on my back, firmly wedged between a log and the ground, completely covered by the ferns. For some reason, it struck me as funny, and I started laughing so hard I couldn't move or talk. Just at that time, Fred turned around to see how I was doing. He couldn't see me, of course, so let out a yell, "Anne." No answer, so he rushed back up the hill, still calling. By then I recovered some of my wits, and waved a leg above the ferns. We both nearly expired with laughter as he helped pry me out from under the log (Morse 1965: 59-60).

Cindy McReynolds, first USFS woman animal packer, Wallowa-Whitman National Forest, Oregon, c. 1970. USFS photograph.

All-woman tree-planting crew, Willamette National Forest, Oregon, 1969. USFS photograph.

Cindy McReynolds, on the Wallowa-Whitman NF, was the first Forest Service woman horse packer. A Reed College student, she was employed beginning in 1970. Her packing skills were used extensively in the Eagle Cap Wilderness. During the same year 100 women were assigned to fires in various capacities such as timekeeping and other clerical work, but not firefighting. That came later.

Opportunities for professional women in the Forest Service began in the 1970s. Among the first was Wendy Milner Herrett, who was employed in the Regional Office soon after she graduated from the University of Oregon in 1970. Herrett began working with Warren Bacon, the landscape management specialist, on the new visual management system for recreational visitors. She recounted,

> One of the things while working in the regional office, they were trying to find a forest to place me on … The first forest that had a vacancy for a GS7 Landscape Architect was the Gifford Pinchot National Forest. It was one of the forests that timber was king. They sent me out to visit

> with the recreation staff, the forest landscape architect, and some others on the forest, and then I went back to the regional office. I later heard the forest supervisor didn't want to hire me, and I'm going, "Whew! That's a relief because I don't want to work there!"... He didn't want to hire me because I was a woman (Herrett 2001: 21-22).

Herrett later became the first woman district ranger in the Forest Service and the first woman forest supervisor in the Region. She noted that one of the reasons for not hiring women in professional positions was that the work often required going into the field with married employees. Not only were the men worried about how well the women would do the job, but also their wives were concerned about possible inappropriate behavior (Herrett 2001). There were concerns as well about professional women in office settings. In at least one instance, women clerical staff on the Umpqua NF around 1980 expressed being very uncomfortable about working for a woman professional.

In 1974, Leslie Wildesen was hired as the first regional archaeologist in the Cultural Resource Management program in the recreation section. Wildesen was the first woman in a professional series, rather than clerical series, in the Regional Office. She did much to shape policy and procedures for cultural resources including a 40-hour training session for cultural resource technicians called REC-7, which continues today (Wildesen personal communication).

Gail Throop, who was a student of Wildesen at Portland State University, did her masters thesis on CCC structures in Region 6, rediscovering and documenting more than 1,000. Throop was hired in 1983 by the Regional Office as the regional historian (Wildesen personal communication).

Gail Kimbell, who became chief of the Forest Service in 2007, began her career in the federal government as a forester with the Bureau of Land Management in Medford, Oregon. She then held a variety of forester positions in Region 6, serving as a district ranger in Oregon and Washington, then forest supervisor on the Bighorn NF, Pike and San Isabel NFs, and Tongass NF. Kimbell became associate deputy chief for the national forest system in the Washington Office, then regional forester for the Northern Region, before being appointed chief.

Linda Goodman started with the Forest Service in 1974 as a receptionist at a ranger station on the Olympic NF. After a number of jobs in the Region, she served as the national director of the Job Corps program in the Washington Office, then deputy regional forester for Region 6, then acting chief of staff for Chief Bosworth, finally coming back to the Region as regional forester in 2002; she retired in 2008.

Minorities hired by the Forest Service were few and far between in the 1960s. That slowly began to change with passage of the Civil Rights Act of 1964. The federal government began a major push to provide equal access and employment for people of color. But in the Region there were strong centers of hatred against African-Americans. The Rogue River Valley was one of those places that had a heritage of not allowing Black people to live or work there.

James Stoops, director of the Fort Vannoy Job Corps Center just outside of Grants Pass, reported that the area still had a small Ku Klux Klan following from the 1920s and 1930s, as well as a growing John Birch Society membership (*Congressional Record*, November 14, 1967). An account by Barbara Roland gave the essence of the problems in Grants Pass:

> I remember when they first set up the Fort Vannoy Job Corps ... The town went completely nuts. Because they were going to be bringing in people from big city ghettos, and they knew that they would be black and brown and whatever. The town went totally crazy. They would not let any of the young men come into town without a person in control of them. They weren't allowed to go into the stores unless they were in a group with someone in control of them. And it was simply because the town was scared to death of the color (Roland 1995: 59-60).

When employees of color, especially African Americans, were stationed in remote ranger stations many people in the small towns, including the regular Forest Service employees, had to face their own prejudices. Fred Cooper noted that the Region organized sensitivity sessions for employees that were initiated and led by

> Jetie B. Wilds, Jr. beginning in 1970. Jetie later became the first Civil Rights Director for the Region in 1975. The objective [of the training] was to help those in the work force, who were primarily white males, understand that minorities and women could be trusted and that they were looking for acceptance in the workplace. Another approach was the more traditional "teaching" of the principles and concepts of the Civil Rights Act and that it was the law of the land (Cooper personal communication).

Land Status and Ownership

NATIONAL PARKS

In 1963, the Secretaries of Interior and Agriculture appointed a study team to investigate the North Cascades. "The study was completed in 1965 and the result was a fine comprehensive report encompassing a history of the area, its uses, its resources and final recommendations for its administration." (Chriswell 1989: 125) The Forest Service proposal was, of course, for continued management by the agency of all the North Cascades area. Chris Chriswell described the changing situation:

> Suddenly politics changed the whole picture. The *Sunset Magazine* came out with a feature article suggesting a National Park ... Governor Dan Evans issued a State recommendation for a National Park and John Nelson of Seattle City Light supported a Park provided their Skagit power projects were fully protected ... Word filtered down that President Johnson wanted the national park created during his administration. The Secretary of Agriculture had to stop supporting our plans and we were cautioned not to make public remarks covering our plans for the area proposed as a national park (Chriswell 1989: 131-32).

After 30 years of wrangling (see Bob Marshall's visit in Chapter 5), North Cascades National Park was established on October 2, 1968, when President Lyndon B. Johnson signed the North Cascades Act. The act created the 684,000-acre North Cascades National Park, as well as Ross Lake and Lake Chelan National Recreation Areas. The act also created the 431,974-acre Pasayten Wilderness on the adjacent Okanogan NF and Mt. Baker-Snoqualmie NF and enlarged the Glacier Peak Wilderness to 573,973 acres on the Mt. Baker-Snoqualmie and Wenatchee NFs.

Oregon Senator Wayne Morse introduced a bill in 1967 that called for the National Park Service to study the High Cascades of Oregon from the Columbia River south to the northern boundary of Crater Lake National Park for national park status. *The Oregonian* for February 2, 1969, reported that the "bill got nowhere, and that's just as well, because an extensive park probably would be too rigid in view of the multiple uses that legitimately are made of the Cascade crest. But [Regional Forester] Connaughton admits that the hint of competition in the Cascades from a respected rival agency did 'further stimulate' compilation of the Forest Service [Oregon Cascades] plan." Later that year, Senators Bob Packwood and Mark Hatfield also suggested similar park status for the Cascades. They also failed.

National Forest Changes

In 1961, for the first time since 1907, a new national forest was created in the Region when the Winema NF was established in south-central Oregon, taking 419,000 acres of the Klamath Indian Reservation that had been sold by the tribe to the federal government and combining it with Rogue River NF lands lying east of the Cascade Range crest. These lands were used to create the new 915,000-acre Winema NF. Another approximately 100,000 acres of Klamath Reservation lands were added to the Fremont NF, while other pieces were sold to private timber companies. New ranger districts were established at Chiloquin, Chemult, and Klamath Falls, which was an older Rogue River NF district office. The supervisor's office was located in Klamath Falls under the direction of Alex Smith.

A new reorganization study was conducted in 1973 at the national forest level, considering the consolidation of several NFs. A news release on April 9, 1973, noted that "Forest Supervisors in the six areas [under study] will head up the studies in their locations. The Supervisors [will make proposals that will be] forwarded to [Regional Forester Ted] Schlapfer and, later, to Forest Service Chief John R. McGuire." The six areas considered for consolidation were (1) Snoqulamie and Mt. Baker NFs, (2) Wenatchee, Okanogan, and Colville NFs, (3) Malheur, Umatilla, and Wallowa-Whitman NFs, (4) Deschutes and Ochoco NFs, (5) Fremont and Winema NFs, and (6) Siskiyou and Rogue River NFs. As a result of the studies, only the Mt. Baker and Snoqualmie NFs were combined, although 35 years later the Wenatchee and Okanogan, Fremont and Winema, and Rogue River and Siskiyou NFs were combined administratively. These combinations were primarily caused by the need to reduce budgets.

In 1974, the Colville NF, which had started in the Region in 1908, but been transferred to Region One in 1943 (see Chapter 5), returned to the Pacific Northwest Region. By the early 1970s, Congress wanted to realign the national forests along state lines as much as possible. For the Colville NF, much of which was inside Washington, this involved splitting the Kaniksu NF, some of which is still in eastern Washington; the Priest Lake RD went with the Kaniksu NF administered by the Northern Region (R-1), while the Newport RD went to the Colville NF now administered by Region 6 (Holstine personal communication).

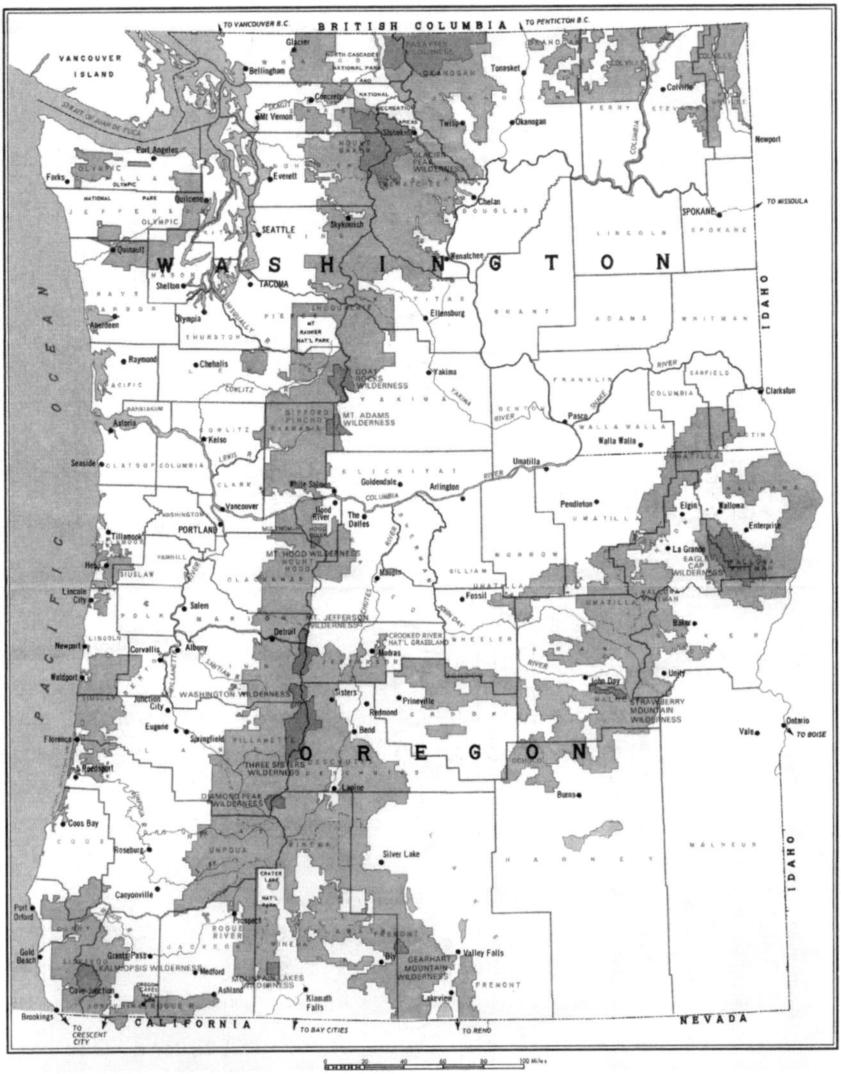

National Forests in the Pacific Northwest. USFS map.

In 1855 the "Indian tribes of Middle Oregon" ceded 10 million acres of their ancestral lands to the United States, reserving approximately 600,000 acres for their sole ownership and use as the Warm Springs Reservation. The treaty specified that the northeastern corner of the reservation was to be "in the middle of the channel of the De Chutes River opposite the eastern termination of a range of high lands usually known as the Mutton Mountains." The first official boundary survey was conducted in 1871 by Tom Handley and R. T. Campbell. However, they began their survey from a mountain ridge farther south, resulting in a reservation much smaller than the tribes had expected in the treaty. This prompted an immediate protest by the Warm Springs people.

Congress responded by authorizing a second survey in 1886 and John McQuinn began his survey using the Mutton Mountains as his landmark and subsequently ended up with a potential reservation over 80,000 acres larger than the first survey. Yet the area was not added to the reservation. This area along the northern and northwestern boundary of the reservation became known as the McQuinn strip. The disputed land was within the old Cascade Range Forest Reserve of 1893, and subsequently within the boundaries of the Mt. Hood NF. The Warm Springs tribes continued their protests. The next few years of the dispute were described by Joshua Binus (2002):

> In 1917, Congress commissioned another study which confirmed the validity of the McQuinn survey. Unwilling to transfer disputed land, the federal government offered the tribes money to compensate them for their loss. The tribes rejected the offer. In 1930, Congress passed legislation allowing the tribes to take the land dispute before the U.S. Court of Claims. Eleven years later, in *Warm Springs v. United States,* the court recognized that minus an 8,000-acre triangular plot in the northeastern corner, the McQuinn line was legitimate. Unfortunately for the tribes, the court still refused to transfer ownership of the McQuinn Strip and offered only financial compensation for their loss (Binus 2002).

On December 8, 1971, Congressman Al Ullman introduced a bill in Congress to finally settle the McQuinn ownership dispute by transferring most of the McQuinn Strip to the reservation. Forest Service Chief John R. McGuire, at a hearing before the House Interior subcommittee in May 1972, said the Nixon administration opposed the bill. But the bill was passed and on September 21, 1972, President Richard Nixon signed it, and the Warm Springs Reservation added all but a small portion of the long-disputed McQuinn Strip area.

Weather Becomes a Problem

On October 12, 1962, a massive windstorm that quickly became known as the Columbus Day Storm hit the Pacific Coast. The storm consisted of the remnants of two typhoons (Emma and Freda) that had crossed the North Pacific, reformed off northern California, and raced northward along the coast. Hardest hit were the coastal forests, such as the Siskiyou, Siuslaw, and Olympic NFs, but others along the Cascades were also battered. All told, 18 of the 19 national forests in the Region were affected by the storm. The 1962 Pacific Northwest annual report noted that 90 percent of the standing trees at the Tahkenitch Campground in the Siuslaw NF had been knocked down by the fierce winds. The wind gauge at Cape Blanco was destroyed in the storm, but wind speeds were estimated at 170 miles per hour. The storm killed 58 people in the two states. Approximately 2 billion board feet of timber were knocked down by the storm on the national forests, and more millions were thrown over on BLM and private lands. The Region estimated that around 70 percent of the downed trees were available for salvage. By the end of 1963, 1,424 salvage sales were let to recover some 1.8 billion board feet of storm-damaged timber.

Another natural disaster hit the Region in 1964. The Christmas Week Flood actually began with a late November hard freeze, followed by deep snow fall in early December. Then just before Christmas, warm Chinook winds quickly melted the heavy snowpack, flooding the already rain-swollen rivers. The frozen ground under the snowpack prevented the melted snow from soaking into the ground. A report from the Willamette NF described the scene:

> Stillness had fallen over the Detroit Ranger Station as Old Mother Nature covered the area with snow … It was December 21 … We had approximately 16 inches of snow … and it started raining. We are all accustomed to rain in Oregon so this was not an alarming fact. However, by 9:00 a.m. on December 22 we all knew that we had problems. Total rainfall by now was 4.61 inches with 3 inches of this being in the last twelve hours. Rainfall for December totaled 31 inches. Whitman Creek, which is the first creek above Detroit Dam, had reacted to the rain and took out approximately 200 to 250 feet of the North Santiam Highway … By 12:00 o'clock on December 22, 1964, Boulder Creek … had washed out the highway. Detroit was now isolated. Roads were closed on both sides … We knew we had lost most of our bridges [on the forest]. The roads were all closed by slides or washouts.

The flood cost the lives of 42 people and the economic damage was estimated to be as much as one billion dollars. The Mt. Hood, Willamette, Umpqua, and Siskiyou NFs suffered the most damage. Numerous campgrounds were destroyed or severely damaged; roads suffered slides and washouts; 126 road and trail bridges were destroyed, and culverts by the hundreds were washed away. Damage to the roads and trails alone was estimated at $12.2 million. Another $2 million of damage was estimated to administrative and recreation sites. Louise Parker, from the Willamette NF, noted that "debris on the Corps of Engineers Detroit Dam in the Willamette National Forest covered an estimated 230 acres. Many of the logs were branded and logging companies would salvage them." (Parker 1965: 16) Much of the damaged timber in the forests was salvaged, but it was many months before the roads and campgrounds could be restored.

Grasslands

The national grasslands were established by the Secretary of Agriculture in 1960. As noted in Chapter 6, in the depths of the Great Depression, the federal government had obtained title to several million acres of eroded and unproductive farmlands. On June 20, 1960, some 3,804,000 acres were designated as 19 national grasslands as part of the national forest system.

This new designation included the 105,925-acre Central Oregon Land Utilization Project on the Ochoco NF in central Oregon (now enlarged to 112,357 acres). Chief Richard McArdle selected the name: the Crooked River National Grassland. Joe Mohan became the first manager of the new grasslands. Much like the national forests, the grasslands allowed management for outdoor recreation, range, timber, watershed, and wildlife and fish. The Crooked River National Grassland, the only national grassland in the Region,

Sheep on the Crooked River National Grasslands, Oregon, 1964. USFS photograph.

contains portion of three rivers—Crooked, Deschutes, and Metolius, as well as the Round Butte Dam and Haystack Reservoir. Along with the Cove Palisades State Park, the grassland is a popular summer recreation site. A February 24, 2003, *Federal Register* notice for improving the grassland stated:

> The Grassland was created after the homesteading era failed in the 1930's. Between 1935 and 1965 the Soil Conservation Service rehabilitated and developed the Grassland by seeding approximately 60,000 acres of abandoned farmland to crested wheatgrass and beardless bluebunch wheatgrass, developing livestock drinking water, fencing and implementing grazing allotment management plans (AMPs). During the 1960's the USDA Forest Service assumed the management of the CRNG.
>
> Sagebrush steppe and western juniper woodland plant associations primarily characterized the historical vegetation of the Grassland. Historically wildfire maintained the balance of these plant associations. It is believed that approximately twenty to forty percent of the land area of the Grassland would have had a major component of western juniper. Today western juniper exists on approximately ninety percent of the land area of the Grassland.

Employment, Youth, and Volunteer Programs

Work programs became important for the nation and the national forests in the 1960s. The Forest Service became involved through the Accelerated Public Works Program (APW), designed to put unemployed men to work (there were no women in the program) on projects to develop or improve national forest resources. The program provided immediate work for over 9,000 men on more than 100 national forests in 35 states. APW projects in the Region included work on developing and improving campgrounds like the Grayback Forest Camp on the Siskiyou NF. Many other APW projects across this Region included construction of camp toilets, telephone lines, a smokejumper center at Redmond, bridges, fire lookouts, and other facilities, as well as planting trees, thinning timber stands, improving fish and wildlife habitat, and constructing or improving roads. By January 1, 1964, "projects totaling $6.3 million had been completed or were committed in Oregon and Washington. The program affects 10 National Forests and 15 counties in Oregon involving expenditures of $3,485,000, and 6 National Forests and 11 counties in Washington, with an outlay of $2,660,000." (*Pacific Northwest Report* 1963: 20) The APW program ended in 1964 when Congress failed to renew it.

A new work program called the Job Corps began in 1964 for unemployed youth ages 16 to 24. Many of the enrollees were Black and Latino. Most of the original Job Corps center sites were run by the Forest Service and the Park Service, in cooperation with the Department of Labor's employment and training administration. The work assigned to the Job Corps enrollees was similar to the old Civilian Conservation Corps in the 1930s—trails, campgrounds, buildings, etc. In the Region, Job Corps centers were under construction in 1964, opening in 1965, at Wolf Creek (Umpqua NF), Camp Angell (Siuslaw NF), Timber Lake (Mt. Hood NF), Cispus (Gifford Pinchot NF), and Curlew (Colville NF). Each camp had an authorized 220 corpsmen (no women at the time), except for Camp Angell which had only 120 young men. Cispus center was removed as a Job Corps Center in 1969 and became the Cispus Environmental Learning Center the following year. The center is now used for outdoor and environmental education of Washington State school children and adults.

George Foreman said at age 16 he was headed for a life of crime in Texas. His life changed when he saw football players Jim Brown and Johnny Unitas in a television ad praising the new Job Corps. Foreman signed up and was assigned to Fort Vannoy on the western edge of Grants Pass, Oregon. His life changed again on November 22, 1965: the Job Corps "boys were all sitting in the dayroom, listening to the radio broadcast of the championship fight between Cassius Clay and Floyd Patterson, when one of them challenged Foreman, 'If you think you're so tough, why don't you become a boxer?' 'I really didn't want to be a boxer ... but I took that challenge that day in the dayroom and here we are today'."(Silva and Nelson 2006: 129) Foreman would go on to win an Olympic gold medal in 1968 and become the heavyweight champion of the world in 1973.

In 1970, a three-year pilot Youth Conservation Corps (YCC) program began; it became fully established in 1974. It was designed to further the development and maintenance of natural resources by America's youth between the ages of 15 and 19. The young male and female YCC members, from all parts of the country and all walks of life, spent the summer months working on conservation projects on the national forests. In the summer of 1972, five YCC camps were established in the Region—two operated by the Snoqualmie NF, one on the Gifford Pinchot NF, and two on the Ochoco NF—employing a total of 152 young people. In 1976, a pilot year-round YCC coeducational program was started at the Cispus Environmental Center on the Gifford Pinchot NF.

The Volunteers in the National Forests Act of 1972 authorized the Forest Service to recruit and train volunteers to help manage the national forests. It has been a highly successful and visible program. Many of the volunteers are retired people who enjoy working outdoors and with the public in a wide variety of capacities, ranging from being campground hosts to assisting with archaeological digs. Often 100 or more volunteer opportunities are available each summer in most parts of the national forest system.

Regional and National Forest Planning

MULTIPLE-USE SUSTAINED-YIELD ACT OF 1960

The Multiple-Use Sustained-Yield (MUSY) Act of 1960 was necessary because many members of Congress and interest groups felt that the Forest Service was giving too much attention to timber harvesting on the national forests—just 15 years after the huge postwar push to open the national forests for needed timber to be used in the national housing boom. One important aspect of the MUSY Act was the creation of multiple-use planning, which brought a number of new specialists such as soil scientists and wildlife biologists into land management decisions.

In 1960, the Region was host to the 5th World Forestry Congress, held in Seattle, was attended by 1,970 members. The central theme of the Congress was the "Multiple Use of Forest and Associated Lands." Chief Richard McArdle delivered the keynote address, noting: "Although 'multiple use' may not be a customary term everywhere, the practice of multiple use has been long established in some intensively managed forests of other countries … In past years many of us have thought that we had enough land in forest in the United States of America to meet all foreseeable needs for wood and other products and services of forest lands. Today we are not so sure. We think our earlier estimates were too conservative."

MUSY helped to introduce or formalize the idea of zoning the forests for single uses, such as timber or recreation, that when totaled gave a forest a mosaic of multiple uses. It was never intended that all forest activities should be available on every acre, although several overlap like timber harvest and road access. National forests began a process to devise single-use unit plans. Other forests took individual single or functional resource plans and merged them into one overall multiple-use plan, and thus allocating different land uses over the entire forest without conflicting overlaps of priority resources. In Region 6, the Deschutes, Willamette, and Umpqua NFs in the mid-1970s devised forest-wide, multiple-use plans using the unit planning process.

In the early 1970s, the Region began a planning project that was called "area planning" to supplement other regional planning efforts, as well as to provide a better tie between forest and regional plans. Wendall Jones, former Regional timber staff, recounted:

> I do not recall what triggered this effort but it came on [Regional Forester] Ted Schlapfer's watch. The Region was divided into several Planning Areas: (1) Eastern Washington, (2) MBS [Mt. Baker-Snoqualmie] and Olympic NFs, (3) the GP [Gifford Pinchot], Mt. Hood, Willamette, and Siuslaw, (4) the Rogue, Siskiyou and Umpqua NFs, (5) the "iron triangle" made up of the W-W [Wallowa-Whitman], Umatilla, and Malheur NFs, and lastly (6) the Deschutes, Ochoco, Winema, and Fremont NFs. The job was completed using a 4 or 5 member team in each Area consisting of one representative from each Forest … and sometimes one or two add-ons for planning and writing skills. The job was completed in early 1974, but I don't know what use, if any, was made of it (Jones personal communication).

Multiple-use planning was a great opportunity for the national forests to gather new data and undertake long-range planning for the Region's natural resources, especially timber. The Gifford Pinchot NF developed a 10-year timber management plan that was approved by Regional Forester Ted Schlapfer on May 7, 1975. A Regional Office news release about the plan noted that, "The plan is the first of its kind to be approved in the Forest Service. … As a pilot project, the Chief of the Forest Service has approved an investment level that will maintain a 404.2 million board foot average annual timber sale program for the period of the plan. If additional funding is made available, this level can be increased in future years."

However, by the early 1970s, after a decade of constructing single resource MUSY plans for the national forests in the Region, it became evident that not all resource plans were equal—the timber plan usually overruled the other resource plans. Many advocates inside the Region and from the timber industry saw that small areas set aside for different management (e.g., RNAs or wilderness) were just fine, as long as these were not too large to "infringe" on the commercial forest lands or reduce the timber harvest levels. The act was never fully implemented, at least not so as to change the agency culture and management. Many employees believed that multiple use was simply a restatement of what the agency had been doing since Pinchot's time, and thus there was no need to change. Multiple uses or benefits were coming from timber production, they believed, including insect and disease control, firefighting, road access to remote areas, hunting and fishing, hiking, and, of course, jobs in the local economy. However, members of the skeptical environmental groups still only saw a road through the forest leading to a clearcut—the same as before the act was passed.

By the late 1970s, the National Forest Management Act of 1976 (see Chapter 8) would establish multiple-use planning over entire national forest areas, mandating public involvement through the act itself and the National Environmental Policy Act of 1969 (Fedkiw 1999).

NATIONAL ENVIRONMENTAL POLICY ACT OF 1969 AND OTHERS

The decade of the 1970s saw a dramatic shift in the environmental attitudes and values of the American public. The National Environmental Policy Act of 1969 (NEPA), signed into law January 1, 1970, mandated the comprehensive analysis of environmental impacts of proposed federal projects. An important part of the act made it mandatory for agencies to seek public participation on projects from the planning stage to the review-of-documents stage. These requirements were quickly incorporated into the many projects already underway in the Region. The law was intended to produce environmental documents that were brief and informative to the decision maker (usually the forest supervisor). But the NEPA requirements rapidly grew, as did the length and depth of environmental impact statements and other planning documents.

The Forest and Rangeland Renewable Resources Planning Act (RPA) of 1974 required, for the first time, that national-level planning and budget preparation be linked with the multiple-use plans. The RPA was described by historian David Clary as "the answer to a bureaucrat's prayer ... which authorized the endless generation of paperwork." (Clary 1986:189) With the RPA program in hand, the Forest Service would go to Congress to obtain the necessary money to implement its program. However, the RPA process never met expectations, even though millions of dollars and many people were assigned to assess the national situation and come up with a reasonably funded plan. In fact, former Chief Dale Robertson said that the best way to fund the agency was to use the budget allocation of the previous year, not start the process anew each year.

The series of congressional acts passed in the 1970s to further protect and enhance the environment included the Federal Water Pollution Control Act of 1972 (which amended the Clean Water Act of 1948), Endangered Species Act of 1973, Freedom of Information Act of 1974, Forest and Rangeland Renewable Resources Planning Act of 1974, National Forest Management Act of 1976 (see Chapter 8), Clean Air Act of 1977 (which amended the Clean Air Act of 1955). These laws and others have been at the root of numerous administrative appeals and lawsuits, some of which have successfully challenged and changed how the federal agencies would interpret the laws, implement regulations, build plans, and manage the federal lands (LeMaster 1984).

Timber Management

Timber harvesting in the Region reached a peak during the era. In 1960 4.1 billion board feet was harvested from the national forests in the Region. The Region's allowable cut was increased in 1962 by "about 125 million board feet, bringing the annual allowable cut to 4.266 billion board feet. Since 1950, the allowable cut has been increased 74 percent." (*Pacific Northwest Report* 1962: 11) The cut rose again to 5.1 billion board feet in 1965, but five years later, had dropped slightly to 4.8 billion board feet, dropping again by 1976 to about 3.9 billion board feet. The 1960 harvest was worth around $86.2 million , while in 1965 the harvest was valued at $98.2 million, rising to $280.0 million in 1976 despite the smaller quantity of timber harvested. (See Appendix E.)

The Oregonian reported on April 4, 1969, that the Region would sell an additional 410 million board feet of timber over the next 15 months. This was part of an effort by the Nixon administration to raise the timber harvest on all the national forests in the country by 1.1 billion board feet during the same period. These sales were intended to involve thinning and salvage volumes. Timber volumes sold did increase in the Region by over 1.1 billion board feet in 1970 but the harvest levels remained almost the same (see Appendix E).

Seed orchards in 1960 reported a bumper crop, in part due to the use of fertilizers. In the late 1960s and early 1970s, the forest genetics program in the Region really started to grow. Tree-seed collection became tightly controlled to ensure that off-site seeds were not used to grow seedlings that were destined for different locations. The seed collection and selection criteria used included latitude, longitude, and elevation of the seed sources. Wendall Jones (personal communication) noted that "progeny plantations were being installed, and seed orchards were beginning to produce. The Region began recruiting forest geneticists to bring professionalism into the program."

The 1960 USFS Pacific Northwest annual report said that permanent inventory tree plots "on a grid 1.7 miles have been established on more than two-thirds of the national forests in the region. The information provided includes data on growth, mortality, vigor, and stand condition, as well as actual volume by species."

In the summer of 1961, some 6,000 Douglas-fir seedlings were planted on 21 acres in the Lewis River RD on the Gifford Pinchot NF. This was a joint operation, dubbed the Youth Forest, cosponsored by the junior clubs of the Washington State Federation of Women's Clubs and the Forest Service. Youth groups such as the 4-H, Future Farmers of America, Boy Scouts, Girl Scouts, and Camp Fire Girls participated in the tree-planting operation. The seedlings were provided by the Wind River Nursery (Morse 1962). Reforestation programs in the Region by 1970 had increased to nearly 72,000 acres, with over 22 million seedlings planted. Knutson-Vandenberg funds provided the major funding for the reforestation work.

New techniques for timber harvest developed during this time were more efficient or less disturbing to the land. Long-span skyline systems had first come into use in the late 1950s. Based on experience, the contractors and the Forest Service made a number of significant improvements to the skyline

Sikorsky S-64E "Skycrane" helicopter dropping logs at the landing, Siskiyou National Forest, Oregon, c. 1970. USFS photograph.

systems over the years. Balloon logging, using Goodyear Aerospace Vee-Balloons instead of spar trees, was tried as early as August of 1964 on the Elliot State Forest near Reedsport, Oregon. Tests by the Pacific Northwest Research Station were conducted on the Willamette NF near Oakridge in 1964 to "study the feasibility and economics of balloon logging." (Forbes 1965: 13)

Helicopter logging was experimented with and generally found to be satisfactory, but expensive. In helicopter-logging operations logs were suspended from heavy lift helicopters and the logs were literally flown to landing areas to be put on trucks. This method was, and still is, used in steep terrain where other systems (especially roads) would be too expensive or environmentally damaging. These many changes in logging techniques triggered the hiring of logging engineers in the Regional Office. Wendall Jones (personal communication) noted that logging engineer "candidates were sent to a Logging Engineer training course at OSU [Oregon State University] before they were certified."

As late as 1965, probably the last attempt to establish sustained-yield units under the 1944 act failed. Secretary of Agriculture Orville Freeman denied a request for the Pacific Northwest Federal Sustained Yield Unit "made of 13 National Forests in … Washington and Oregon and portions of Idaho, California, and Nevada." The effort was spearheaded by the Save Our Logs Committee from Everett, Washington, which formed in 1964 to prevent the export of logs from the national forests. Action to restrict log exports came four years later, in April of 1968, when Secretary Orville Freeman "determined that export controls were needed in Western Washington and Oregon to help protect purchasers dependent upon National Forest timber … Later in 1968, an amendment to the foreign aid bill by [Oregon] Senator Wayne Morse required domestic processing of all but 350 million board feet of timber sold each year on Federal lands west of the 100th meridian." (*Pacific Northwest Report* 1968: 23) This action by Congress would adversely affect a number of timber companies that were shipping raw logs from the national forests to Japan.

At a congressional hearing on November 26, 1968, Chief Edward Cliff testified about the contents of a report that he had previously sent to Morse. The chief described the study, which was titled "Saving Timber Values, a Program to Salvage Tree Mortality on National Forests in Western Washington and Oregon": "All or part of 13 national forests [including NW California], involving 16 working circles, are included in the study. There are 9.2 million acres of commercial forest land on the national forests in this region. Over 7.3 million acres are classified as timber key value, that is, lands managed primarily for timber production ... Almost two-thirds of it is Douglas-fir. This inventory of timber presently provides an annual cut of 2.9 billion board feet. This is about 7.5 percent of the Nation's annual output of all species." Chief Cliff also explained the dangers of increasing timber harvest levels by simply removing the standing old growth:

> [A]llowable cuts could be increased by shortening rotations. However, much of such increase from shorter rotations would be derived merely from faster liquidation of the residual old-growth timber accompanied by falloff in levels of annual cut at the end of the rotations. The faster the liquidation, the greater degree of falloff. The falloff would carry into the next rotations because the backlog of old-growth would be gone and the cut would be essentially reducing our principal at a faster rate and speculating on future growth.

In 1969 the Region and the Oregon Department of Environmental Quality agreed to a cooperative plan to manage smoke from timber harvest debris or slash, which was extended to all the national forests in the Region the following year. A news release on February 16, 1970, documented the intent of the new plan: "Regional Forester Charles A. Connaughton said new regulations on disposal of timber harvest debris also will reduce the hazard of wildfires, which can cause extensive air pollution and do serious damage to water, wildlife, and other resources ... [The woody material would only] be burned when the smoke can be dispersed without getting into populated areas."

Landscape management started in this era. The program was an effort to "soften" and "sculpt" the standard clearcuts and roads to make them blend into the forest background. Clearcuts had to resemble natural openings in the forests rather than large geometric blocks of cut timber with exposed soil and rocks. Warren R. Bacon led the effort under the direction of Ronald M. Walters, who was in charge of the landscape architecture program in the Regional Office and remembered: "Warren Bacon ... did much of the development work. He also did most of the training for the timber folks. It was tough getting it implemented because of the ingrained way of doing business in R6 ... the public response to [against] clearcutting had already solidified." (Walters personal communication)

Although intensive forestry took on more importance with the adoption of many new forest practices and procedures, certain intensive-forestry practices, especially clearcutting, became a problem. Regional Forester Charles Connaughton wrote that, ""the toughest problem facing the forestry profession today results from a major segment of the public not realizing commercial forest lands can be managed without destroying its utility and appearance. Consequently, much of the public lacks confidence in foresters as stewards of the land." (Connaughton 1966) In his autobiography, Connaughton wrote that "I could see this revolt against clear cutting coming and … [the] article [quoted above] pointed out to foresters the need for clear cutting but also the need to reconcile public viewpoints … The admonition of this article proved true and a ground swell of anti-clear cutting grew nationwide and culminated in a Congressional hearing on the subject." (Connaughton 1973: 103)

In the late 1960s, there was controversy over the management of Montana's Bitterroot NF, when residents became concerned about the scenic and reforestation problems being caused by clearcutting and terracing on steep slopes. On the other side of the country, in West Virginia, legal challenges were brought in the late 1960s by the Izaak Walton League against clearcutting on the Monongahela NF (*Izaak Walton v. Butz*). The agency strongly defended the practice, but to no avail. The suit was decided in 1973 against the agency and the judge issued an injunction against any clearcutting in West Virginia. The Forest Service appealed, but lost, and the injunction was spread throughout the 4th Circuit of the Court of Appeals. Injunctions based on similar cases were soon issued in Alaska and Texas, and were clearly going to be applied across the rest of the agency. The clearcutting issue caused an extensive internal review of Forest Service management. The situation would be resolved partially by Congress with the National Forest Management Act of 1976 (see Chapter 8).

Wendall Jones, former R-6 timber director, recalled that in the early 1960s the Siuslaw NF in a new timber management plan recommended that a partial cutting program be put into effect on the forest (personal communication). Partial cutting instead of clearcutting occurred in the 120-year-old stands of Douglas-fir that had regrown after the big coastal fires of the 1800s. However, once timber harvesting using this method began it became obvious that the remaining trees were prone to "wind throw"—that is, trees were blown over in severe wind storms. Areas that had no partial cutting were not affected by the winter storms.

Other examples of using alternative methods to clearcutting involved heavy and light selective harvesting, smaller patch cuts, and even cutting narrow strips of trees within larger forest stands. Pacific Northwest research staff experimented with several forest types. For example, Wendall Jones (personal communication) reported that on the Siuslaw NF at the Cascade Head Research Area researchers established plots "to determine the extent of damage to the soil and tree roots as a result of partial cutting in spruce-hemlock stands and logging with rubber tired skidders. Workable small cable systems were not yet available for logging thinning. Research discovered that root damage, especially on hemlocks became the entry point for root rot that would probably result in windthrow problems in the future and rotting of the butt log. Recommendation from research—clearcut, treat slash and plant."

With the 1970s, came a concern from Pacific Northwest Research Station researchers and university scientists for the increased need to preserve more old growth as habitat for birds, mammals, and invertebrates. There was a growing constituency in the 1970s and 1980s that old growth was needed for esthetic and recreation reasons. One question became a major one in the Region—how much old growth should be protected? This was followed by another question—what or who defines old growth? Related questions arose, such as how many acres of old growth existed and where were they located? The environmental groups argued that only 5 percent or less remained of the original ancient forest in the Region and elsewhere in the agency. The Forest Service in turn argued that perhaps as much as 20 percent remained, if all the wildernesses, research natural areas, wild and scenic rivers, and other special areas that didn't allow timber harvesting were taken into account.

The Pacific Northwest Research Station's Web site describes the typical old-growth forest in the Pacific Northwest as

> a conifer forest, dominated by large, old trees. In the Pacific Northwest, the most common type of old-growth ecosystem is forests dominated by Douglas-firs and western hemlocks, generally 350 to 750 years old. The youngest old-growth forests are 200 years old, and the oldest are about 1,000 years old. The Pacific Northwest also has old-growth forests dominated by Sitka spruce and western hemlock, along the Pacific Coast, and at higher elevations in the Cascade Mountains, true fir, and hemlock old-growth forests. Among all the forests of the world, the Pacific Northwest old-growth forest is unique because of the size and old age of its trees, the accumulations of biomass (weight and density of living organisms), and the climate, with its wet, mild winters and dry, warm summers (www.fs.fed.us/pnw/).

PESTICIDES AND HERBICIDES

Further problems with forest management came about in regard to the use of pesticides. The Federal Pest Control Act of 1947 had declared it the policy of the government to protect all forest lands, regardless of ownership, from destructive forest insect pests and diseases, and authorized the Secretary of Agriculture to work out cooperative agreements with other forest landowners to carry out control measures. In 1962, Rachel Carson published *Silent Spring*, an indictment of insecticides and related chemicals. The book, widely opposed by the chemical industry, eventually led to modifications of existing practices relating to the approval of all pesticides.

The annual cooperative aerial surveys in 1965 covered some 65 million acres of federal and private forest lands and indicated that some 1.4 million acres were infested by insects. The Douglas-fir bark beetle, in the aftermath of the Columbus Day Storm of 1962, killed an estimated 2.6 billion board feet of timber in western Oregon by 1965. Almost 66,000 acres of forest were treated for tussock moths in 1965. DDT mixed with fuel oil "achieved a tussock moth kill ranging from 93 to 100 per cent, with an overall average of 98 per cent." (*Pacific Northwest Report* 1965: 7)

Several huge outbreaks of tussock moth affecting an area of about 500,000 acres occurred in 1973 in southeastern Washington, northeastern Oregon, and small portions of Idaho. The only known insecticide for tussock moth control or eradication was DDT. However, by this time DDT use had been prohibited by the Council of Environmental Quality, based in part on Carson's studies showing damage caused in the environment by use of the insecticide. "We were caught really between a rock and a hard place, knowing that the only way we could positively control it was to use DDT." (Schlapfer 1984: 48)

Yet there was one loophole—if there was an emergency of national significance, DDT could be used. The Forest Service, Bureau of Indian Affairs, BLM, dean of the OSU School of Forestry, Forest Service research, and state forester's office pushed for emergency use. Their petition was denied. A year later, the petition was accepted after about 1.2 million acres were defoliated. Regional Forester Ted Schlapfer worked with state officials and the BLM to conduct the project in the summer of 1974. The results of this aerial DDT spray project were mixed as some of the insects were already dead as they had run out of food sources. This was the last approved aerial spraying of DDT in the forests of the U.S.

Schlapfer said that the tussock moth control issue did have some positive aspects: "One of the real positive things that came out of that was a lot of effort put on by agency leaders, including Forest Service folks, saying, We don't want to do this again. We have got to find alternative solutions to controlling tussock moth." It generated a lot of support for additional research money. That research effort went into … trying to find a biological control for the tussock moth; and indeed, they did find a breakthrough in developing a way of disseminating *Bacillus thuringiensis* [*B.t.*], which is a biological agent that infects the tussock moth with a virus." (Schlapfer 1984: 50)

By 1964 many special forest products were being harvested from national forests in the Region, including Christmas trees (over 3 million), Christmas boughs (1.4 million pounds), floral greens (over 8 million bunches and bales), cedar products like shakes and poles (32.5 million boards and posts), cordwood or firewood (almost 91,000 cords) seed cones (538,000 bushels), and cascara and conks (2.1 million pounds).

Recreation and Wilderness Management

Recreation was beginning to share with timber a dominant place in forest use. Camping made up about 25 percent of the normal day use, followed by motorized travel, water sports, fishing and hiking, and organizational camping (e.g., Scout and church camps). Camping and picnicking facilities built by the CCC had served well during the 1940s, when there was relatively little use; but by the 1960s they were showing signs of wear and tear. New campgrounds were built, some with modern refinements such as flush toilets, community-kitchen shelters, pumping stations to clean out holding tanks for recreational vehicles, and even fish-cleaning stations.

The Regional architects provided standard plans for numerous buildings, including recreation facilities, lookout structures, offices, barracks, and residences. Work was progressing during 1965 on the North Cascades Highway route from both ends of the roadway. The two roads were connected at Rainey Pass in September 1968, where "a great celebration was held there with everyone arriving in 4-wheel drive rigs." (Chriswell 1989: 123) Yet work was still needed to finish the roadway to make it into a highway. The beautiful scenic highway was a cooperative effort with the Washington Highway Commission, Bureau of Public Roads, and the Forest Service.

Recreation use in the Region reached 9,573,700 visits during 1960, increasing to 13.4 million visits by 1965. The method of calculating forest visitors was changed in 1965 to "recreation visitor days" or RVDs (one RVD equals 12 hours of use by one person or one hour of use by 12 people). Using the new method, 23.6 million RVDs were recorded in the Region.

To help finance recreation management, the Forest Service, in 1962, initiated a fee system charging one dollar per day for use at selected sites in the Region. Until this time, no fees had been charged for recreation use of the national forests. In 1965, the Land and Water Conservation Fund was established, which formalized the payment of a fee for use of campgrounds. For payment of $7, a Golden Eagle annual permit would allow entry into all federal areas where a fee was charged. Senior citizens were given permits for free. By 1965, the fees collected amounted to $112,626. Summer home fees, from updated home/cabin valuations, required every five years, were increased by an average of $50 per year on 2,700 summer homes in the Region in 1967. Remote homes were increased around $30, while high-value lots along lakes or streams increased to $225.

The Waldo Lake area recreation plan on the Willamette and Deschutes NFs was approved by Chief Edward Cliff on April 16, 1962. This decision affirmed Regional Forester Herb Stone's approval of a revised plan for the area under the framework of the high mountain policy of 1962 that essentially provided protection for all high mountain areas, while allowing restricted development and logging. The Waldo Lake area had been under administrative appeal by several recreational groups affiliated with the Federation of Western Outdoor Clubs. Sierra Club director Mike McCloskey believed that the decision was based in part on the fact that the Forest Service listened to outside groups and changed its policies. He noted that this was quite a remarkable change in the early 1960s era (McCloskey 2005).

The Cape Perpetua Visitor Center was opened in early August 1967 with Chief Ed Cliff and USDA Assistant Secretary John Baker attending the ceremony. Another special visitor in late June 1968 was First Lady "Lady Bird" Johnson. She, USDA Secretary Freeman, and Chief Cliff visited the Region, spending an afternoon at the Tollgate Campground and overnight at Timberline Lodge. On the Olympic NF during 1968, the Seal Rock Campground was expanded by the purchase of an additional 660 feet of waterfront along Hood Canal. This purchase enabled campers to have public access to about half a mile of tidelands where they could gather oysters and dig clams. These may be the "only productive oyster beds in the entire National Forest System." (*Pacific Northwest Report* 1968: 20)

The Wild and Scenic Rivers Act of 1968 created a new series of categories for management of rivers or parts of rivers within the national forests and elsewhere. The three river categories were defined as wild, scenic, and recreational; wild rivers were those in need of the most protection (with the least development) and recreational the least (with the most development). Only one river, the Rogue (to be administered by the Forest Service and the BLM) was designated as wild and scenic by the original act, with the Skagit and Illinois designated as "study rivers" for possible inclusion as wild and scenic rivers. The studies were to be completed by 1978.

Ski areas continued their growth. By 1961 and 1962, winter sports areas had more than a million visitors. Increased demand was met by new ski areas that opened in 1962 at Crystal Mountain (Snoqualmie NF) and Anthony Lakes (Wallowa-Whitman NF). At Timberline Lodge, the new Magic Mile

Mazama hike on Mt. Hood, Mt. Hood National Forest, Oregon, 1963. USFS photograph.

chairlift went into operation, more than doubling the number of skiers. The Mt. Ashland Ski Area (Rogue River NF) opened for business in 1963. Mission Ridge (Wenatchee NF) was built in the mid-1960s, opening in the 1966 ski season. The following year saw the opening of two new ski areas—Mt. Hood Meadows (Mt. Hood NF) and Alpental (Snoqualmie NF). Other downhill ski areas, such as Crystal Mountain, were developed in the Region using private funding.

In 1967-68 came the first reports of recreational snowmobile use in the Region. Snowmobile use in 1968 was estimated at 66,200 visitors. The Region responded to this new influx of winter sports users by marking some 600 miles of snowmobile routes with black and orange signs. Approximately 2,800 miles of roads that typically get snowed in were also authorized for use by snowmobilers.

Congress passed the National Trails System Act of 1968 that established a system of national scenic, national recreational, and state and metropolitan trails. The Pacific Crest Trail was designated as one of two national scenic trails in this act, and another 14 trails were identified in other parts of the nation for possible inclusion in the system.

WILDERNESS

In 1960, another administrative wilderness was established in the Region—the Glacier Peak Wilderness on the Mt. Baker and Wenatchee NFs. This was one of several wildernesses in the national forest system that was established prior to the Wilderness Act of 1964. The Wilderness Act culminated several decades of political lobbying by various environmental organizations, especially the Sierra Club and the Wilderness Society (Roth 1988b). In addition to legally defining wilderness, the Act also determined that wilderness areas must be designated by Congress. Initially, the Forest Service opposed the bill, since administrative wilderness areas had already been established as early as 1924 in many parts of the country. The agency also thought that to have Congress establish wildernesses would be unnecessary and redundant since the Multiple Use-Sustained-Yield Act of 1960 provided for recreation, including wilderness recreation, on the national forests. Congress won the battle.

With passage of the Wilderness Act, the chief of the Forest Service no longer had the authority to designate wildernesses on the national forests. It was believed by many private foresters and Forest Service employees that wilderness designation was a single use under their understanding of the MUSY. The wilderness and other congressional designations or forest zoning soon would affect many areas within the national forests. It did not take long until "multiple use ran afoul of multiple demands." (Clary 1986: 163)

The Wilderness Act created only a few new wildernesses, as most had already been established as administrative wildernesses, wild areas, or primitive areas by the Forest Service. For Region 6, the areas designated in the 1964 Act were as follows:

Diamond Peak (Oregon)	35,440 acres
Eagle Cap (Oregon)	216,250 acres
Gearhart Mountain (Oregon)	18,709 acres
Glacier Peak (Washington)	458,105 acres
Goat Rocks (Washington)	82,680 acres
Kalmiopsis (Oregon)	78,850 acres
Mt. Adams (Washington)	42,441 acres
Mt. Hood (Oregon)	14,160 acres
Mt. Washington (Oregon)	46,655 acres
Mountain Lakes (Oregon)	23,071 acres
Strawberry Mountain (Oregon)	33,004 acres
Three Sisters (Oregon)	196,708 acres

FRENCH PETE

In 1957, the Three Sisters area on the Willamette and Deschutes NFs had been established as an administrative wilderness by the Secretary of Agriculture. This was 20 years after the Three Sisters Primitive Area had been designated, but a 53,380-acre area on the west side of the original wilderness was eliminated by the 1957 decision—it became known as the French Pete area. In the winter of 1967-68, Michael Kerrick, district ranger of the Blue River RD, snowshoed in to take a look at the French Pete area in which an 18.5-million-board-foot timber sale with 11 cutting units was proposed for early 1968. This sale was to be the first timber sale in the 19,200-acre French Pete Creek drainage, the last drainage in the area to be developed.

However, several environmental groups found out about the proposed timber sale and began to use it as an opening wedge to attack the Secretary's 1957 decision. The Eugene-based Save French Pete Committee was formed in early 1968 by local people who proposed that the area be left in its natural state, with no roads or management activities, noting that the valley was only one of three roadless valleys 10 miles or more in length remaining in western Oregon.

Kerrick and his staff felt that the sale, which had been laid out in 1964, was not up to current Forest Service standards so far as silvicultural and some environmental aspects were concerned. The timber sale was postponed and, on the advice of Kerrick, then canceled. In June 1968, Kerrick submitted a new management plan for the French Pete area. The Forest Service had worked this out in their multiple-use plan, as directed by the Secretary's 1957 decision and the regional forester's high mountain policy statement of 1962, which limited and applied restrictions to timber harvest in the mountains.

French Pete Creek drainage, Willamette National Forest, Oregon, c. 1970. USFS photograph.

The Save French Pete Committee prepared an alternative plan called the French Pete Intermediate Recreation Plan. It seems to have been the result of collaboration between Brock Evans (the Sierra Club's Northwest Representative since 1967) and Richard Noyes from the Save French Pete Committee. They wanted French Pete to be "an area from which public roads and commercial timber harvest was excluded, of sufficient size to accommodate a variety of outdoor experiences without crowding." This plan resembled the 1930s "semi-wilderness" plan of Regional Office recreation planner Fred Cleator for nearby Pamelia Lake and Marion Lake. Cleator's plan had been based on the assumption that management degrades, rather than enhances, habitat and that such natural events such as tree disease and insect infestation will cure themselves as they have for centuries.

A student group from the University of Oregon, calling itself Nature's Conspiracy, demonstrated in November of 1969 at the forest supervisor's office in Eugene. Forest Supervisor Dave Gibney talked with the demonstrators but dismissed them, which only tended to infuriate the protestors. This was

Save French Pete Creek protest at Willamette NF supervisor's office, Eugene, 1969. USFS photograph.

one of the first, perhaps *the* first, anti-Forest Service public demonstration ever held (Williams 1993). Phil Hirl reported that during this event, "I ... remember rocks hitting my office window, and felt uneasy about leaving the office. Sometimes you had to walk through a group of people, 'hippies,' so that I chose not to wear my uniform unless I was headed out early for the field." (Hirl personal communication)

Secretary of Agriculture Clifford Hardin asked that implementation of the Forest Service plan be postponed, and requested that the affected parties make suggestions and comments by January 17, 1970. The chief then appointed a special task force, and on June 2, 1970, the chief upheld the French Pete management plan. Also in 1970, several events occurred that changed the course of the French Pete controversy. Senator Robert Packwood endorsed the Willamette's Intermediate Recreation Area Plan, and in late 1969 brought it to Congress as a bill. Zane Smith, forest supervisor after Dave Gibney, spent time repairing the public relations damage, putting off the French Pete decision until the Willamette land use plan could be completed. This placed the issue in the political arena, where it remained for the next three years until the French Pete area was added to the Three Sisters Wilderness by the American Wilderness Act of 1978 (Rakestraw and Rakestraw 1991 and Williams 1993).

ROADLESS AREA REVIEW AND EVALUATION (RARE)

One provision in the Wilderness Act of 1964 directed the Forest Service to study all existing primitive areas (those primitive lands and forests that were eligible for wilderness designation) within 10 years for possible inclusion in the National Wilderness Preservation System. As early as 1967, "the Forest Service manual directed Regional Foresters to review and report by 1969 [later changed to 1972] on roadless areas that might have wilderness potential." (Roth 1988b: 37) After a 1970 court decision in Colorado, the Forest Service began a national study of such roadless areas (sometimes called "de facto wilderness").

The highly controversial Roadless Area Review and Evaluation (RARE) process begun in 1971 included all roadless areas anywhere on the national forests. There was great acrimony and dissent over the methods of identifying and evaluating the existing roadless and undeveloped areas over 5,000 acres in size or smaller if located adjacent to existing wildernesses or primitive areas. The roadless areas were quickly identified and mapped by personnel at the national forest level, but many were not as carefully identified as they could have been if more time had been allocated to the process.

The inventory across the national forest system identified 1,449 roadless areas containing 55.9 million acres that were studied for their wilderness potential. The final report was published in 1973, with 274 of the roadless areas (12.3 million acres) selected for possible wilderness designation by Congress. In Region 6 the RARE inventory identified 255 roadless areas (5,519,146 acres), 112 in Washington and 143 in Oregon. Only 16 areas (371,920 acres) were recommended for wilderness status and another 11 areas (600,980 acres) were recommended for further study. Areas that had high timber values were not generally considered for wilderness. (This trend of eliminating resource-rich roadless areas would continue through to the present.)

The decision became immediately embroiled in controversy. As a result of a lawsuit in California (*Sierra Club v. Butz* or the "Conti Decision") in 1973 over a roadless area that had not been selected for wilderness designation, the assistant Secretary of Agriculture and the chief of the Forest Service ordered a new study of all roadless areas, called RARE II, in the summer of 1977. The agency had to reevaluate the roadless areas from updated RARE (now referred to as RARE I) inventories, as well as adding unroaded areas that had been missed (see Chapter 8).

NATIONAL SCENIC AND RECREATION AREAS

A new scenic area was approved by the Secretary of Agriculture in 1963—the 130,000-acre Hells Canyon-Seven Devils Scenic Area. The scenic area encompassed parts of the Wallowa-Whitman NF in Oregon and the Nezperce and Payette NFs in Idaho and extended for some 22 miles along the Snake River north of Homestead, Oregon.

There had been much discussion over the years about whether the Forest Service or the National Park Service should manage the Oregon dunes area (see Chapter 6). The Forest Service became the "winner" as the 37,237-acre Oregon Dunes National Recreation Area on the Siuslaw NF was dedicated on July 15, 1972. Chief John McGuire, Senators Mark Hatfield and Bob Packwood, Representative John Dellenback, as well as Governor Tom McCall were present at the dedication.

An unusual purchase of land occurred in the Snake River Canyon with backing by Congress in the early 1970s. A Regional Office news release on May 10, 1973, announced that 8,108 acres had been purchased on the Oregon side of the river under eminent domain for the preliminary estimate of $1,457,714. The "Declarations of Taking" had been filed in federal court. Another 6,980 acres of private land in Idaho, also along the Snake River, were acquired in a similar fashion. This may have been the only time that eminent domain was used to expand the national forests in the Region. The news release continued:

> The United States started negotiations to acquire these lands about a year ago. The owners of these lands indicated that the livestock ranching had become financially marginal in this area. Several of the landowners sold their stock and indicated their intent to subdivide their ranches into recreational homesites. Subdivision activity has been started on three of the ownerships in both Idaho and Oregon after the Forest Service and the landowners were unable to reach agreement on the price of these lands. ... Testimony on the [1973 congressional] appropriation centered on the need to guarantee the "primitive" integrity of the canyon.

The 622,000-acre Hells Canyon National Recreation Area was established here on December 31, 1975. Dedication of the area was on July 31, 1976, at Hat Point on the Wallowa-Whitman NF. Speakers at the dedication included Representative Al Ullman, Senator Bob Packwood, Governor Bob Straub, and Forest Service Chief John McGuire. The Nezperce and Payette NFs in Idaho and the Wallowa-Whitman NF in Oregon have jurisdiction over the area.

CULTURAL RESOURCES/HERITAGE MANAGEMENT

The National Historic Preservation Act of 1966 declared it to be national policy that "the historical and cultural foundations of the nation should be preserved as a living part of our community life and development in order to give a sense of orientation to the American people." Cultural resource management (CRM), now called heritage management, became the catch-all term for such work. Reports recommended that more CRM surveys needed to be conducted prior to road construction and timber sales, and that the Forest Service needed to conduct an education program to provide awareness of resource values and laws, especially in view of illegal collecting by citizens and Forest Service employees alike.

By 1977, there were cultural resource guidelines for all 19 national forests in the Region. Leslie Wildesen developed the "yellow" or "lemon-aid" book that was used for more than a decade to train cultural resource technicians on the proper ways of doing the CRM work (Wildesen personal communication). The CRM program and growing numbers of site reports yielded important American Indian discoveries in spite of the fact that many university archaeologists had thought that the Region would not yield any important or significant archaeological finds. Initial historic preservation efforts focused on the remaining CCC structures in the Region, where over 1,000 CCC structures were still standing, though many were in rough shape.

Yet the cultural resource data gained has not been without internal controversy. Most of the early opposition came from Forest Service managers who viewed the CRM program as expensive and time consuming. Some district rangers believed there were few, if any, benefits. Also, found sites would limit management options, especially road construction and timber sales. Initially, some archaeologists thought that every obsidian flake or rusty tin can found was "significant" and thusly had to be preserved. The CRM technician on the Umpqua NF referred to this as the "cult of the sacred flake." That was to change under the leadership of regional archaeologist Jim Keyser, through a program of using mitigation and avoidance (Keyser personal communication). Sometimes when a site was found, an alternative route for a road or location of a timber sale overcame the problem. In extreme cases, sites could be excavated fully or covered by gravel before the ground-disturbing project began. The costs for the CRM site inventory work and mitigation were paid for by the benefitting program—usually timber. By the middle of the 1980s, district rangers could see the positive benefits and began to emphasize the CRM program.

Engineering

Phil Hirl wrote about changes to the engineer system in the Region during the 1960s: "Regional Engineer Ward Gano started a move to make all Forest Engineers professional civil engineers. Well over half were foresters, though many took forest engineering at school. By 1970 that goal had been accomplished on all Forests." (Hirl personal communication) Part of the plan to get more civil engineers was simply to hire them directly from engineering schools at various universities. These new civil engineers were professionals but in many cases had little or no training in forest engineering needs.

> The new CEs [civil engineers] influenced how the FS work was done. They were the first big influx of a different professional skill. Many of then went to work on Ranger Districts … For the most part it worked pretty well. They got involved in sale layout and how roads should fit with planned [timber] harvest and the ground conditions and learned forest practices. Because they were getting higher pay scale there was some resentment from the foresters…We also hired a number of civil techs, who became the backbone of the engineering organizations, particularly in road building activities … Many of these people were good public contacts because they were visible as FS employees to

> the visiting public. Most of them were proud to do so (Hirl personal communication).

The Forest Roads and Trails Act of 1964, as well as earlier acts, authorized construction, reconstruction, and easements for roads on or across national forest lands that were not considered part of the forest road system. Of particular interest were easements on roads entering and crossing national forests from private and other public lands. The 1964 act stated in part that "the construction and maintenance of an adequate system of roads and trails within and near the national forests and other lands administered by the Forest Service is essential if increasing demands for timber, recreation, and other uses of such lands are to be met … The Secretary is authorized … to grant permanent or temporary easements for specified periods or otherwise for road rights-of-way."

During this era, a large road construction program was ongoing in the Region. The roads, built by private contractors who bid on projects, were funded by appropriated or public funds provided by Congress through the forest road and trail program. Reconstruction work was generally about 100 to 130 miles per year, while new construction was some 30 to 60 miles per year. Many of these were gravel and paved roads built with two lanes. Some of the roads were used for campgrounds and other recreation projects.

During fiscal year 1960, more than 1,700 miles of roads were constructed in the Region, according to the Regional Office yearly report. This brought the total forest road miles in the Region to around 26,750 miles. The Forest Service worked cooperatively with the Bureau of Public Roads, which awarded some $3.6 million to build an additional 40 miles of roads. In addition, 108 miles of trails and four bridges were constructed or reconstructed. This brought the total trail mileage in 1960 to more than 18,000 miles.

Five years later, in response to the major damage caused by the Christmas Week Flood of 1964, a total of $5.9 million was spent on restoration of the forest transportation system in the Region. This money was expended on rebuilding 2,389 miles of roads, 43 bridges, 256 miles of trail, and four trail bridges. In addition, 1965 also saw the construction of 1,520 miles of new roads, 973 miles of reconstructed roads, and 28 new bridges. Total cost was $51.5 million that was "accomplished by National Forest timber purchasers as part of their timber sale contracts." Acquisitions of permanent rights-of-way "resulted in 241 new easements for raw land right of way, 173 easements for existing road, and 146 title approvals on old right-of-way easements. Negotiations proceeded on 25 new share-cost road agreements." (*Pacific Northwest Report* 1965)

This was also when the road system was redesignated for administrative purposes to be more in line with state and county highway systems. Instead of simple forest roads, in 1972 they became known as arterial, collector, or local roads. Arterials were main forest roads that were used to connect the lower class of roads (collector and local roads) to major state or federal highways. Local roads were often forest access logging roads that connected to more primary collector roads, which in turn connected with the major arterial roads. In many cases the arterials were paved, while the collectors were graveled and the local roads either graveled or dirt.

Regional Forester Charlie Connaughton "returned from a field trip [in the mid-1960s] over some forest roads and wrote a brief new policy that said, 'Hereafter, all roads in Region 6 will be designed and built so as to eliminate slips and slides.' This soon became known as the 'No Slips, No Slides Policy'… Although slips and slides were never completely eliminated, the intent of the Regional Forester's policy was willingly carried out." (Unruh 1990: 558)

New housing facilities at various ranger stations in the Region amounted to more than $2.8 million in 110 contracts. The largest single building contract was let in 1961 to construct 11 administrative buildings, including employee housing, at the Chemult Ranger Station on the new Winema NF. The engineering division in 1964 "designed standardized prefabricated wood buildings required for the four [Job Corps] centers in Region 6. These plans had wide use throughout by the Forest Service and other agencies, and earned a Presidential citation for the Region." (Grefe 1990: 280) Designing these Job Corps sites was quite challenging and time consuming.

> We designed and built small towns, buildings, water supply, sewage collection and disposal. Some of these were fairly sophisticated at sites far removed from city facilities. We could not just hook [up water and sewer] to [a] city system, we had to build our own. Well at Angel [Job Corps camp] we did hook up to city system at some point … Buildings were not routine when designed for 250 people, in addition to housing we had to build mess halls, meeting and recreation buildings. And this was all in a year or so (Hirl personal communication).

Aerial survey work continued unabated across the Region during the era. *The Oregonian* on August 17, 1969, ran a story about four women who were working on aerial survey controls for the Regional Office. It was unusual in the late 1960s for women to be employed doing field work, a job traditionally filled by men. "What the girls are doing is running surveys to establish controls necessary to make accurate maps from aerial surveys. They're also proving that 'women can do the job as well as men and are an untapped labor force in surveying,' according to Cecil A. Navin, chief of field service unit … Division of Engineering, Region Six, U.S. Forest Service. … Next goal for B. [Beulah] J. Jones is to qualify as a registered land surveyor after two years of supervisory experience. There are more than 600 men on the rolls. No women. Yet." She did qualify.

Range Management

Grazing in the Region continued on the 730 range allotments. By 1960, there were 93,290 cattle and 137,800 sheep grazing on national forest ranges in the Region. Revegetation was conducted on 9,000 acres, another 5,000 acres were sprayed for sagebrush control as well as noxious weeds and poisonous plants, and 60 acres were improved through water spreading. The Regional program also involved the building of some 220 miles of fences and 390 stock water developments. Range analysis work was completed on about half of the allotments. In addition, on several "key areas, permanent Parker 3-step condition [assessment method that was developed in 1950 by the USFS] and trend transect plots have been established which … will indicate the condition of the range and whether it is improving or deteriorating." (*Pacific Northwest Report* 1960: 11)

In 1963, the Region was host to the 15th annual meeting of the American Society of Range Management, Pacific Northwest Section, in Portland. Discussions included the use of herbicides on rangelands. The following year saw a concentrated effort to control sagebrush and other undesirable vegetation on 13,600 acres of Forest Service rangeland on the Snoqualmie, Okanogan, Mt. Hood, Ochoco, Siskiyou, and Willamette NFs. Spraying of 2,4-D was accomplished by use of helicopters, followed by seeding of grasses that included mixtures of wheatgrass, orchard grass, timothy, and meadow foxtail. Grazing animals were prohibited on the sprayed sections for two years to allow the grasses to grow.

A report in *The Oregonian* on October 31, 1965, noted that the oldest grazing permit was for the Stearns Cattle Company on the Umatilla NF that dated to 1906. The cattle grazing permit had been obtained by the Brosnan family, who had previously run horses for many years on the forest. John Brosnan obtained the permit from his father, but changed it from cattle to sheep in 1928. Jerry Brosnan obtained the permit in 1954, changing it back to cattle. William (Will) Vernon held the second-longest grazing permit in the

Region. His father, Thomas Vernon, had obtained a grazing permit in 1907 for portions of the Fremont NF. Four years later the permit was changed to Will Vernon, who still held it in 1965.

By 1965, there was a shift in grazing use as cattle grazing increased to 119,249, while sheep grazing reduced to 92,383. Cattle grazing was up slightly by 1970, but sheep numbers were still down. The last sheep were grazed on the Malheur NF and Okanogan NF's Horseshoe Basin in 1967. The Regional annual report noted these changes: "The decline in sheep numbers on the National Forests reflects a continuation of a grazing shift from mountain ranges to private pastures, coupled with decisions of livestock operators to convert their permits from sheep to cattle, largely for economic reasons." (*Pacific Northwest Report* 1967: 20)

A total of 203,421 animals were permitted in the Region in 1970. Grazing fees paid by 1,019 ranchers amounted to $280,805, while their stock grazed over 6.8 million acres of national forest lands. Grazing improvements in the Region this year amounted to almost 26,000 acres of range seeding, around 3,900 acres of brush and weed control, 245 miles of fence construction, 66 cattle guards, 103 spring developments, 181 ponds and reservoirs constructed, and 17 miles of stock driveways improved. Regional Forester Ted Schlapfer spoke about the grazing program of the early 1970s:

> We worked hard to create coordination between users of National Forest forage … and worked very closely with BLM, Soil Conservation Service, [and] livestock people in the development of coordinated resource plans. I spent a lot of personal time in that activity working … with John McKeen, Director of Fish and Game … Because integrating wildlife with domestic livestock, elk and cattle for example, are very important [there was]…the potential of a lot of conflict … Livestock grazing, in my judgment … would be an easy one to let go because it didn't have a lot of conflict around it, which usually gets the administrator's attention (Schlapfer 1984: 47).

Wildlife and Fish Management

As in the previous decades, wildlife management tended to focus on huntable and trappable species. Cooperative agreements were worked out with the state game commissions. During the summer of 1961, 24 fishers (a medium-sized mammal of the weasel family) were reintroduced in the Mountain Lakes Wild Area on the Winema NF and in the Minam River drainage on the Wallowa-Whitman NF.

By the end of 1965, most forests in the Region had completed comprehensive wildlife management plans. These plans were developed in cooperation with the game departments in both states. The same year saw the first hunting of mountain goats in Oregon with five tags issued for a hunt in the Eagle Cap Wilderness (Wallowa-Whitman NF). The State of Oregon also issued for the first time 300 tags for the taking of non-native wild turkeys on the easterly slopes of Mt. Hood. In 1972, four mountain goats were trapped on the Olympic NF, transported by truck and helicopter, then released on the Gifford Pinchot NF in the Mt. Margaret area. The following year, another four goats were released in the same area.

There were growing public concerns over unique wildlife that lived or nested on national forests around the country. In the West, the imperiled status of spotted owls, marbled murrelets, grizzly bears, caribou, Pacific salmon, bull trout, and wolves caused increasing concern. In 1968, the Winema NF stopped a timber sale that would conflict with the nesting habitat of the rare greater sandhill crane. The same year, on the Deschutes NF, at Crane Prairie, efforts were made to protect 27 osprey nests. The nation's first Osprey Management Area was dedicated at Crane Prairie on August 8, 1970, in a cooperative effort by the Forest Service, the Oregon Game Commission and the Bureau of Reclamation. The area encompassed 10,600 acres. Chief Ed Cliff, Governor Tom McCall, and Congressman Al Ullman attended the dedication ceremony.

The Christmas Week Flood of 1964 had devastating impacts on fisheries in the Region. Untold numbers of fish were washed out to sea with the flood. The state hatcheries were especially hard hit. Louise Parker noted that "over two million spring Chinook and 340,000 steelhead fingerlings were lost at the Willamette Hatchery near Oakridge when a slide cut off the water supply. Another 148,000 spring Chinook were lost in the hatchery on the South Santiam River near Foster. Loss of the salmon brood will be apparent in the runs of Coho salmon in 1967 and Chinook runs in 1968-69." (Parker 1965: 61)

SALMON AND THE BOLDT DECISION

Many Indian tribes in the Northwest traditionally depended on salmon fishing. As early as 1937, tribal members had obtained a federal court injunction to protect them against state interference with their traditional salmon fishing rights as set forth in the Medicine Creek Treaty of December 26, 1854. But the State of Washington began an effort to stamp out what it considered illegal fishing. The campaign to reassert Indian fishing rights began in 1964 with "fish-ins" on the Nisqually River and nearby Puyallup River (the upper portions are in the Mt. Baker-Snoqualmie NF). Washington State law enforcement officers made a series of raids and stings to entrap the Indians who were fishing illegally according to state regulations. The situation escalated; gunshots were fired.

In September 1970, the federal government, on behalf of the tribes, filed suit against the State of Washington to force the state to comply with the treaties that specifically obligated the state to allow salmon fishing by Indians. Most of the treaties negotiated by Territorial Governor Isaac Stevens during the mid-1850s had included language that allowed the treaty Indian tribes to fish at "all usual and accustomed grounds and stations" in the future. The trial (*U.S. v. Washington*) was held in U.S. District Court Judge George Boldt's court. On February 12, 1974, Judge Boldt affirmed the right of several western Washington tribes to continue to harvest salmon as they had for thousands of years.

> [T]he court finds and holds that every fishing location where members of a tribe customarily fished from time to time at and before treaty times, however distant from the then usual habitat of the tribe, and whether or not other tribes then also fished in the same waters, is a usual and accustomed ground or station at which the treaty tribe reserved, and its members presently have, the right to take fish … sharing equally the opportunity to take fish at "usual and accustomed grounds and stations"; therefore, non-treaty fishermen shall have the opportunity to take up to 50% of the harvestable number of fish that may be taken by all fishermen at usual and accustomed grounds and stations and treaty right fishermen shall have the opportunity to take up to the same percentage of harvestable fish, as stated above. (*U.S. v. Washington,* 1974)

The court decision noted the differences between fishing rights written into Indian treaties from the 1850s and state fishing regulations, concluding that Indians have a right to fish, not a mere privilege. Today, the court decision remains unchanged and the tribes are equal partners with the State of Washington in management of the salmon resource. The tribes have also been active in trying to restore salmon runs on the Yakama Reservation and the Umatilla Reservation in Oregon. Both of these reservations adjoin national forest lands.

In *Sohappy v. Smith*, a case similar to the Nisqually case involving the fishing rights of the Yakama Tribe, Judge Robert C. Belloni in 1969 ruled in favor of the tribe and the treaty signed in the 1850s. The Boldt and Sohappy decisions would have implications for management of fisheries in the Columbia River system for decades to come. Because of dwindling native fish runs—basically salmon species and steelhead—on many rivers throughout the Pacific Northwest, especially in the Columbia Basin, in recent years forest planning rules have been put in place to try to maintain the native fish runs (see Chapter 9). In addition, the affected tribes have become active participants in the management of the national forests concerning riparian protection and related fisheries protection.

Watershed Management

The Region's national forests are estimated to supply 45 percent of the total water runoff in Washington and Oregon, as well as drinking water for at least 151 cities in the Pacific Northwest. The Corps of Engineers and the Bureau of Reclamation had some 50 water storage projects proposed for location partially or totally within the forest boundaries in the Region by 1963. At the same time, more than 30 major dams and hydroelectric projects relied on water from the Region's national forests. A reconstruction project at the Clear Creek Dam on the Tieton RD (Snoqualmie NF) was completed in 1964. The 46-year-old concrete dam was deteriorating. The reconstruction allowed users to once again irrigate, fish, boat, and have adequate drinking water.

The Federal Power Commission issued a license for the High Mountain Sheep Dam to Northwest Power in February 1964, despite objections by organized citizen groups and the Department of the Interior, which challenged the granting of the dam license because of expected serious environmental impacts, especially on salmon and steelhead. The dam was to be built about one mile upstream from the mouth of the Salmon River and was to be 670 feet tall, with a reservoir extending 58 miles to the south (Brooks 2006). The license was upheld by an appeals court, but overturned by the Supreme Court in a 1967 decision written by Justice William O. Douglas. Douglas interpreted the Federal Power Act to require the consideration of more alternatives, including no development.

> The objective of protecting "recreational uses" means more than that the reservoir created by the dam will be the best one possible or practical from a recreational viewpoint ... The importance of salmon and steelhead in our outdoor life as well as in commerce is so great that there certainly comes a time when their destruction might necessitate a halt in so-called "improvement" or "development" of waterways. The destruction of anadromous fish in our western waterways is so notorious that we cannot believe that Congress through the present

> Act authorized their ultimate demise ... The test is whether the project will be in the public interest. And that determination can be made only after an exploration of all issues ... including future power demand and supply, alternate sources of power, the public interest in preserving reaches of wild rivers and wilderness areas, the preservation of anadromous fish for commercial and recreational purposes, and the protection of wildlife (*Udall v. Federal Power Commission* 1967).

The dam was never built. The Hells Canyon Wilderness, established in 1975, within the Hells Canyon National Recreation Area on the Wallowa-Whitman, Boise, and Nezperce NFs was another outcome of the battle over the High Mountain Sheep Dam. This Supreme Court decision would also have implications for later federal environmental impact statements and the range of alternatives that must be considered.

From 1963 to 1965, the Region established four "barometer" watersheds that were to serve as pilot test areas for evaluating watershed management practices: the 148,000-acre Green River watershed (Snoqualmie NF), the 130,000-acre Entiat River watershed (Wenatchee NF), the 87,000-acre upper Clackamas River watershed (Mt. Hood NF), and the 85,000-acre Umatilla River watershed (Umatilla NF).

The Bull Run watershed, which provides drinking water for the City of Portland, came under increasing scrutiny beginning in 1971. Joe Miller, a resident at the edge of the watershed, led the effort to stop logging and all recreation in the watershed. The battles lasted for a decade; Miller sued the Mt. Hood NF for allowing regular management activities—including building the Lolo Pass Road that provided access on side roads to the watershed—to continue in violation of the Bull Run Trespass Act of 1904. The court ruled against the Forest Service in 1976, stopping most activities.

Fire Management

By 1963, the Region was relying less on aerial applications of borate and water to fight forest fires and more on retardants that have commercial fertilizer as a base. It was hoped that fertilizer would provide a boost to vegetation growth on burned-over areas once the fires had been put out. The 1963 Pacific Northwest annual report noted that "as a comparatively new element in the air attack program, three highly mobile helitack crews were in action last season and proved to be excellent fire suppression units." (*Pacific Northwest Report* 1963: 26)

Fire crew on the Mitchell Creek Fire, Wenatchee National Forest, Washington, 1970. USFS photograph.

The Redmond Air Center opened on August 29, 1964, at the airport outside of Redmond, Oregon. It "is the home base for smokejumpers, air tankers, air cargo handling, and a fire suppression crew that may be flown anywhere in the West." (*Pacific Northwest Report* 1964: 22) About one-third of the total cost of $580,000 was paid from the Accelerated Public Works funds, which helped construct two 25-person barracks, a 100-person mess hall, a paraloft building, and a residence. Pioneer smokejumpers Earl Cooley and Francis Lufkin attended the opening celebration.

There was a move to have aerial detection become the dominant method for spotting fires. Back in 1948, there had been 717 lookout houses and towers in the Pacific Northwest, but the number of lookout structures was rapidly reduced so that by 1964, for example, 105 lookouts (out of 390 remaining) were not used. Aircraft were employed to take up the slack and this saved the Region money. "The Mount Hood Forest, for instance, saved at least $10,000 by discontinuing 22 lookouts." (*Pacific Northwest Report* 1964: 23)

Retardant drop on a fire, Fremont National Forest, Oregon, 1966. USFS photograph.

But lookouts were still being used during this period. A story appeared in *The Oregonian* on July 7, 1969, about Pat Lake, who had at the time served 21 years as lookout on the Winema NF atop Yamsay Mountain. She said that life as a lookout has been interesting:

> [S]leeping in a sleeping bag on a narrow bed, carrying every drop of water up long stairs, running to the 'out-behind-house' in icicles or low temperatures, and living with mosquitoes, flies, ticks, woodchucks, squirrels, cougars, bobcats and badgers. … without television, telephone and electric lights. It's washing clothes and ones self in a dishpan. … My tower has been struck by lightning and I've seen balls of fire marching down the lightning net, I've been snowbound and I've had run-ins with bears and badgers, but there's nothing like it.

Research Work on the National Forests

Research work began to shift from field research to the laboratory, and Congress provided funding for laboratory construction. In 1961, the Forest and Range Experiment Station, working closely with the forestry department of Oregon State University, built a Forest Products Research Laboratory at Corvallis. The 14,490-acre H. J. Andrews Experimental Forest (or simply "the Andrews") on the Willamette NF served as a demonstration forest for the lab. Between 1948 and 1986 over 600 studies were published regarding research activities in the H. J. Andrews Experimental Forest, but research showed some changes in emphasis during this period. As before, there was close cooperation between federal and state research agencies, and particularly with Oregon State University. The first studies in the Andrews in the late 1940s through the 1950s had tended to focus on logging and road system designs, as well as forest regeneration techniques. By the 1960s, the research emphasis shifted to watershed studies with the examination of the effects of logging methods on water quality and timber yield. A series of studies in hydrology involved stream flow in unlogged as compared with logged areas and other studies of soil movement in logged areas, especially concerning water quality, water temperature, and oxygen levels (Geier 2007).

Establishing new research natural areas (RNAs) became an increasing priority for research in the national forests in the Region. A dozen RNAs were established in this era to preserve undisturbed natural features for scientific and educational purposes:

- Canyon Creek RNA on the Malheur NF in 1960 as an example of virgin ponderosa pine;
- Olallie Ridge RNA in 1963 on the Willamette NF to provide examples of mountain meadow and true fir;
- Gold Lake Bog RNA in 1965 on the Willamette NF to preserve subalpine bogs and rare plants;
- Bull Run RNA in 1966 on the Mt. Hood NF to study a mix of true fir, hemlock, and Douglas-fir at middle elevations;

Bernie Carter measuring grass seed, Wallowa-Whitman National Forest, Oregon, 1964. USFS photograph.

- Sisters Rock RNA in 1967 on the Gifford Pinchot NF for Pacific silver fir research;
- Pataha Bunchgrass RNA on the Umatilla NF in 1968 as an example of transition bunchgrass vegetation;
- Rainbow Creek RNA on the Umatilla NF in 1968 as an example of grand fir, western white pine, and western larch stands;
- Wildcat Mountain RNA in 1968 on the Willamette NF for researching noble fir;
- Wolf Creek RNA in 1969 on the Okanogan NF to study bitterbrush-bunchgrass community;
- Ashland Creek RNA in 1970 on the Rogue River NF to study ponderosa pine and ponderosa pine mixed with Douglas-fir;
- Bluejay RNA in 1971 to exemplify ponderosa pine and lodgepole pine interactions with bitterbrush and needlegrass on the Winema NF;
- Mill Creek RNA on the Mt. Hood NF in 1971 that was chosen for the forest-grassland transition.

By 1972, 48 such RNAs had been established in the Region. (See Appendix I.)

On December 22, 1974, Congress established the Cascade Head Scenic Research Area (CHSRA) on the Siuslaw NF. This unique area in the fog belt near the Pacific Ocean is home to more than 350 species of wildlife. The 9,670-acre CHSRA includes the western half of the Cascade Head Experimental Forest that had been established in 1934, several prairie headlands, the Salmon River estuary, and adjacent private lands. The legislation establishing the area includes the following statement: "In order to provide present and future generations with the use and enjoyment of certain ocean headlands, rivers, streams, estuaries, and forested areas, to insure the protection and encourage the study of significant areas for research and scientific purposes, and to promote a more sensitive relationship between man and his adjacent environment." The YWCA in 1979 sold 200 acres of land, including more of the Salmon River estuary, to the Forest Service and this became part of the CHSRA. In 1980 the entire area was designated a Biosphere Reserve as part of the United Nations Biosphere Reserve system.

Growing "superior" (i.e., fast-growing and disease-resistant) trees became a major research effort in the 1960s and 1970s. Wendall Jones reported that the Region during this era, in addition to using superior seed sources, developed "stocking level control … [that] must be started when the trees are very young by pre-commercial thinning, later entries to remove more stems [trees] commercially, and then even later to prepare the area for natural seeding (shelterwood) or clearcut and start over." (Jones personal communication) Over the next few decades, there were many lessons learned about planting "off-site" trees like eastside ponderosa pines in westside Douglas-fir areas, as well as latitude, longitude, and elevation differences (such as planting low elevation seedlings in high elevation sites), as well as microclimates that can affect tree survival and growth.

During the 1970s, the focus shifted to ecosystems, involving studies of nutrient cycling, energy flow, and the basic forest community. The Andrews was selected in 1969 as one of the intensive study sites of the Coniferous Forest Biome Project of the International Biological Program (IBP). During the field seasons of 1971 and 1972, 19 reference stands (plots) were established as part of the IBP studies. These reference points enabled researchers to focus on limited areas to investigate species diversity, density, biomass, leaf area index, structure, and forest succession. Research activities grew rapidly in this decade as a result of increased support from the National Science Foundation through the IBP and grants for individual projects.

In the aftermath of the 1970 Entiat Fire, scientists found that portions of the Entiat Experimental Watershed (Wenatchee NF) had burned. "After their initial sadness at seeing years of [research] work wiped out, watershed scientists decided to convert a scientific disaster into scientific knowledge. And thus, Operation Phoenix was born. … the program became one of evaluating the effects of a destructive fire on a watershed." (*Pacific Northwest Report* 1970: 16)

Chapter 8

Turbulent Times 1976–1989

The middle 1970s through the end of the 1980s was an era marked by the passage of a variety of new environmental laws. The population in the Pacific Northwest was rapidly increasing and becoming more urbanized. Suburbs grew, while the inner cities declined. The U.S. was still reeling from the Vietnam war and social protests in the cities. The federal government was viewed by many, including President Reagan, as the problem with society.

By the end of the 1970s, there was a new wave of national concern about the conservation of natural resources that was fundamentally different from the Pinchot conservation theme. The new way of thinking was coined "environmentalism." The scientific forestry methods of the turn of the twentieth century, once thought to be the backbone of progressive forestry and conservation, were increasingly viewed by many parts of the urban public as conservative, overly production oriented, and not sensitive to resources other than timber. It took the Forest Service years to adjust. So-called "timber wars" were evident throughout the Forest Service, but especially rancorous in the Pacific Northwest. Tensions rose to the point of violence against the Forest Service and the timber industry.

A rather unexpected natural event occurred in the spring of 1980, when Mount St. Helens awakened and on May 18, 1980, exploded in a fury, scattering tons of ash across Washington and dust around the world. After months of negotiation, as well as threats to have the National Park Service take over Mount St. Helens, the Mount St. Helens National Volcanic Monument was established in 1982.

Regional Foresters

Richard Worthington was appointed regional forester in 1977, succeeding Ted Schlapfer. Worthington had worked in various national forests in Region 6, was promoted to the regional forester's staff in the Pacific Southwest (California) Region in 1970 as assistant regional forester, becoming deputy regional forester in 1971. In 1972 he returned to the Washington Office where he became director of the division of timber management. He returned to R-6 as regional forester in June of 1977. Worthington faced the continuing RARE II controversy even after the publication of the final environmental impact statement (EIS) in 1979 and the continuation of the regional and forest planning efforts under the mandate of the National Forest Management Act (NFMA). In addition, the Regional Office put together a team to prepare an EIS on the proper use of herbicides in forestry. In 1982 Worthington retired, and was succeeded by Jeff Sirmon.

Sirmon had served as an engineer in various positions from 1958 to 1963, when he was transferred to R-5, and in 1969 to the engineering division of the Washington Office. In 1974 he became deputy regional forester for the Intermountain Region (R-4) and was promoted to regional forester in 1980. He was appointed as regional forester for R-6 in 1982. One significant problem that he had to address was the so-called "timber depression," when high interest rates at the national level collapsed the timber market. Sirmon also faced numerous legal actions against the RARE II study and the new Regional Guide for planning (1984). During the winter of 1982 and 1983, Sirmon ordered a study of the national forests in the Region under the direction of the Asset Management Program (of which more details are given below) to see which forests or portions thereof should be sold. Sirmon also supported an effort in 1985 to transfer or interchange certain BLM and Forest Service lands in Oregon. During his time as regional forester, in 1983 and 1984, there were efforts to resolve the roadless area question through Congress, and after months of negotiation, the Oregon and Washington Wilderness Acts were passed in June of 1984. Protesters from Earth First! occupied his office for a short time. Sirmon transferred to the Washington Office in the fall of 1985 as the deputy chief for programs and legislation. He was replaced as regional forester by Tom Coston.

Charles "Tom" Coston worked in the Washington Office in 1975 before returning to the Northern Region (R-1) in 1977 as deputy regional forester for resources. Two years later Coston was appointed as regional forester for R-1. After his appointment in the summer of 1985 as regional forester in R-6, Coston unexpectedly retired after a few months and was replaced in early 1986 by Jim Torrence.

James Torrence started with the Forest Service on the Ochoco NF. He served in the Washington Office and several regions, including as the forest supervisor of the Mt. Baker NF, deputy regional forester in R-6 in 1977, regional forester in R-2 in 1984, before coming back to R-6 as regional forester in the spring of 1986. Torrence faced the controversial proposal to have portions of the Columbia River Gorge established as a scenic area. In addition, his tenure saw the terrible fire season of 1987, the publication of the first NFMA forest plans in 1987, and the Oregon State Wild and Scenic

Rivers Act of 1988. During 1987-88 there were Regional efforts to address through EISs the very controversial questions of spotted owl habitat and the management of competing and unwanted vegetation through herbicides and other methods. Torrence retired in August of 1989 and was replaced by John F. Butruille, former R-6 deputy regional forester for resources (see next chapter for a biography).

Regional News

During 1977, another new youth employment program was begun—the Young Adult Conservation Corps (YACC). It had similar aims to the YCC program (see Chapter 7), and was intended for young adults (both male and female) between ages 16 and 23. The Forest Service provided many opportunities for enrollees to work on important projects in the Region. The first two YACC camps in the nation were opened on the Gifford Pinchot NF at Camp Mt. Adams on December 14, 1977, and at Camp Hemlock three days later. By 1980, more than 3,700 enrollees were hired in the Region; this program was short-lived, because funding was eliminated on September 30, 1981.

Regional Forester Jeff Sirmon in 1982 became embroiled in dealing with the mistreatment of Hispanic or Latino workers who were working on reforestation contracts. There was an idea among the Hispanic community that whenever a contract was awarded to a Hispanic contractor the agency would call the Immigration and Naturalization Service (INS). On the other hand, Wendall Jones reported that "there were a number of instances when the INS folks accused FS people of participating in a warning system via portable radios to tell contractors the INS was in the area. I don't think … that was ever proven, but it did not help our relations with the INS." (Jones personal communication) In other cases it was suspected that competing reforestation companies reported on companies they suspected of hiring undocumented workers. Reports in *The Oregonian* and other newspapers fanned the flames of the Hispanic community. Sirmon, through a news release on May 13, 1982, in part acknowledged that some employees were overzealous in dealing with Hispanic workers:

> Sirmon asked National Forest Supervisors to be certain Forest Service employees know and are guided by official directions and guidelines only; that employees understand they are not immigration officers, but as Forest Service employees they are to cooperate with and assist INS upon request through proper channels; that only top administrators have authority to involve the Forest Service in INS activities. The

> Forest Service values its positive relationship with the Hispanic community in this Region too much to allow it to deteriorate through misunderstanding or through misguided activity of a few employees.

Another news release on November 2, 1983, told the story of the Region becoming more active in overseeing federal contracts where illegal immigrants were suspected to be employed. The Region "is taking renewed action to assist the Oregon State Bureau of Labor and Industries (OSBLI), the Department of Labor (DOL), and the Immigration and Naturalization Service (INS) in controlling contractor hiring of illegal aliens." Sirmon said that "Forest Service personnel are being directed to intensify their efforts at notifying the OSBLI and the INS of suspected illegal activities. Although these agencies have the primary responsibility for enforcing the laws … they need our help." Seven new contract specifications were put in place to ensure that every contractor was aware of the new direction.

Charles "Chip" Cartwright worked as a district ranger on the Conconully RD on the Okanogan NF from 1979 to 1982. He was the first African American district ranger in the Region and in the Forest Service. He served as district ranger on the Gifford Pinchot NF in 1982-83. Later he was the first African American forest supervisor on the Jefferson NF in Virginia and the first African American regional forester in the Southwestern Region.

A training program for new employees was instigated by Dick Worthington. He noted that in his travels around the Region, he found employees with as many as four to five years experience who still did not know the mission of the agency. He described the situation in his oral history: "If a guy is going to work for us, he ought to know what our job is and where our finances come from." (Worthington 1984: 45) In shades of things to come, he also was confronted

Charles "Chip" Cartwright, first African American District Ranger, Gifford Pinchot National Forest, Washington, 1983. USFS photograph.

with a new type of employee—one who didn't see things quite the same way as the old Forest Service: "One time there was a delegation of young people, sincere as heck, who came to see me. They said we didn't understand—He said, 'one of the reasons I came to work for the Forest Service was to stop this awful harvesting of timber, that's bad.' I said, 'It depends on whose eyes you look at it through … If you want to stop the Forest Service from cutting timber, you should leave the Service and work for somebody else.'" (Worthington 1984: 45-46)

A new division of planning was announced on September 14, 1982, and Allan Lampi was chosen to become the director. Previously, the position had been grouped with the regional programming and budget staff, but with the growing importance of land use planning under the National Forest Management Act of 1976 (NFMA), a new unit within the Regional Office was needed to coordinate efforts. Many new technical experts were hired in the planning shop during the late 1970s and early 1980s. New, highly qualified employees, such as Alan Fox as regional economist and Arnie Holden as regional sociologist, were hired to coordinate NFMA planning related to National Environmental Policy Act (NEPA) analyses in planning documents. NEPA specialists were employed to oversee the quality of plans and documents coming from the forests in the Region. It was a time when the planning shop was coordinating with many of the Regional Office specialists in order to produce quality NFMA plans on schedule. Outside experts, usually from universities, were brought into the planning process to provide background documents, new procedures, computer programs, and training of employees in the Regional Office and on the national forest planning teams. Relatively speaking, there were abundant dollars for the planning effort.

There were many opportunities for national forest planning specialists to work on a temporary basis (called "detailers") in the Regional Office for a variety of projects and EISs. So many of the detailers stayed at the Imperial Hotel in downtown Portland that it was often referred to as the "bunkhouse." A number of forest-level planners and other specialists would try to arrange details before Christmas so that they could shop for gifts in the "big city." The Regional Office was an exciting place to work.

An outgrowth of Reaganism called the Asset Management Program began in 1982. The purpose was to identify "federal lands and other real properties no longer needed for achieving the objectives of the federal agencies having jurisdiction over those properties." The program created three categories under which all federal property was to be classified: retain in public ownership under federal management, study further, and sell. A preliminary study in the

Woman Forester, Pacific Northwest, c. 1980. USFS photograph.

Region was completed by March 15, 1983, with the following results: areas to be retained, 24,054,000 acres; land to be further studied, 617,000 acres; and parcels for sale, 1,972 acres. There were tales told of real estate agents contacting every Forest Service office trying to get permission to list the properties, but little actually occurred.

The Forest Service in 1985 adopted the Data General (DG) computer system, which electronically linked all agency locations—Washington Office, research stations, regions, national forests, and ranger districts. The Forest Service was the first federal agency to have all employees linked electronically. Yet this progress was not fully embraced by Forest Service managers for several years. At least one forest supervisor in Region 6 declared that he didn't want a computer on the desk of his staff as typing was for the typing pool. Yet the first DG system was introduced in the Region on the desk of Regional Forester Jeff Sirmon. The first DGs on the forests in the Region were installed on the desks of the forest supervisors; they quickly gained widespread use, especially the e-mail function, which instantly connected everyone in the agency (the World Wide Web was still in its infancy). Within a couple of years, the e-mail system was used to start a new organization whose membership included a number of Forest Service employees in the Region—the Association of Forest Service Employees for Environmental Ethics (see discussion below). Ten years later, the Forest Service adopted an IBM/UNIX-based system to replace the Data General.

Regional Forester Jeff Sirmon on March 27, 1985, announced that the Region would reduce 700 to 900 employees before the end of the year. He said, "[W]e anticipate a significant reduction in workload over the next five years as a result of a large volume of timber being returned to the government under the 'Timber Contract Modification Act' (buy-back) of 1984." About 60 percent of the reductions came as a result of the act, while the remaining reductions came from program and budget reductions. The news release noted that about 9,000 people had been employed in the Region in 1975, but that a decade of reductions had already lowered the number of employees to around 7,000.

Nevertheless, a series of new programs were developed at the national level of the agency in the late 1980s and early 1990s. The "challenge cost-share program," established by Congress in 1986, provided the means for the Forest Service and the private sector to share management and financial costs for projects on the national forests. The Region, along with other federal and state agencies and nonprofit organizations—like Ducks Unlimited, Rocky Mountain Elk Foundation, and many others—carried out several thousand cooperative wildlife habitat enhancement projects.

BLM/Forest Service Interchange Proposals

On January 30, 1985, the Reagan administration announced a proposal to interchange/exchange or consolidate lands between the Forest Service and the BLM in order to save money. An eight-member joint Forest Service and BLM team was formed, headed by Jim Space from the Regional Office and Paul Vetterick from the BLM. Regional Forester Jeff Sirmon and BLM State Director William G. Leavell announced their preliminary recommendations on May 7, 1985, followed by public hearings in June. The most noticeable proposal was to move 3.6 million acres of BLM's O&C lands in western Oregon to Forest Service administration, while shifting 300,000 acres of national forests to BLM management. In addition, the proposal was to transfer 312, 000 acres in Washington to Forest Service management. On February 18, 1986, the two agencies produced a national 250-page legislative environmental impact statement on a revised proposal under which the BLM was to transfer only 1,085,000 acres to the Forest Service in Oregon, while also reducing the national forest exchange to about 27,000 acres.

Former Associate Chief George Leonard noted the Forest service's opposition to the plan: "[T]here was simply no significant constituency for government efficiency, [but there] were strong constituencies for maintaining the status quo." Especially strong opposition came from the Association of O&C Counties in western Oregon, as this group believed that any such proposal in Congress might give lawmakers the opportunity to rework the lucrative O&C revenue-sharing formula. The administration proposed legislation to institute the interchange, but could not get the proposal introduced in Congress.

However, several recommendations for combining Regional and BLM State Office staffs were implemented over the next few years, including co-locating the BLM State Office with the Regional Office and sharing staff in selected programs. Also, the Forest Service took over from the BLM administration of surface claims, especially mining, on the national forests in the Region.

Law Enforcement

In the late 1970s and early 1980s, the illegal growing of marijuana on the national forest lands caused numerous management problems. Many of the national forests responded to this problem and other lawlessness by hiring law enforcement specialists, who worked closely with other federal, state, and local authorities. By the mid-1980s, 15 of the 19 national forests had special agents who developed criminal cases, with around 50 law enforcement officers who patrolled the forests. Around 900 Forest Service employees in the Region had the authority to issue notices of violation that were often vandalism related. Law enforcement issues in the Region included firewood theft, cutting of evergreen boughs, timber theft, theft of small trees, marijuana gardens, and acts of vandalism. In 1984, 171 marijuana gardens were discovered and 9,737 plants were destroyed. Cooperative efforts with local law enforcement agencies helped discover many of the gardens through helicopter surveys of the forests. Criminal cases on the Region's national forests increased during the 1970s and 1980s, in part as a result of having more law enforcement personnel. During 1984, for example, 123 cases were prosecuted, up from 83 the year before. Financial losses were estimated at $754,487 in 1984.

For at least 20 years, law enforcement personnel had been assigned to the respective manager—a district ranger, forest supervisor, or regional forester. At times the independence of the law enforcement officer was called into question. After much acrimony, early in the 1990s law enforcement was "stove-piped" to the chief of the Forest Service—that is, law enforcement and special investigations officers were no longer responsible directly to local district rangers or forest supervisors. Instead, they reported to their Regional counterparts in law enforcement, who in turn reported to the Washington Office without going through the chain of command of Forest Service line and staff officers. In some cases this new law enforcement paradigm strained relationships between local law enforcement officers and the staff of the forests.

Opposition to Forest Service Policies

By 1985 there was a rising tide of outrage by western landowners about the alleged mismanagement of BLM and Forest Service administered lands. Calling itself the "sagebrush rebellion," the critics got the ear of the new Ronald Reagan administration through the office of Secretary of the Interior James Watt. The rebels wanted to see federal lands given over to local control or sold into private hands. While this effort was much ballyhooed across the country, the massive sale of millions of acres of federal lands never materialized.

Other special interest groups became more vocal during this period, and in some cases destructive. The Sierra Club and more moderate environmental groups had, during the 1960s and 1970s, relied on letter-writing campaigns, influencing and applying pressure on Congress, then lawsuits (McCloskey 2005). By the 1980s, new fringe groups, such as Earth First!, carried out the most serious protests through sabotage, or in the words of its proponents, "ecotage" or "monkeywrenching." In Region 6, there were a number of instances of violent protests. Several on the Willamette NF were particularly troubling to both the Forest Service and the local community (Kerrick 1985).

In the spring of 1985, Earth First! members came from around the United States to the Willamette NF to take action against the Willamette Industries logging operation. After the Corvallis-based Cathedral Forest Action Group tried forest road blockades that proved ineffective, Mike Jakubal from Earth First!, believing that tree sitting would be a more effective alternative, he made the first Earth First! tree sit in the United States on May 23, 1985. This tree sitting lasted less than a day; he came down that evening and was arrested by a Forest Service law enforcement officer, as were several protesters at the base of the tree. The timber sale continued, but the sale areas were modified. A month later a group of activists arrived at Willamette NF and set up as many as a dozen tree-sitting platforms in the Squaw/Three timber sale on the Sweet Home RD in the Willamette NF. On July 10 all the trees with platforms were cut down except for Ron Huber's. He remained in his tree for the next 10 days until two Linn County deputies, who were lifted by a crane boom high into the tree, wrestled him off his platform. These incidents were the first of many tree-sitting episodes in almost every national forest in the Region, slowing but not generally stopping timber sales and the cutting of old-growth trees.

An unusual group of Forest Service professionals formed on the Willamette NF because the new Data General computer system made it easy to send messages anywhere in the agency. Led by Jeff DeBonis, a timber sale planner on the Blue River RD, the group called itself the Association of Forest Service

Employees for Environmental Ethics or AFSEEE. Much of DeBonis' work was planning timber sales in old-growth forests. He felt that his and others' comments about environmental concerns were not being heard. He sent an eight-page, single-spaced letter to Chief Dale Robertson about the sales of the old-growth timber. In an interview, DeBonis spoke about the start of AFSEEE:

> "We were set up to produce timber outputs," he says. "The system rewarded those who met board-foot production targets"… His concerns about overcutting were refueled when he was transferred to the Willamette in 1988 … Then there was the Ancient Forest Seminar he attended in Eugene early in 1989. In a two-page report to his superiors, he rather audaciously concluded, "We as an agency are perceived by the conservation community as being an advocate of the timber industry's agenda. ... I believe this charge is true." (McLean 1990: 31)

The organization started a newsletter called *Inner Voice*, which received wide attention in the agency and from outside professionals and environmental groups. By 1990, the organization was becoming more vocal in management decisions. A study of the AFSEEE membership by the University of Idaho found that most members were from Regions 5 and 6 and that the membership identified themselves professionally as foresters, followed closely by wildlife and fish biologists. These younger professionals (sometimes referred to as "ologists") often did not have the same background and training as the line/staff on forest ranger districts, which put them at odds with management decisions (Brown and Harris 1991). The issue of using the new computer system's e-mail and that of free speech was tossed about in all levels of the organization.

> Meanwhile, at Willamette National Forest headquarters … staff members have been tussling with some landmark questions about dissent from within, free speech, conflict of interest—on a surprisingly positive level, while communicating rather openly, if not benevolently, with Jeff. Explains Forest Supervisor Mike Kerrick with remarkable patience, "All this is a natural outgrowth of growing diversity within the Forest Service. We are, after all, trying to encourage change within the organization, and Jeff represents a very different view of change." (McLean 1990: 66)

The organization changed when Jeff DeBonis left. There was a slight renaming of the AFSEEE to FSEEE. The current organization, according to its Web site, has "grown to be the largest forest protection organization in the country, with more than 500 Forest Service employee members providing the technical expertise on which we rely, and 12,000 citizen members who provide political and financial support."

Mount St. Helens Eruption

Probably one of the most dramatic events in the history of the Region began in the spring of 1980 when Mount St. Helens "awoke" from over a century-long "sleep." In late March the area under the mountain experienced a growing number of shallow earthquakes. Forest Supervisor Bob Tokarczyk issued an order closing the mountain to the public above timberline on March 25. Geologists believed that the mountain, long dormant, was getting ready for a major volcanic event. It was not understood exactly what the extent or duration of such an event might be.

In the nineteenth century smoke and ash had been observed near the top of the mountain several times, but nothing could have prepared the Gifford Pinchot NF or the population of Washington State for the upcoming drama. Even Ken Johnson, St. Helens district ranger, commented that he and his employees were unconvinced of the danger. He said, "I think the mood of a number of employees and their families [was]: It's a joke—there wasn't anything going to happen." (Beals et al. 1981: 13) But others didn't take the situation so casually. On March 26, Tokarczyk convened a meeting of state and federal agency representatives, as well as TV, radio, and newspaper reporters, to discuss the situation.

A private airplane carrying a Portland news crew observed fresh ash deposited on the top of Mount St. Helens on the following day. Forest Service employees and the public were evacuated from a circle one and a half miles in diameter around the mountain on the same day.

> The evacuation of the Ranger Station itself … would be conducted … in a more urgent atmosphere, because by then there was no doubt as to the volcano's eruptive intentions … Kathy Anderson was one of the District employees … [who] later described her experience: "I had some bags packed at home … So I kept thinking of another thing to grab … what I ended up grabbing was … the rifle and French shotgun … because I figured … what if we get stranded? I'd at least be able to shoot something to eat … I thought: What if trees blow down? So I grabbed the chainsaw … Then I thought … I better take some beer … I threw these three things in the truck … and raced back to the Station." (Beals et al. 1981: 17)

Within days a small hole appeared on the snow-covered dome indicating that there was major underground magma movement to near the top of the mountain. On March 30, 93 smoke and ash eruptions were observed coming from the top of the mountain. Two craters near the summit merged into one larger crater on April 1. By mid-April an ominous bulge appeared on the northwest flank of the mountain. It continued growing for more than a month. The Forest Service designated Mount St. Helens a geologic area on April 9. Restrictions were put in place to prevent airplanes from flying over the mountain and landing was prohibited at elevations above treeline, but because the top of the mountain was owned by Burlington Northern Railroad (part of the old Northern Pacific land grant), there was little that could be done to stop helicopters from landing there.

Earthquakes were increasing in number and intensity. The USGS geologists strongly suggested that the immediate area be evacuated. The state imposed a red zone (restricted to scientists, law enforcement, and rescue personnel) around the north side of the mountain. Although Mount St. Helens was somewhat quiet, the growing bulge on the north flank concerned scientists. Forest Supervisor Tokarczyk decided to close additional areas around the mountain. The red zone and a blue zone (restricted to property owners with permits, but logging could continue during daylight hours) were enlarged. Forest Service public affairs specialist Jim Unterwegner was concerned that people did not seem to be taking the volcano seriously: "One of the problems we're having ... is that people in the Northwest have never faced this kind of thing before ... I feel the general public really doesn't understand the danger."

Washington State Governor Dixie Lee Ray and local government officials were pressured by residents, summer cabin owners, and logging outfits to make the restricted areas smaller than recommended by the USGS. In the weeks that followed, the mountain was once again quiet. Many people believed the danger had past and the governor ordered several of the road blockades removed to allow residents to take belongings from their homes. Harry R. Truman refused to leave his beloved Mount St. Helens Lodge, perched on the edge of picturesque Spirit Lake. His stubborn stand in the face of the unstable volcano attracted many newspaper and other media reporters. "Seismic activity, once as great as 40 events per hour, decreased on consecutive days from May 15 to 17. Geologists of the USGS hoped it was a sign of decreasing danger." (Beals et al. 1981: 32) A group of property owners signed waivers and were escorted on May 17 to retrieve possessions from their summer homes in the Spirit Lake area. Another trip was planned for the following day.

Then on Sunday May 18 the unimaginable happened—the mountain blew up! The huge explosion at 8:32 a.m., heard in many parts of Washington and northern Oregon, blasted 1,314 feet of summit into the atmosphere, buried picturesque Spirit Lake, and flattened many thousands of acres of prime timber and recreation land. The ash column reached 80,000 feet into the atmosphere in 15 minutes, spread across the U.S. in three days, and around the globe in 15 days. The area looked as if a nuclear bomb had exploded—there was nothing left standing for miles and the once green landscape now resembled a moonscape with only undulating mountains and hills. An estimated 57 people were killed in the blast, including volcanologist David Johnston and the crusty 83-year-old Harry R. Truman. The deaths were the first recorded from any volcano in the lower 48 states. Only Mt. Lassen on the Lassen NF in R-5 during 1914 had ever experienced a volcanic eruption during recorded history.

Cities such as Spokane were buried by several inches of volcanic ash, requiring residents to turn on car headlights by noon. Rivers such as the Toutle were a muddy flow of ash, trees, boulders, and wrecked buildings speeding downstream. Supervisor Tokareczyk signed an occupancy and use order on the afternoon of the eruption, closing the forest to the public to ensure public safety. Public Affairs Officer Jim Unterwegner moved the volcano information center from the supervisor's office to the Shilo Motel in Vancouver. "That

Mount St. Helens eruption on May 18, Gifford Pinchot National Forest, Washington, 1980. USFS photograph.

action not only cleared, in part, the Forest Service office of reporters, but it also provided the media with better communication facilities. The telephone company installed a number of phones for the use of reporters, as well as for the various agencies using the facility." (Beals et al. 1981: 39)

One of the visitors not long after the eruption was President Jimmy Carter. Just days after the big eruption, Bob Tokareczyk (2005) received a phone call from Air Force One then flying over the Mississippi River asking if it would be possible for the president to have a volcano briefing in three hours. Of course Tokareczyk said yes.

> President Carter arrived with his entire cabinet along with the Governors of Washington, Oregon and Idaho at the Gifford Pinchot supervisors office in Vancouver, Washington. The forest hosted this group with presentations by Forest Supervisor Bob Tokareczyk and staff [as well as Regional Forester Dick Worthington] and other agencies involved with volcano safety and Columbia River cleanup. The U.S. Geological Survey briefed the group about the volcano's activity. The following day Forest Supervisor Tokareczyk guided President Carter to the volcano in the President's helicopter with the red [war] phone. They were followed by the President's cabinet in a Chinook helicopter. Due to cloud cover the volcano was not visible and after a trip around Spirit Lake the two helicopters returned to the Portland Air Base (Tokareczyk personal communication).

Supervisor Bob Tokareczyk and President Jimmy Carter at Mount St. Helens briefing, Gifford Pinchot National Forest, Washington, 1980. USFS photograph.

For weeks, the media visited the forest, talking with the staff, USGS geologists, and state officials, and members of the public wanted to visit too. The Gifford Pinchot NF and the Regional Office were quick to respond. Visitor information trailers were set up along I-5 at the Ridgefield exit (opened July 4) and another at Lewis and Clark State Park (opened on July 12). By the end of the month, approximately 138,000 people had visited the volcano information sites.

Pressure was quickly mounting from various Seattle environmental groups to establish a Mount St. Helens National Park. As the mountain was still steaming, the Forest Service initiated a land trade with the Burlington Northern Company for their land on the mountain in exchange for national forest land elsewhere. With these activities in motion, it was not surprising that instead of a new national park, Congress passed the Mount St. Helens National Volcanic Monument act, which was signed into law by President Jimmy Carter on August 26, 1982. This 110,000-acre national monument is managed by the Gifford Pinchot NF. This was the first national monument administered by the Forest Service that had its origins in Congress rather than by proclamation of the president.

Eruptions continued at Mount St. Helens for years. Only in 2008 was the mountain declared quiet. Other mountain peaks in the Cascades have, since the coming of the European settlers, experienced some low levels of volcanic activity. Mt. Baker, Mt. Rainier, and Mt. Hood have been seen steaming near their summits for decades, but no eruptions have been recorded on these three peaks since the late 1800s. An ominous ground rising (a matter of inches) has appeared near the South Sister in the Three Sisters Wilderness. Seismic monitoring by the USGS on many of the Cascade peaks, because of the Mount St. Helens disaster, has become routine.

National Forest Planning Laws (NFMA)

The Forest Service began major (and costly) land use planning efforts in the 1970s and '80s. Some of these efforts were simply updates of the multiple use plans from the late 1960s and early 1970s. Others were designed to replace earlier plans with more comprehensive forest plans. This idea of forest planning would come to a head with the plans written under the National Forest Management Act of 1976 (NFMA).

Because of the lawsuits and congressional hearings over the clearcutting issue in the early 1970s, there was considerable debate in Congress about how much latitude the agency should have in any new act to address the inadequacy of the Organic Act of 1897. Those on one side wanted to see a very open law that would give Forest Service managers great flexibility in management by just fixing the wording of the Organic Act. Others in Congress and from special interest groups wanted to limit the agency's flexibility by providing very tight, prescriptive controls. Former Chief R. Max Peterson noted that: "[I]t became obvious to most that neither Congress nor anyone else could possibly write management prescriptions that would fit the many physical situations on National Forests. ... This led to a recognition that the legislation would have to set forth a process rather than specify answers." (Peterson 1988: 31)

The result was the National Forest Management Act, which was signed into law on October 22, 1976, and amended the RPA act of 1974. The implementing regulations, created for the most part by a committee of scientists from outside the agency, took almost three years to become final. As with the wording in the act, there were many debates about how specific to make the regulations. The new regulations required a long-range planning process to be carried out by a team of interdisciplinary specialists. Regional plans for each Region were written to help guide the planning efforts at the national forest level, with the Region 6 draft plan and draft EIS released to the public on July 20, 1981. The Regional Plan was renamed the Regional Guide, then a few years later all mention of the guide was dropped.

At the same time, an intensive NFMA forest planning effort began on each of the national forests in the Region. The Forest Service hired many new specialists, many of them women, including public affairs specialists, economists, archeologists, sociologists, geologists, ecologists, and operations research analysts. In the early 1980s, each national forest was required to use the computer FORPLAN, or forest planning model, to test alternatives under consideration in the plan. The FORPLAN model was written by Norm Johnson from Oregon State University and tested at the Forest Service computer and planning facility at Ft. Collins, Colorado.

> Numerous training sessions on NFMA—called the Eisenhower Consortium—were given to the forest teams in the Region. The consortium was "linked strongly to the linear program geeks at Colorado State … [and] a collection of colleges and universities who were loosely banded together to partake of grants and contracts; I think they were mostly located in the Rocky Mountain States. I think as time went on, more and more of that course's teaching was done by FS people as academics found it to be less and less fun. It was a national contract out of the WO [Washington Office], specifically WO-LMP [WO Land Management Planning section]." (Holden personal communication)

New resource inventories were contracted or old ones updated. New land use allocations were discussed internally and applied in the forest plans. The Deschutes NF and Wallowa-Whitman NF plans were considered to be pilot plans for the Region, intended to move the planning process along as quickly as possible so that other forests could learn from their efforts and mistakes. One stumbling block came from an unexpected quarter—review of every step of the process by the Department of Agriculture, especially by John Fedkiw, who served as Department plan reviewer. His intent was noble, but his review of the planning documents for each forest was frustrating. The biggest problem was that the standards were quickly changing as the forest planning teams learned from mistakes and produced better materials.

Problems cropped up as well over land allocations—not only the names but also the management prescriptions for each. Then there was the problem that arose when two forests had different land allocations for an area that straddled the forest boundaries. Which should be dominant—unroaded recreation on one side of the mountain on one forest, or general forest including timber management on the adjacent forest?

Initially, the Regional planning shop believed that minimal direction was best, as each forest could come up with solutions that would fit their circumstances. This approach was quite different from that of other Regions, which imposed standardized definitions, land allocations, mapping standards, and even criteria for development of FORPLAN computer models. The Deschutes NF and the Wallowa-Whitman NF on the same day made presentations to the Washington Office about their draft forest plans. One forest said that they used economic figures calculated from the 1980 RPA report, the other that they had not used the recommended RPA figures because their situation was so different. The Washington Office reviewers must have been amazed. Region 6 was going it alone, but that was about to change.

Then in 1983 Regional Forester Jeff Sirmon imposed a pause on the forest plans in order to come up with regional solutions to roadless area designation and other issues (see the section on RARE III later in this chapter). Sirmon described the problem:

> Forests were going their own way, and the regional industry watchdog planning group was watching and challenging everything we did. They were finding lots of differences between Forests on identical issues … I immediately called a halt to all Region 6 planning (I called it a pause), got a new planning director, and directed Forest Supervisors that they were staff to me as far as LMP's [land management plans]

> were concerned since I had to sign all the plans and could not tolerate differences on identical problems on two or more Forests without a justification I agreed to. I started frequent meetings with the Supervisors and their planners to accomplish the degree of uniformity in definition, processes, and procedures needed for success (Sirmon 1990: 682-83).

One protracted problem that first appeared in the public involvement activities of the Umpqua NF plan was the question: What is the "no-action alternative"? Douglas Timber Operators wanted the no-action alternative to only show what the earlier 1978 plan had proposed. The forest believed that the no-action alternative should incorporate changes in direction since the 1978 plan. The Region compromised with two alternatives: the no-action alternative with updated inventories and direction and another no-change alternative with whatever earlier plans had called for. There was also strong pressure from the Department of Agriculture to include an alternative that would maximize timber harvest, especially through elimination of the non-declining even-flow restrictions that would quickly eliminate the standing old-growth forests. Non-declining even-flow was the idea that forest production would always remain the same over many decades, no matter what the economic conditions and demand were at the time. Sustained yield, on the other hand, also looked ahead but was better suited to the demand principal where production could vary but no more could be harvested than could be grown. In addition, for the Oregon forests, there was an agreement with the State of Oregon to include another alternative that would meet the wood needs as outlined by the 1976 Beuter Report (Beuter et al. 1976). Many of the forests in the Region redefined the maximum timber alternative as the state

Land use planning public meeting, Siuslaw National Forest, Oregon, 1987. USFS photograph.

Shelves of Region 6 forest plans, Regional Office, Portland, Oregon, 2008. Author's collection.

alternative. Other forests simply made more FORPLAN computer runs that had similar parameters. Sometimes FORPLAN runs made it into alternatives, while at other times they were background computer printouts. Often each national forest analyzed as many as a dozen alternatives in the draft plans. What had been envisioned as a process taking only a couple of years to complete draft forest plans soon became a decade. The Deschutes NF plan was published in early January of 1986 and was the first draft environmental impact statement in the Region. Most of the NFMA plans were completed by 1990.

However, the completed NFMA forest plans were not greeted with great fanfare or relief from the old battles over resources, especially timber supply. The agency tried to integrate the science of forestry, range management, and other resource disciplines, along with public desires, into these comprehensive plans. In doing so, they had to ignore a number of thorny resource issues that were outside their control. These issues, which carried over to the 1990s, have in most cases been handled outside the NFMA planning process: spotted owls, marbled murrelets, grizzlies, wolves, cougars, caribou, steelhead, salmon, roadless areas, old growth or ancient forest, Pacific yew, clearcutting, monoculture, air quality, international impacts including climate change, herbicide use, and so on. Some of these remain unresolved today.

Timber Management

Timber harvest levels in the Region during this period were very high. In 1977, the total harvest was 4.342 billion board feet; this fell to a low of around 2.3 billion in 1982, then rose to the highest level in 1987 with around 5.6 billion board feet cut. At the end of the period, in 1989, 5.2 billion board feet were harvested. The value of the timber harvested rose from $423 million in 1977, then dropped to around $192 million in 1982, then spiked to over $857 million in 1989. The average dollar per million board feet during the period was $97.44/million in 1977, rising to $163.96 in 1989. (See Appendix E.)

The 311,000-acre J. Herbert Stone Nursery, located in Central Point, Oregon, opened in 1977. The nursery serves the Forest Service, BLM, tribes, and other federal, state, and local agencies in the western United States. Over a hundred species of evergreen and deciduous trees and shrubs for reforestation and watershed-restoration work have been grown here. The reforestation program in the Region during 1980 topped out at 115,000 acres, dropping to below 75,000 acres by 1985, then rising again to over 100,900 acres in 1988.

The Forest Service experimented with a helium-filled balloon and tethered four-helicopter mix, which was referred to as a "helistat," to transport logs from remote areas, especially roadless areas. The balloon would lift the heavy log load, with assistance from the helicopters, then move to a landing site where the logs would be unloaded and placed onto trucks. One project using the helistat was planned on the Wind River Experimental Forest. But before this could take place, the innovative system crashed at Lakehurst, New Jersey, with the loss of one pilot on July 1, 1986. The project was not renewed and the Wind River experiment was cancelled.

PESTICIDES AND HERBICIDES

The use of herbicides, especially the chemicals 2,4,5-T and 2,4-D, came under fire during Regional Forester Richard Worthington's administration. The national forests in the Region had been spraying vast areas that had been reforested after timber harvest, arguing that neither herbicide had a major environmental or human impact except on targeted brush and grass species. The Forest Service noted that the herbicides could be applied to northwest forests under EPA label guidelines to reduce competing vegetation in young tree stands, as well as along roads and other locations where unwanted vegetation, including invasive plant species, were present.

Herbicides and other pesticides have been used since the early 1970s to effectively and cheaply kill off competing vegetation in new tree stands. In

Old-growth Douglas-fir with Dick Swartzlender and Jim Hunt, Umpqua National Forest, Oregon, 1981. Author's collection.

previous decades, the most effective ways to remove or kill off unwanted vegetation were by pulling, cutting, digging, back-pack spraying, or in some cases direct injection of a poison into the competing tree. Ground-spraying equipment using the herbicides was tested in many locations, but on steep ground and in heavy brush this proved inadequate. Aerial spraying from either bi-planes (used extensively on farms) or helicopters proved to be the preferred method to apply herbicides over large areas of the forests in the Region, especially west of the Cascades. Many new restrictions were put in place for any aerial-spraying operation, including setbacks from lakes, streams, and human habitation, as well as accounting for weather conditions including wind speed and direction. Wind drift of the spray was always a top consideration.

There were accidents, including wind drifting the spray over populated areas, as well as accidental spills into creeks (the issue is still around today, with reference not only to herbicides but use of fire retardants). Wendall Jones (personal communication) recounted his experiences with an apparent herbicide spill on the Siuslaw NF:

> I cannot pin down the year, but it must have been between about 1970 to 73, that the spill catastrophe occurred on the Alsea district, upstream

> from the State Fish Hatchery. This drainage was being very carefully managed because of the hatchery downstream. Apparently the hatchery folks suddenly saw many fish dying. Immediate reaction was to do water sampling. The USFS was doing brush spraying upstream, very carefully avoiding direct application to any streams. The percentage of 2,4,5-T in the hatchery water was the equivalent of about a 55-gallon drum of the stuff being dumped directly in the hatchery. Contractors could account for all of their barrels and containers of herbicide. I don't recall that anyone or any group were ever brought in for interrogation. But this "accident" really put the pressure on use of herbicides on the Siuslaw. It was not immediately shut down, but was a limited program after that.

In 1975, Carol Van Strum and others living in the Alsea River drainage near the Siuslaw NF noticed that, after aerial spraying near their homes, various garden plants died while chickens, geese, and ducks were born deformed. Van Strum began researching the 2,4,5-T and 2,4-D herbicides that were being sprayed. She helped form the organization Citizens Against Toxic Sprays (CATS), determined to halt herbicide use in the forests. CATS filed a federal lawsuit against the Forest Service in May 1976, claiming an environmental impact statement for the herbicides was insufficient.

U.S. District Judge Otto Skopil Jr. ruled against the Forest Service in 1977 in the use of 2,4,5-T and Silvex containing 2,4,5-T and 2,4-D, sometimes called Agent Orange, on the Siuslaw NF. A couple of months later, on March 3, 1977, Regional Forester Ted Schlapfer extended the ban to all national forests in the Region. The Region completed an environmental impact statement on the use of herbicides in 1978, which was approved by Judge Skopil. The judge essentially said that if the Forest Service followed label instructions, as well as protecting streams from contamination and making efforts to protect people and domestic animals, spraying could continue. Besides, there were no viable alternative methods to control competing vegetation at that time. A new Regional EIS on using methods other than chemical sprays to control vegetation was completed in 1980. This new EIS took some of the heat off the herbicide issue, but required closer monitoring of all types of hand pulling and cutting, mechanical methods, use of grazing animals, as well as herbicides, insecticides, fungicides, and other chemicals. In 1979, the EPA banned the use of 2,4,5-T (Schneider 1989).

Another anti-herbicide group, Northwest Coalition for Alternatives to Pesticides (NCAP) also filed a federal lawsuit to end the use of 2,4,5-T. Similar lawsuits were filed against the BLM in western Oregon. On January 27, 1984, the federal district court found that the Forest Service was in violation of the NEPA, but remanded the case to allow the district court to expand its injunction to include both aerial and ground spraying over a larger territorial area. In 1983, the EPA and Dow Chemical both agreed the herbicide should be withdrawn from the market for spraying on forest areas. The use of herbicides continued across the Region, but for the most part different chemicals were used on fewer acres and in fewer locations. Former Regional timber director Wendall Jones (personal communication) recounted some of these changes:

> There was some pressure from the RO [Regional Office] after I went to the Mt. Hood NF to get some brush control with herbicides started on the forest. My silviculture assistant did not feel we needed to use them, but could rely on mechanical methods, or some hand treatment. I agreed with him. My position was that herbicides were very necessary in the management of the Siuslaw, and to a lesser extent the Siskiyou and Umpqua. Starting up an herbicide program on the Mt. Hood would be met with rigorous opposition by the Portland area enviros, and we didn't need to do that and jeopardize the use of herbicides on these other Forests. I was able to get the RO folks to back off using that argument. In later years the Ceanothus brush, that we were being pushed to treat to release young DF [Douglas-fir] trees, turned out to be protective cover for the increased elk herds who were turning to DF as a source of food in winter.

Another comment on the changing need for chemical herbicides was reported in 1989: " 'I don't foresee ever having to use chemical herbicides on this forest again,' said Tom Turpin, who oversees the reforestation projects at the Siuslaw National Forest here. 'We have proved that we can manage without chemicals, and we've seen that in many cases what we're doing now works better and is less expensive.' " (Schneider 1989) In the same article, Michael Ferris from the Regional Office public affairs office said, "We don't want to go back to doing business as usual … We were wrong to use chemicals the way we did." (Schneider 1989)

During the summer of 1987, an epidemic outbreak of western spruce budworm was discovered on more than 900,000 acres of the Mt. Hood, Deschutes, and Willamette NFs. The moth larvae ate the new growth of needles and buds on the evergreen trees, eventually killing the infected trees. Eradication efforts proved too little, too late for many thousands of acres of forest.

In 1979, one Asian gypsy moth was detected in Oregon at Lake Oswego near Portland. Following this discovery, small outbreaks of the defoliating insect, which can attack evergreen trees (unlike the European gypsy moths that only attack deciduous trees), were found and successfully treated in Gresham, Salem, and Corvallis. Then, in the summer of 1984, the largest single outbreak of Asian gypsy moths ever found west of the Mississippi River was discovered near Pleasant Hill at the edge of the Willamette NF. The moths were found on over 200,000 acres of forest land in Lane and Douglas counties. Timber industry and state officials insisted on aerial spraying with chemical insecticides, while environmental groups, led by Lane County Commissioner Jerry Rust, were opposed. Rust and others recommended trying *Bacillus thuringiensis* (*B.t.*):

> The issue went to court, but before a judge could rule, the state elected to aerial spray with *B.t.*, if enough supplies were available. Some 140 people were assigned to the project, including Forest Service, BLM, USDA Animal and Plant Inspection Service, Oregon Departments of Agriculture and Forestry... In the spring of 1985, 227,000 acres were sprayed in Lane County, including portions of the Lowell Ranger District and the BLM Eugene District, with 190,000 acres sprayed in 1986, and just a fraction of these acres in 1987. During the summer of 1988, only a handful of moths were detected in the sprayed areas. This was the largest eradication program using *B.t.* ever attempted in the U.S. and the outcome was even better than expected, which surprised even the environmental groups (Rakestraw and Rakestraw 1991: 150).

TIMBER BIDDING

Allegations of collusion and fraud dogged the timber industry in the early 1980s. In one instance, on October 15, 1980, the Bald Knob Land and Timber Company of North Bend, Oregon, was suspended from bidding on national forest timber for 60 days. The suspension was based on a U.S. District Court conviction of one of the company partners "for conspiring to defraud the United States in bidding activities relating to the Sucker Creek and Tellim Lumber timber sales on the Siskiyou National Forest." The company also had violated many contract provisions on previous timber sales.

Bidding for national forest timber reached an all-time high in 1979 and 1980, just before a wood-products depression hit the timber industry. Because of very high interest rates, the new-home and home-remodeling markets became very depressed, with the demand and price for lumber products falling

Log truck carrying alder logs, I-5 near Roseburg, Oregon, 1989. Author's collection.

to lows that had not been experienced for decades. Timber companies could not economically harvest the timber that they had purchased at high prices. In some locations around the Region, timber companies would fell the trees, then let them lie in the forest, even trees for which they had already paid a deposit, hoping that the market would turn around to make harvesting profitable. Some smaller companies went bankrupt, while other larger companies tried a variety of tactics to avoid paying the high prices that they had bid. The Roseburg Forest Products Web site noted the changes to the company during this era, which at the time had the largest sawmill in the world:

> In the early 1980s, the timber industry suffered a downturn as the housing market slumped. The downturn hit Roseburg [Lumber Company] particularly hard. Because it had contracted to cut government timber while prices were high, it was liable for significant losses when prices fell over the next few years. During the late 1970s, Roseburg signed contracts with the United States Forest Service worth $383 million for 1.3 billion board feet. The company's average cost worked out to about $300 per 1,000 board feet. By 1983, however, the going price per board foot was only $120. Roseburg stood to lose a huge amount of money if it harvested the wood at this price, and this was not the only lumber company in this position. Scores of companies in the Northwest were affected, and total possible industry losses were estimated at $2 billion, according to *Forbes* magazine (January 30, 1984). While a legislative bailout stalled, President Reagan acted to grant relief to the industry by negotiating a five-year extension on the unprofitable contracts. At Roseburg, business picked up rapidly, and

> by 1984 the company had opened a new melamine plant in Dillard, Oregon. In 1985, the company went through a series of organizational changes. A new company was formed, Roseburg Forest Products, which then [later] bought the assets of Roseburg Lumber.

The new company left all the high-priced sales with the old company, probably assuming that if there was no bailout the old company could declare bankruptcy and possibly wipe out the older, expensive timber sales. Chief Max Peterson, USDA Assistant Secretary John Crowell, and the Office of Management and Budget argued against any bailout, but the timber industry began a successful lobbying campaign. The resulting Federal Timber Contract Payment Modification (Timber Relief or Buyout) Act of 1984 was a "godsend" for timber companies in the Region that still held large amounts—some 18.3 billion board feet—of high-priced federal timber. Some of the companies had bid over $1,000 per thousand board feet for standing timber. Under the act, timber companies could give back to the Forest Service and BLM up to 55 percent of their outstanding timber sale contracts, but they had to pay a charge ranging from $10 to $30 per thousand board feet. When the fees had been paid, the contracts were extended with a provision for extra monitoring and restraints on future timber sale bidding (Wiener 1987).

Across the Forest Service in the mid-1980s there was considerable discussion and acrimony over the issue of below-cost sales. In most Forest Service Regions there was a negative cash flow—the federal government was paying more for preparing and managing the sale that was received when the timber was sold (Gorte 2004). Many environmentalists and libertarians seized on this alleged "mismanagement and waste of government funds" (O'Toole 1988). Region 6, however, only had a few sales that were in deficit. These tended to be timber salvage sales where the volume was low and trees heavily damaged. In some cases, the forests added nearby "green" trees that were undamaged as an incentive. This too caused outrage from the environmental community but was welcomed by the timber industry.

In some cases, the below-cost sales were tied to an accounting system called "purchaser road credits," established in the National Forest Roads and Trails Act of 1964. Basically, the term means that the cost of a sale is reduced by the amount that a timber purchaser has to pay to construct a road system to cut and move the logs from the forest. Beginning in fiscal year 1999, the Congress eliminated purchaser road credits for future timber sales.

REFORESTATION FUNDING CHANGES—THE K-V ACT

The Forest Service and the Bureau of Land Management use both congressionally funded appropriations and moneys from special trust funds to reforest harvested areas. The most important of these funds during the late 1970s and through much of the 1980s came from the Knutson-Vandenberg (K-V) Act of 1930, which put a portion of timber sale receipts a special trust fund to pay for reforesting the area from which the timber had been cut. The "K-V funds" were originally designed to provide a constant source of funding for reforestation work after a timber sale was completed and to eliminate the backlog of reforestation and timber-stand improvement work. In forest areas destroyed by natural causes, such as wind storms or fires, for which K-V funds have not been collected, reforestation work is funded through annual congressional appropriations. The 1980 Reforestation Trust Fund is another smaller source of funding.

The K-V Act prohibited spending more money for reforestation and other related activities on a sale area than had been collected from timber receipts for that sale area. The reforestation projects eligible for K-V funding included such activities as growing trees for planting, planting trees, sowing seeds, removing weeds and other competing vegetation, and preventing animals from damaging new trees. The act was amended in 1976, under the National Forest Management Act, to allow the Forest Service to pay for non-timber-related activities, such as creating fish and wildlife habitat or improving recreation facilities on the sale-area lands, from these funds. There have been a number of criticisms of the K-V Act, with most coming from the General Accounting

Helicopter loading fertilizer, Willamette National Forest, Oregon, 1980. USFS photograph.

Office (GAO) and the Department of Agriculture, as well as the environmental community. In 1987, the USDA Office of Inspector General criticized the direct expenditure of K-V funds on Forest Service overhead, such as furniture, rent, and other office supplies.

> After the enactment of the 1976 amendment to the K–V Act, FS management encouraged "imaginative and innovative management practices" in the use of K–V funds … liberal interpretations were employed in the actual allocation of costs to K–V funds. For example, the purchase of furniture was funded from K–V because it was used by timber and silviculture personnel whose salaries were all, or primarily, funded by K–V monies. Also, the vehicles, utilities, supplies, rent, magazine subscriptions, books, uniforms, etc., purchased or used by these personnel were considered legitimate direct assessments to K–V funds … This practice was employed at the RO, FSO [forest supervisors office], and RD levels of the FS. There appeared to be no established limitations as to how far this concept could be carried.

Despites these criticisms, the availability of the K-V funds greatly enhanced the replanting of trees after harvest as well as related timber-stand improvements, but these funds decreased greatly after the timber harvest reductions in the 1990s.

Engineering

The administration of the forest roads systems changed in the mid-1970s to mid-1980s. By 1976, the forests in the Region inventoried and placed all the temporary logging roads into the road system. This nearly doubled the miles on the inventoried transportation system to around 90,000 miles. At the time the road numbering system was rather confusing as the road numbers were based on which township the road started in, much like the BLM road numbers today. Beginning in 1978, all roads in the Regional road system were assigned maintenance levels, ranging from one to five. Level one roads were those that were closed (16,000 miles), level two were usable for high-clearance vehicles (52,000 miles), while levels three to five were for passenger cars (20,000 miles). This generated the need for a new road numbering system.

Then, four years later, in a somewhat controversial decision, the road numbering system changed again. The major arterials now had a two-digit number, collectors were assigned a four-digit number that tied to the arterial number, while local roads had the same number as the collector roads but with an additional three digits. For the most part, older road names like the

Robinson Lake Road on the Willamette NF were changed to a number. Another example on the Malheur NF illustrates the changing system: the old Prairies-Drewsey Road became road 14, at the next level down a collector road from road 14 became 1450, then a local road was numbered as 1450-485. Several newspaper writers complained about removing the road names, but the new number designations continued. In response, the Region printed a brochure, "Getting Around on Forest Roads," in 1985, in part, to address the renaming concerns for the public.

Near the end of the 1980s images from Landsat satellites were beginning to be used in the Region. However, better photography from space would have to wait another 15 to 20 years. Regional Engineer Dave Trask was appointed in 1983 to a national team to look into a new computer system for road design. "The national engineering Road Design System (RDS) had gone through evolution since its inception in the late 1950s [see Chapter 6], and it, too, was undergoing major changes in the 1980's. I believe some of the significant modernizations were the incorporation of aerial photography/digitizing (DTIS) and interfacing with road design and the Local Interactive Digitizing and Editing System (LIDES) that allowed field offices to have small stand-alone computers with peripherals to do low-standard road design and logging system analysis." (Frost 1990: 564)

Range Management

During the late 1970s and into the 1980s, grazing use varied considerably; in 1979, roughly 160,000 cattle and horses were authorized, with an additional 80,000 sheep and goats. However, in 1983, the number of authorized cattle and horses increased to almost 275,000 animals, but by 1988 the number was again at 160,000. In the same year, sheep and goat grazing was reduced to 50,000. The total animal unit months (AUMs) during the same 10-year period was lower in 1979 and 1980, then rising in 1981 through 1985, falling in 1986, then rising considerably in 1987, then falling to the average of around 725,000 AUMs in 1988. By 1988, there were 742 active grazing permits in Oregon national forests and another 214 in Washington forests. Two years later the number of active permits declined significantly to 601 in Oregon and 93 in Washington.

Recreation and Wilderness Management

Recreation use on the national forests in the Region was increasing from an estimated 34 million RVDs in 1981 to an estimated 36 million RVDs in 1989. The following year there was a jump in recreation use to over 43 million RVDs. Wilderness visitation, hiking, camping, driving, swimming, boating, fishing, and many other activities were having an impact on the ability of the national forests and rangers districts to cope. Money for recreation projects was small, but many recreation improvements were made through timber sale funds (like K-V funds) within timber sale areas. In Region 6, trail management was transferred from the engineering division to recreation in 1982.

The landscape management program was gaining headway, especially when tied to improving the visual impact of clearcut logging practices. Warren Bacon in the Regional Office was instrumental in designing methods and procedures, as well as training, to implement a landscape management program for the Region and across the country that

> involved identifying areas of the forest that were of varying levels of landscape or scenic diversity (variety classes) and viewer sensitivity (sensitivity levels along roads, trails and use areas). Measurable objectives were then described to guide activities (timber harvest, power lines, etc) in these areas. They were termed visual quality objectives (VQOs). The VQOs included retention where activities existed but were not evident, partial retention where activities were evident but were not visually dominant, modification where activities could be dominant but borrowed from natural openings and patterns sufficiently to appear natural...The VMS was later adopted by the Washington Office for use in all Regions ... Regional Siliviculturist Asa D. "Bud" Twombly teamed up with Warren to develop a handbook on managing vegetation called the "timber chapter." (Bacon personal communication)

The 277,000-acre Columbia River Gorge National Scenic Area was established on November 11, 1986, after years of discussion and intense debate, sometimes quite heated. The legislation directed the Forest Service to acquire lands in the scenic area through outright purchase or exchange. A Region 6 news release commented on the new scenic area: "We look forward to working with the Governors of Washington and Oregon, the Columbia Gorge Commission, and the six counties involved in developing management direction for the area," noting that Regional Forester James Torrence was "appointing a special management team to begin immediate implementation of this legislation ... Art DuFault, Forest Supervisor of the Winema National Forest ... has been selected to head that team as the National Scenic Area Manager."

The legislation also provided that the Forest Service expend several million dollars for an interpretive center at The Dalles (the Gorge Discovery Center at a cost of $6 million) and a convention facility near Stevenson (Skamania Lodge at a cost of $5 million). Other provisions in the legislation called for spending $2.8 million to restore part of the Columbia River Highway, another $10 million for recreational facilities, and finally $10 million to enhance the economic development opportunities in conjunction with the new scenic area. Additional monies were contributed by the states and counties. Many of the residents and some of the counties involved agreed with the designation and subsequent management plans; however some vocal residents, including an organization called Columbia River United, believed that the scenic area restrictions were overly restrictive and interfered with landowner and county development rights. The controversies are expected to go on for years.

The Newberry Crater National Natural Landmark (NNL) was designated in 1976. The NNL program is administered by the National Park Service, but management is by the land owner—in this case the Deschutes NF. Newberry Crater is the only NNL located on the national forests in the Region. The NNL program, established on May 18, 1962, was designed to recognize and conserve important natural areas. To be considered for NNL status, a site must be one of the best examples of a natural region's characteristic biotic or geologic features and the Newberry Crater certainly qualifies for this important distinction.

Columbia River National Scenic Area from Skamania Lodge, Washington, 2006. Author's collection.

There was an effort by various environmental groups, beginning in 1987, to designate many Washington and Oregon rivers into the National Wild and Scenic Rivers System. After several months of wrangling about which rivers or segments would be included, bills were introduced, which were strongly supported by the various environmental groups, mostly supported by the congressional delegations, mostly opposed by the timber industry, and strongly opposed by several groups. One of the groups in strong opposition was the Alsea River Alliance in Oregon, a loosely organized assemblage from the Alsea River drainage; another was the American Inholders Association, which has generally opposed all wild and scenic river bills.

After considerable congressional discussion, pressure from interest groups, and the strong influence of Oregon Senator Mark Hatfield, the Omnibus Oregon Wild and Scenic Rivers Act was passed by Congress and signed into law on October 28, 1988. It was the first state-wide wild and scenic rivers bill ever enacted by Congress, and added 40 river segments to the National Wild and Scenic Rivers System, most of which had already been designated for protection in the forest plans. Several of these rivers have more than one classification, such as the Skagit River on the Mt. Baker-Snoqualmie NF and the Wenaha River on the Umatilla NF, which have all three categories—wild, scenic, and recreation. A number of residents along the segments were not very happy, as at least some viewed the designations as another, unwanted layer of government management. (See Appendix H.)

"America's great outdoors" program, a presidential initiative, was designed to encourage cooperation between the Forest Service and the private sector in developing and improving recreational facilities and opportunities. Another popular program, managed in conjunction with other federal agencies, was the "scenic byways" program, which has designated about 7,700 miles of roads and highways in the national forests for recreational use. The Region began the program in the mid-1980s; Phil Hirl (personal communication) noted that "we had several Scenic Byway routes well before National and then State SB [scenic byway] Programs were formulated about 1992 [see Chapter 9]." Hirl continued, "Region 6 had several scenic roads that we promoted in the early '80s. Elkhorn Drive was an early one, another was Blue Mountain Drive from Heppner to Elkhorn Drive … Another was Aufderheide Drive on [the] Willamette … Century Drive on the Deschutes was another. Mountain Loop near Darrington on [the] Mt Baker/Snoqualmie was one. Most of these were on Forest Service and County roads."

ROADLESS AREAS AGAIN

The draft RARE II report was published on June 15, 1978. Nationally 2,919 roadless areas were located, using new criteria. The final EIS was produced on January 4, 1979. Regional Forester Dick Worthington described what happened in Region 6:

> Director of Recreation, Tony Skufca, who was overseeing it—Tony did a headsup [sic] job of keeping the integrity of that program. We wound up [with] … a very minimum acreage being recommended for wilderness; a lot less than anybody thought. So we had a meeting where each supervisor presented what his [recommendations] were … I don't remember the acreage but it was less than 100,000 acres … I knew that wasn't going to fly. … We ended up raising the acreage up some, not much, but some … Then Tony and I took this to Washington D.C. and we had to present it in front of the chief [and] to [Rupert] Cutler, the assistant secretary. We were sitting there and Cutler arbitrarily more than doubled the figures … We wound up with much less acreage than the timber industry thought we were to come out with, and so they embraced it. … The Forest Service wound up being criticized by the preservationists. I was attacked—the ex-timber chief was out to do away with wilderness, which was not true. The reason that Region 6 had such a low amount was the high value of timber on these [roadless] areas (Worthington 1984: 46-48).

In the final RARE II tally for Region 6, 34 roadless areas totaling 634,401 acres were recommended for wilderness designation by Congress: 372,507 acres in Oregon, 256,944 acres in Washington, and 4,950 acres in California (on the

Regional Forester Dick Worthington at RARE II news conference, Portland, Oregon, 1979. USFS photograph.

Siskiyou NF). In addition, 32 areas comprising 660,500 acres were proposed for further planning, and 230 areas totaling 4,285,000 acres were proposed for non-wilderness status. The results pleased neither the environmental community nor the timber industry.

There were major protests against the RARE II process and outcome. The logging industry began the "yellow ribbon" campaign and held log truck rallies, letters bombarded the congressional staffs, and the agency was inundated with petitions, but they did little to sway the Region. The environmental groups tried similar tactics, with similar results. Regional Forester Worthington retired in early 1982 with the wilderness battles still affecting him. The consummate "timber beast," he wrote a letter to the forest supervisors and directors in the Region on December 30, 1981, in which he expressed his feelings,

> Many wilderness demands are entirely unethical. They conflict with any sound, esthetic or biological value; yet they are trumpeted as being in the public interest. In reality classification of land as wilderness means locking up large acreages for the use of very few people. It is entirely contrary to basic good government when carried beyond minor amounts. The issue is not really yes or no, but how much ... The wilderness extremists have sold the public a bill of goods that the only alternative to unethical land use practices is wilderness. The cultists who advocate locking up now or lose forever are using unethical tactics to prevent unethical practices (Roth 1988a: 27).

The RARE II decision, much like the earlier RARE decision, was appealed and taken into court. A decision by the Ninth Circuit Court in 1982 (*California v. Block*) ruled the RARE II EIS inadequate. As a result USDA Deputy Assistant Secretary Douglas MacCleery directed each national forest to reevaluate the roadless areas in the forest planning process as potential wilderness. Regional Forester Jeff Sirmon, on October 17, 1983, also directed each of the national forests in Region 6 to reinventory and reevaluate, for the third time, every roadless area in their forest plans.

First, roadless reinventories had to be made in the Region as many changes had been documented in the original surveys. In the four-year window from 1979 to 1983 these surveys found that a number of roadless areas had been split by roads or reduced in size because of previous road building or timber harvesting along their edges. On some forests, road planning and timber sales had been hurried to make roadless areas ineligible for wilderness status as they would no longer be at least 5,000 acres, adjacent to existing wildernesses, or next to national parks. Thousands of timber projects were planned in the

Region for the released areas. Yet these drastic changes did not happen, in many cases, since the process for planning timber sales and road construction projects takes several years, and contractors are generally given three to five years to complete the projects (Roth 1988a and 1988b).

When the wilderness bills were passed for most states, it meant that the timber harvest levels had to be adjusted downward, especially in Region 6. The National Forest Management Act planning was also underway and this would in turn further reduce the timber harvest levels. These reductions caused the timber industry and timber-dependent communities and counties to howl in protest. Log trucks carrying anti-Forest Service signs appeared on many city streets. Pressure groups and organizations, such as the Douglas Timber Operators in Roseburg (established in 1968), tried to turn the tide of timber harvest reductions through the use of public letter-writing campaigns, telephone calls to congressional delegations, and rallies of log trucks in public locations to portray and publicize the devastating economic effects on local communities. Yet the planning process continued and timber harvest reductions were made.

Discussions about the "release" language for those areas not chosen to be wilderness took months of very heated debate both in Congress and in the public arena. The timber industry position, led by U.S. Senator James McClure of Idaho, attempted to get "hard" release language into the bill that would make it impossible for these areas to be considered for wilderness designation in the future. The House, led by Morris Udall of Arizona and John Seiberling of Ohio, wanted "soft" release language, which would allow reevaluation of these areas in the future. After lengthy negotiations, Oregon Senator Mark Hatfield carried the day with a compromise that was basically soft release language with a few words to appease the timber industry—essentially, the modified language allows reevaluation of those roadless areas that are still eligible for designation as wilderness in the next NFMA planning process (in 10 to 15 years). In spite of the valiant efforts of all involved to finally settle the roadless area question, these state wilderness acts, because of the release language, did not completely resolve the issue.

For Oregon and Washington, the roadless areas of RARE I and RARE II were finally decided in 1984. The Oregon Wilderness Act (PL 98-328) was signed into law by President Ronald Reagan on June 16 and the Washington Wilderness Act (PL 98-339) on July 3, adding almost 1.9 million acres (40 new wildernesses) to the existing 2.8 million acres of wilderness in the Region. In quick succession, many other state wilderness bills became law.

Environmental groups were disappointed that many areas they had wanted for wilderness were not included. New efforts were made to delay or halt timber harvest in existing roadless areas outside the wildernesses. Some of these were legal and constructive, such as working with the Forest Service to develop a consensus on management. Others were illegal and destructive actions at road and timber harvesting sites, such as tree sitting, sabotage of road and logging machinery, blockading of roads by protesters sitting in the roadways or chaining themselves to trees or gates, and driving steel and ceramic spikes (which could not be found by metal detectors) into trees to stop chain saws, causing injuries in both the forests and the mills.

Further complicating the situation was the fact that during the RARE I and RARE II studies the Regional Office had not allowed the timber harvest for each national forest to be reduced as harvesting and other development activities were not allowed in the roadless areas during the study periods. The Forest Service had expected the RARE I study to be completed quickly, so that the restriction of harvesting in the roadless areas would not be a management problem. But by the time the final EIS for the RARE II study had been implemented, almost a decade had passed and the forest lands that were available for development during this time period were being over-cut. Jack Usher, the Region 6 director of timber management stated:

> For more than a decade the allowable cut [for each national forest in the Region] has included timber volumes and anticipated tree growth in some of these roadless areas, even though the actual cutting has been in areas where roads are already established. ... We have assumed the political controversy would be resolved quickly and no one area would be heavily impacted by this intense harvesting [outside the roadless areas] ... But when it goes on for a decade, we think we may be perilously close to violating our soil and watershed protection requirements in some of our roaded areas (Usher 1984).

Because of all these problems, and the threat of a lawsuit by environmental groups, Regional Forester Jeff Sirmon announced in late January 1984 that timber sales in the Region would be reduced by 880 million board feet over the next two years. The reductions would come from planned timber sales from the released RARE II roadless areas that were in public contention or were likely to be very controversial. "Privately, regional Forest Service officials hope a reduction in Northwest timber sales [which are the largest in the nation] will spur Congress toward prompt resolution of the wilderness issue." (Sadler 1984)

In the 1980s there was discussion among the wilderness community about both naming and numbering of trails and other features in wildernesses. The issue spilled over to wilderness maps. Purists felt that existing trails should be obliterated and there should not be any identification of trails on maps so that a wilderness visitor could get the true experience of wandering the wilderness without any guide. Wiser heads prevailed and the trails, maps, and names remain.

Wildlife and Fish Management

In 1984 a federal court ruled that the Forest Service had failed to protect fish and wildlife habitat on the Mapleton RD of the Siuslaw NF because it had not done the proper NEPA work on assessing environmental damage and long-term consequences of risk of landslides. The court issued an injunction that temporarily shut down timber sales on the Siuslaw NF, which caused howls of protest from the timber companies and local citizens along the coast. The ruling also meant that there had to be better analysis in the future by the agency of the long-term consequences or worst-case scenarios in forest planning in the Region. The Endangered Species Act of 1973 provided for protection of rare, threatened, and endangered animal and plant species. It established federal procedures for identifying and protecting plants and animals in their native, critical habitats and declared broad prohibitions against taking, hunting, harming, or harassing the listed species. The intent of the act was to restore endangered species to a level where special protection would no longer be needed. Implementing this act would have serious consequences on all types of management of the national forests, especially timber and road construction programs, during the 1980s and 1990s.

Fish enhancement projects, such as stream rehabilitation, followed decades of efforts by the Forest Service and the state fish and game departments to rehabilitate many streams on the national forests. Before the 1970s, much of the stream rehabilitation took the form of removing obstacles to salmon and steelhead passage. Debris in some cases was blasted from stream beds, large rocks were removed, and creeks straightened. A Regional Office news release on February 14, 1986, noted the change of approach: "Several years ago, many resource managers were clearing streams of logs and other debris on the assumption that this would improve fish passage. But fish biologists have learned that logs and boulders can be used to direct water flow, creating habitat that allows for gravel accumulation and rearing pools."

Fish enhancement projects were many and varied across the Region, with nearly $2 million spent in 1985. One fish enhancement project on the Siskiyou

NF, as announced in a Regional Office news release in late 1985, was described as "placing log weirs on two tributaries of the Rogue River and blasting a fish ladder around a falls ... On the nearby Powers Ranger District ... [f] orty-five fish structures were installed on Rock Creek, principally log weirs and boulders, but dynamite was used to create four rearing pools ... Workers on the Mt. Baker-Snoqualmie National Forest this year [1985] completed new projects on six Puget Sound rivers. ... [and over the past four years] 900 structures of various kinds have been installed."

Another Regional Office news release on February 14, 1986, announced a fish enhancement program for steelhead and salmon: "The Forest Service recognizes the need for direct habitat improvement on coastal streams and Columbia River tributaries. It's taking action to attack the root cause of declining fish numbers from within National Forests by improving and enlarging spawning and rearing habitat for anadromous fish."

The northern spotted owl (*Strix occidentalis caurina*) lives in dense stands of mature and old-growth coniferous forest in western Washington, Oregon, and northwestern California. Although some owls have been located outside of old-growth areas, especially in California, studies show that over 95 percent nest in old-growth trees, where they hunt flying squirrels, rabbits, and wood rats. Beginning in the early 1970s, Eric Forsman, a Forest Service research biologist, recognized that the northern spotted owl needed forests that had never been logged or burned for hundreds of years for their habitat. Other scientists estimated that one pair of owls needed hundreds of acres of forest habitat to survive. When roads were cut through old growth, the spotted owl disappeared from the immediate area. The spotted owl was regarded as an indicator species, one whose numbers reflected the health of other species and the forest ecosystem. The westside national forests in Oregon and Washington contain an estimated 4.1 million acres of suitable owl habitat and another 1.5 million acres on the BLM and Park Service lands. During the late 1970s and 1980s, Forest Service and BLM employees and contractors spent many nights in the Region's and BLM's forests calling/hooting for spotted owls in their efforts to estimate their population.

The National Forest Management Act of 1976 required the Forest Service to maintain "viable populations of existing native and desired non-native vertebrate species in the planning area." In 1985, an appeal was filed with the Secretary of Agriculture by the National Wildlife Federation, the Oregon Wildlife Federation, the Lane County Audubon Society, and the Oregon Natural Resources Council, arguing that the Forest Service had been negligent in assuring the spotted owl sufficient habitat to survive.

The Regional Office set up a task force to address the issue. After consulting with a number of in-house and university wildlife specialists, the Regional Forester decided in a draft EIS published in 1986 that 400,000 acres should be set aside to protect 400 pairs of owls. Later that same year, the Region bumped up the acreage to protect the northern spotted owl from decline and extinction by withholding from road building and timber harvest 2,200 acres of forest for each of the 512 known nesting pairs of owls. The immediate impact was an outcry both from the forest products industry and from environmentalists. The industry complained that the measure went too far and would cost thousands of jobs, and environmentalists claimed that not enough was being done to protect the species. The final EIS on spotted owls, released in July 1988, recommended a protection buffer ranging from a high of 2,700 acres per pair on the Olympic NF surrounding the Olympic National Park to a low of 1,000 acres in the Klamath Mountain province.

Meanwhile, in December 1987, the U.S. Fish & Wildlife Service (FWS) declined to list the northern spotted owl under the Endangered Species Act. Environmentalists sued. In the litigation, the agency admitted that it had taken into account economics and politics in its decision not to list though, by law, these issues are not to be considered. In June 1989, the FWS, in a position reversal, proposed listing the owl as threatened under the Endangered Species Act (see Chapter 9). Loggers and mill workers blamed environmentalists and the owl for the jobs lost in subsequent years, while environmentalists blamed mechanization and log exports.

Mineral Management

The Three Sisters pumice claims in the Three Sisters Wilderness on the Willamette and Deschutes NFs again came into public discussion in the 1970s and early 1980s. In 1976 the U.S. Pumice Company applied for patent, largely to clear the way for a court settlement. The court reached a decision in 1981 that 790 acres were invalid claims and 670 acres were valid. Final settlement came in 1983, when Congress appropriated $2 million for purchase of the patented claims (Rakestraw and Rakestraw 1991).

Watershed Management

In the 1980s, as the acid rain issue was splashed across the nation's newspapers, the Regional Office joined with the Environmental Protection Agency (EPA) to collect water samples from the many lakes in the Region. A news release on August 30, 1985, discussed the Regional effort to collect water samples from "195 lakes ... Of this number 83 lakes are in National Forest wilderness (51 in Washington, 32 in Oregon). This effort is part of a larger National Surface Water Survey (NSWS) involving 888 lakes in 10 western states ... The NSWS is an EPA water sampling program and is designed to quantify the chemistry of lakes and streams throughout the United States." Water samples were drawn by EPA personnel using helicopters in non-wilderness lakes, whereas in wilderness areas Forest Service employees collected samples while on foot or on horseback. The sampling began in late September in Washington and mid-October in Oregon. Rick Ross, director of range and watershed in the Regional Office, was in charge of the Forest Service portion of the operation.

Fire Management

The Siskiyou Smokejumper Base at Cave Junction was closed at the end of the 1981 fire season, but continued for several years as an emergency base. Many operations were consolidated at the Redmond Air Center.

On August 30, 1987, more than 1,600 lightning strikes were recorded striking southwestern Oregon and northwestern California, mostly on the Siskiyou NF. Several of the strikes sparked the nearly 150,000-acre Silver Fire complex, which burned an estimated 230 square miles. The Silver Fire continued burning through October and was finally controlled on November 9. The Silver Fire burned in a mosaic pattern typical of many forest fires—some areas were heavily fire damaged while other areas were barely singed. Region 6 spent more than $26 million and the Oregon Department of Forestry an additional $10 million to fight this and other fires that year.

As a result of the terrible fires that spread through Yellowstone National Park and adjacent national forest lands in the summer of 1988, the Forest Service and the National Park Service received considerable public pressure to change their policies of letting some fires burn naturally (the so-called "let-burn" policy). After much public and scientific debate about the proper role of fire in the environment and after viewing the subsequent "rebirth" of the park and adjacent national forests, the agencies have modified their policies; they now put out fires more quickly, but still allow some natural fires to burn under strictly controlled conditions.

Research Work on the National Forests

Growing "superior" trees, that is trees that are straight and tall with no obvious defects, became a major research effort in the 1970s and 1980s. "[F]ield foresters were advised to select, mark, map, and register the locations of trees with genetically superior characteristics ... If I'm picking for tree improvement, the cones are destined for the Dorena Tree Improvement Center [for testing] in the Umpqua National Forest near Cottage Grove, Oregon ... [If I pick] for reforestation purposes only ... that seed goes to either the Ben [Bend] Pine Nursery in Bend, Oregon, or the Wind River Nursery in Washington." (Sherwood 1990: 33-35) From these three sites, superior seeds were grown to seedling size, then planted in special plantations called seed orchards or planted into clearcut units on the forests in the Region. A new genetics program was established to create disease-resistant trees, provide genetic controls over planting stock, and carry on grafting experiments (e.g., Dorena Tree Improvement Center).

Further research into the basic ecosystem occurred in the 1980s because of a National Science Foundation-funded Long-Term Ecological Research Program on the H. J. Andrews Experimental Forest concerned with identifying the factors influencing the decay of logs in the forest and the role of decaying logs in the forest ecosystem. Previously, there had been a great deal of speculation on the role of downed logs in providing nutrients, and this was the first

Douglas-fir seedlings at Wind River, Gifford Pinchot National Forest, Washington, c. 1976. USFS photograph.

USFS Dune Patrol, Oregon Dunes NRA, Siuslaw National Forest, Oregon, 1979. USFS photograph.

systematic study. The downed log study involves 500 logs, 20 feet in length, representing four westside species—Douglas-fir, Pacific silver fir, western hemlock, and western red cedar—to be studied as they decay over a 200-year period. The study is managed by scientists from Oregon State University and the Forest Service's Pacific Northwest Research Station (Rakestraw and Rakestraw 1991).

The area ecology program began in the late 1970s and early 1980s to identify the characteristics of understory (plants such as shrubs, flowers, and grasses) and overstory (trees) in all intact ecosystems in the Region. Miles Hemstrom, regional ecologist in the Regional Office, coordinated studies in all parts of the Region to establish baselines of information about major and minor vegetation types and abundance in the forests of the Pacific Northwest. Area ecologists were assigned to the major ecotypes, where they spent several years doing on-the-ground research to identify plants and trees. The results of these studies have been printed in many Pacific Northwest Research Station publications.

Research lessons learned were passed on to national forest management line and staff at occasional field seminars at the H. J. Andrews and at other locations around the Region. Cooperation and collaboration between the research scientists and the national forests has become common in the Region, resulting in a more rapid transfer of knowledge and technology. In one instance, Willamette and Umpqua NF staffs were given a two-day presentation by government scientists Jerry Franklin, Chris Maser, and others on what had been learned in the years of research at the Andrews. Within days of the session in the summer of 1986, the two forests started applying the lessons to ground operations, especially those dealing with leaving more woody material on the forest floor after logging operations.

Chapter 9

Coping with New Directions 1990–2009

The focus of the 1990s and into the early twenty-first century has been managing natural resources for sustainability, the key to restoring and conserving the health of the land. Sustainable resource management became part of the Forest Service's commitment to landscape-level management across ecosystems and ownerships, in accordance with sound science. Chief Dale Robertson in 1992 introduced the foundation for ecosystem or ecological management, based on the interrelationships of the land, air, water, plants, animals, and people. It was the logical conclusion to earlier management ideas—"new forestry" and "new perspectives." Although the ecological ideas had been talked about for decades, this was the first concerted effort to apply the principles to the whole national forest system.

Probably the most significant change in the Region in the 1990s was the reduction in the number of employees. The efforts, referred to as "realignment" impacted the ranger districts and national forests most heavily. Decreasing timber sales affected foresters and lack of projects and money affected engineers. For many employees, the phrase "get out the cut" was replaced by "survive the cut."

Regional Foresters

John Butruille began his Forest Service career on the Roosevelt NF in Colorado and it included three assignments in Colorado and Wyoming. He came to Region 6 in 1975 to lead the planning effort for the Cascade Head Scenic Research Area on the Siuslaw NF. He then moved to the Regional Office in 1976 to head up the R-6 effort in the Resources Planning Act. He became forest supervisor for the Tongass NF in Region 10 in 1979 and was transferred to the Washington Office in 1981. He was appointed deputy regional forester for resources in R-6 in 1984, leaving in September of 1987 to become the director of recreation management in the Washington Office. Butruille returned to Region 6 as the new regional forester on August 28, 1989, and faced many challenges. Getting forest plans in the Region completed was a top priority. The draft environmental impact statements for the proposed forest plans were all released by late 1987. Timber supply, old growth, spotted owls, and salmon issues were prominent during Butruille's tenure. The salmon situation became a major issue in 1991 with a listing of several Columbia and Snake River species as threatened or endangered by the National Marine Fisheries Service. Butruille retired in 1992 and was replaced by John Lowe.

John Lowe began his Forest Service career in 1963 as a forester on the Ava RD of the Mark Twain NF in Missouri. Subsequently, Lowe was a district ranger, after which he spent time in the Regional Office in Milwaukee. He came to Region 6 in 1977, and served as the assistant forest supervisor for the Willamette NF and then in 1981 as the forest supervisor on the Umatilla NF, both in Oregon. In 1985, Lowe became the deputy director for timber management in the Washington Office, then deputy regional forester for resources in R-6 in 1987. Lowe became the new regional forester on July 4, 1992. He too faced many challenges and difficult decisions. Getting the final National Forest Management Act forest plans implemented was a top priority. Many of the same issues that had faced his predecessor were still prominent: timber supply, old growth/ancient forest, spotted owls, salmon, the Pacific yew tree (source of taxol, used for fighting certain forms of human cancer), and ecosystem management issues. Concerns about spotted owls, marbled murrelet, and salmon stocks intensified external and internal debates. Major efforts went into resolving water-related concerns such as riparian management and anadromous fish habitat. The issue of old-growth-dependent species came to a head in 1993-94. John Lowe, after his retirement in 1996, was replaced by Robert Williams.

Robert Williams entered the Forest Service in 1957, working as a forester on the Payette NF in Council, Idaho, then worked at several forests in various positions including district ranger, becoming supervisor of the Beaverhead NF from 1974 to 1981. He spent the next two years on the legislative affairs staff in Washington, DC, then returning to the West to be the forest supervisor of the Gifford Pinchot NF from 1983 to 1990. For the next five years, he was the deputy regional forester for the Alaska Region, then he returned to the Pacific Northwest Region as the associate regional forester in 1995, becoming regional forester from 1996 until his retirement in June 1999. Williams dealt with many difficult, long-term issues during his tenure as regional forester, among them the changing management of the national forests. The transition from a highly specialized and efficient timber production organization to one that places entire ecosystems and changing public values at the forefront of management was critical. Williams oversaw many changes in personnel and activities that had to occur with management under the direction of the Northwest Forest Plan, and the plan worked effectively in many ways for much of Williams' tenure, although the forests were never able to meet the anticipated timber targets. There were unsuccessful attempts from all quarters to dismantle the plan. At the same time, concerns about most of the salmon and steelhead stocks have intensified external and internal debates, as well as the Forest Service's relationships with other agencies. Upon his retirement in late 1999, Williams was replaced by Harv Forsgren.

Harv Forsgren began his career with the Forest Service in Alaska in 1978, then worked as a fisheries biologist on the Sawtooth National Recreation Area and the Sawtooth NF in Idaho and on the Mt. Hood NF. Beginning in 1991, he was the national fisheries program leader in Washington, DC. In June of 1998, Forsgren became the national director for the wildlife, fish, and rare plants program. Forsgren came to Portland as the regional forester for the Region in January of 2000, and became involved with a number of contentious issues including the aftermath of the Northwest Forest Plan, the Interior Columbia Basin project, a number of threatened and endangered animal and plant listings, fire management, and payments to states. Forsgren was reassigned as the regional forester for the Southwestern Region (covering the states of Arizona and New Mexico) in 2002, and was replaced by Linda Goodman, the first woman regional forester in Region 6.

Linda Goodman joined the Forest Service in 1974 as a receptionist at a ranger station on the Olympic NF. Goodman worked in a variety of administrative and managerial posts, including five different forests in the Pacific Northwest Region. She was the national director of the agency's Jobs Corps program. She was the Pacific Northwest deputy regional forester from August 1999 to October 2001. Goodman then moved to the Washington Office to be the acting chief of staff for Forest Service Chief Dale Bosworth. In 2002, she returned to the Northwest to take on the regional forester post. Goodman's top priorities were restoring and rehabilitating fire-damaged forests and implementing proposed changes in the Northwest Forest Plan. She was challenged by the fact that the plan had not come close to meeting the projected timber sales volumes. She retired in April of 2008 and was replaced by Mary Wagner.

Mary Wagner began her career with the Forest Service in 1983 as a forester on the Sawtooth NF in Idaho. She held various positions with the agency, including deputy forest supervisor and district ranger on the Humboldt-Toiyabe NF in Nevada and district ranger on the Ashley NF in Utah. She served as the forest supervisor on the Dixie NF in Cedar City, Utah. In 2004, Mary became the first national director of wilderness and wild and scenic rivers in Washington, DC. She took the assignment as the deputy regional forester for the Intermountain Region before becoming the regional forester for the Pacific Northwest Region in October 2008. At the beginning of her tenure, she was faced with continuing controversy over salmon and fire issues, trying to meet timber harvest levels mandated by the Northwest Forest Plan, wilderness status for many areas around Mt. Hood and other locations in the Region, and implementing the economic stimulus package for the Region.

Downsizing

In fiscal year 1990, there were 7,560 permanent employees in the Region, along with 3,929 temporary and seasonal employees. The vast majority of permanent employees tended to be technical (3,705) and professional (2,161). The majority of the technical and professional employees were working as foresters (1,066) and engineers (270), although the Region had a total of 164 disciplines (job skills) represented in the work force. At the height of the 1990 summer field season, there were 13,046 permanent, seasonal, and temporary employees in the Region.

But as timber revenues decreased and other sources of funding dried up, there were reductions in the workforce. A number of the employees who lost their jobs were able to find other jobs in the agency, some left government service to work in the private sector, while others were offered directed reassignments and at least a few were given reduction-in-force notices (RIF) that terminated their employment with the Forest Service. Siuslaw NF Supervisor Wendy Herrett tried different methods to keep employees on the forest, especially wage grade employees, who suddenly were no longer funded: "You can try to put them on details and that sort of thing to keep them employed. But I didn't see that it was going to get better in the long range. And so I thought the only tool I have to really be able to reduce that work force is a reduction in force ... So I was the only one in the region that did, I think, a reduction in force [RIF]. Some other forests started the process and didn't do it … it was about eight people [that] actually lost their jobs ultimately." (Herrett 2001: 99)

A memo on June 25, 1991, from Willamette NF Supervisor Darrel Kenops explained how the downsizing was going to impact the supervisors office: "My expectation is that SO [supervisors office] staff will coordinate for the forest as a whole, a look at all aspects of staffing and funding in specific program areas with the understanding that funding will be significantly reduced in timber related programs … This is a budget driver process!" Reductions for the SO were suggested as follows: 19 positions in administration; two in planning; two in public affairs; two in recreation; two in fire and lands; three in timber; three in soils, water, recreation and wilderness; and 10 in engineering. This totaled 43 positions in the SO alone, with many more coming from the ranger districts. The Region began a "surplus list" of employees who were no longer needed or funded. If there was any hiring on other national forests, the first step was to check the surplus list for qualified people. Most employees on the list eventually found other work either inside the Forest Service or with other employers. This process was repeated on every forest in the Region, especially in the aftermath of the timber harvest reductions caused by the Northwest Forest Plan of 1994.

By 1996, the permanent work force in the Region had been reduced to 5,687 positions, with the largest unit being the Willamette NF with 423 positions (the Regional Office had 418 positions) while the Olympic NF had only 168 positions. Four years later in FY 2000, there were 4,685 permanent positions in the Region, and that number was further reduced to 3,646 employees by FY 2007. The overall reduction from 1990 until 2007 was an average of 55.4 percent, with the Olympic NF taking the hardest hit with a 78.7 percent reduction, while the Regional Office had the least impact with a 22.3 percent reduction in permanent positions. Most of the reductions came from forestry, engineering, planning, and clerical support staffs, but other professions were affected as well. When virtually all of the agency's human resources and financial management staffs moved to the new Albuquerque Service Center in 2005, a number of employees left the Regional to staff the new unit, while others retired.

Regional News

The Regional Office moved in 1992 to the Robert Duncan Plaza at 333 S.W. First Avenue, where it remains. Currently, the building is shared with the Bureau of Land Management and the U.S. Army Corps of Engineers.

The Washington Office designed new training courses for regional employees in this period. The 1900-01 Forest Plan Implementation/NEPA course was designed around 1990 by the environmental coordination section to give forest-level employees a better handle on how to implement NEPA by writing EIS and environmental analysis (EA) reports on forest-level projects like timber sales. The training sessions were run by Rhey Solomon and Mikel Shilling from the Washington Office and taught by experts from R-6 and R-10. There was an attempt to introduce a national 1900-03 course to deal with social and economic issues in NEPA, but after several years of trying, the course never completely came together. In the mid-1990s, a new course was developed for forest-level managers, entitled Building High Performing Teams, originally designed and taught around the country by a Blanchard & Associates, a consulting group. It was taught by experts from around the Region.

The Forest Service experienced legal challenges on a wide variety of projects. For example, between 1999 and 2002, the Forest Service nationally was involved in 729 cases litigated by the federal courts. Of this number, the agency won 58 percent of the cases, lost 21 percent, and settled 18 percent, while almost 4 percent were withdrawn. The total cases decided by federal judges or panels of judges instead of the court cases during the same period were 575, with the Forest Service winning 73 percent of the time. The most cases came in the 1994-2002 period, with the highest point reached in 2000. "Litigation varied greatly by Forest Service region. … Region 6 (Washington and Oregon) experienced the most litigation. Region 6 represents 12.8% of total National Forest System (NFS) acreage but accounted for 22.8% of all cases." (Keele et al. 2006: 198) Of the total cases nationally, the majority dealt with logging operations, followed by planning and timber salvage, and were mostly based on NEPA, NFMA, and ESA.

In the early 1990s, several national forests in the Region responded to demands of people who had lost their jobs in the tough economy to live in the forests rather than on the streets; in the Northwest, this often meant people who had been laid off from the timber industry. For example, the Umpqua NF in May of 1992 set up the "first campground in a national forest set aside exclusively for the homeless. … the new residents began assembling a patchwork of tents, tarpaulins and small trailers in an abandoned [rock] quarry pit about 40 miles southwest of Eugene, Oreg." (Egan 1992) Other forests in Oregon used a similar approach when homeless people overstayed the regular campground limits, a maximum of 14 days. The homeless campground idea fizzled after several years when the economy improved.

During this era the operation of several national forests in the Region was administratively joined. The Fremont and Winema NFs were combined in 2002 as the Fremont-Winema NF with headquarters in Lakeview. Then the Rogue River and Siskiyou NFs were combined in 2004 as the Rogue River-Siskiyou NF with headquarters in Medford. Finally, the Okanogan and Wenatchee NFs were combined in 2007 as the Okanogan-Wenatchee NF with headquarters in Wenatchee. In addition, several forests in the Region began sharing staff and specialist work. Wendall Jones (personal communication) noted that the sharing "was done on the Ochoco/Deschutes, and Willamette/Siuslaw, and probably others [e.g. Olympic and Mt. Baker-Snoqualmie and Gifford Pinchot and Mt. Hood] … By 2007, the Mt. Hood, Willamette and Siuslaw shared the same Forest Engineer."

A program involving the public with forest management issues began on May 29, 2007, when Chief Gail Kimbell signed the "proof of concept charter."

The Colville NF is one of only three national forests in the country (the others are the Shasta-Trinity NF and the Allegheny NF) that has been accepted into this novel program. Since the beginning of fiscal year 2009, the Colville NF has had the opportunity to engage with the public to initiate this program locally. This is a new business model that suggests that, through collaboration with the public and with a predictable budget, the forest can increase outcomes while reducing costs. During the 10-year implementation period, the forest will not have to meet specific goals set by the Regional Office. New goals will be set through collaboration, partnerships, and stewardship contracts. The first public meetings on the goals were held in August and September 2007.

Back to Work Program

With the devastating national economic circumstances in late 2008 and into 2009, President Obama and Congress passed legislation called the American Recovery and Reinvestment Act (ARRA) or Economic Stimulus Package to "jump-start" an economic recovery. Chief Gail Kimbell noted in a speech on March 20, 2009: "We will focus on projects that are ready to go, and we will begin our work promptly and complete it within 2 or 3 years for most activities." The projects are designed to be labor-intensive to get people back to work in rural communities. In the Pacific Northwest Region, Lisa Freedman has been assigned to help guide the new program and coordinate with the Washington Office. By June 2, 2009, 55 ARRA recovery and investment projects had been approved. These projects, many of which are carried out through the two states by funds provided by the Forest Service, are summarized by state as follows (see www.fs.fed.us):

Oregon: $115,354,700 for fuel reduction projects, weed treatments, youth employment, salmon habitat and watershed restoration, road maintenance and paving, and bridge work.

Washington: $30,175,000 for fuel reduction projects, capital improvements, road maintenance, and bridge replacement.

National Changes in the Forest Service

With the new Clinton administration in 1993, there was also a new USDA Under Secretary, Jim Lyons, who had a forestry degree, worked with the Society of American Foresters, and served as a staffer in Congress from 1986 until his appointment. His new position had oversight of the Forest Service and the USDA Natural Resources Conservation Service. Lyons, who was often at odds with the Congress and the Forest Service, was quite familiar with Forest Service activities when he came into office. Near the end of Clinton's second term, Congress cut off funding for Lyons because of his speaking out against cutting old-growth trees and related personal issues, but he remained on the job until the end in January of 2001.

Lyons was the first assistant secretary to have a hands-on approach to agency matters, all the way down to the ranger district level. To the frustration of the Region, Lyons would, at times, give "direction" to district rangers without having gone through an approval process through the Washington Office or the Regional Office. Elton Thomas offered a story about how Lyons was directly involved in decision making on the Wenatchee NF in the summer of 1994:

> I specifically remember one occasion when Jim Lyons arrived on the Wenatchee Forest ... he wanted a helicopter flight to view the fires ... when he and [Forest Supervisor] Sonny [O'Neal] returned from the flight, Sonny just shook his head and told me Jim was giving directions so fast that it was obvious he didn't know which end was up ... later that fall and early winter as we were forming and implementing our restoration and rehabilitation plans, which included a proposal to salvage log some of the burned area, Sonny told me that Jim Lyons would frequently call him personally and give him direction on the approach Lyons wanted Sonny to take (Thomas personal communication).

In the mid-1990s a proposal was made by Chief Jack Ward Thomas to reconfigure the regions to better reflect the huge ecosystems that they monitor and manage. The internal Forest Service "reinvention effort" resulted in a December 6, 1994, report (*Reinvention of the Forest Service: The Changes Begin*), which outlined how the agency would accomplish its goals. The most controversial portion dealt with the elimination of several Forest Service regional offices by combining services in other locations. Region 10 (Alaska) and Region 1 (Northern) would have been eliminated. The two national forests in Alaska would have combined with the western portions of Washington and Oregon, while the eastern parts of the two states were to join with Region 4 (Intermountain). The national forests in northern Idaho, Montana, and North Dakota were to be transferred to Region 2 (Rocky Mountain, which was to be renamed the Northern Plains Region). Other adjustments to the regional boundaries were proposed, based roughly on large national ecosystems. In Congress and the various states, the proposal fell on deaf ears and was "dead on arrival."

In the 1990s the new "wise use" or "property rights" or "county supremacy" movement replaced the previous "sagebrush rebellion." The words the activists used were not very different from earlier efforts, but county commissioners from Nye County, Nevada, and Catron County, New Mexico, put new emphasis on local control over federal land. There was also a rash of bombings and threats to Forest Service employees and facilities. However, following the Oklahoma City bombing of the Murrah Federal Building on April 19, 1995, as well as the events of September 11, 2001, this violent extremism cooled.

In 1993, an initiative was put in place to "reinvent" government through more efficient, cost-effective, responsive ways of working. The Forest Service's enterprise initiative was designed to integrate market mechanisms into everyday agency operations. Employees created their own small businesses serving internal customers, operating in a self-determined, self-motivated way to accomplish Forest Service work more effectively and efficiently. One such team in the Pacific Northwest is the Forest Resource Enterprises (FRE), led by Steve Rheinberger and Jerry Smith, which is, according to the USFS Web site,

> in the business of management of historically underutilized forest resources in the most economic and cost effective ways possible. We analyze the economics, marketing, management, and sustainability of forest commodities such as small diameter timber and special forest products. We seek out new and emerging markets associated with underutilized forest products and make these markets accessible to natural resource managers and interested members of the public sector. … Our management philosophy is geared toward producing a sustainable and continuous supply of a variety of forest resources, economic, and employment opportunities to local communities.

The entire Forest Service, along with the rest of the federal government, was shut down briefly on November 14-19, 1995, then again in mid-December through January 6, 1996, because of problems with the budget. The Interior Appropriations Bill for FY 1996 had been held up by the Republicans, in part over issues dealing with Alaska's Tongass NF plan. A continuing resolution had been used several times to fund operations, but the issue came to a head when Congress did not pass a new resolution. A press release from the Regional Office on December 19 said, "Along with the agency nationwide, the Pacific Northwest Region … will shut down at the close of business today for lack of funds … During the shutdown, emergency activities such as fire suppression, law enforcement, and response to natural disasters will continue." Other activities were also allowed to continue, such as "protection of resources, human safety, and government property." Job Corps Centers, which were funded under separate appropriations, continued operating. Other cooperative agreements or contracts and permits with private concessionaires continued, as well as administration of existing contracts. Most "non-essential" employees, including those in Region 6, were sent home until the budget issue was resolved on January 6. It was the longest partial shutdown of the federal government ever recorded.

Service First Initiative

The "Service First" initiative was inspired by the need for better customer service for the public, as well as to reduce management costs. Authority for the BLM and Forest Service to work together on a trial basis was provided initially by legislation in late 1997. The legal authority was expanded with the Service First initiative that was authorized on October 11, 2000, and that legislation was amended in 2006 to include the National Park Service and Fish & Wildlife

Centennial Celebrations

The Forest Service in 1991 celebrated the centennial of the establishment of the first forest reserves.

The second centennial in 2005 celebrated the moving of the management of the forest reserves from the General Land Office to the Bureau of Forestry, the renaming of the Bureau of Forestry to the Forest Service, and the 1905 Forest Congress. The national effort was kicked off in Washington, DC, with the Centennial Forest Congress held exactly one hundred years after the 1905 Forest Congress that had been organized in the same city by Gifford Pinchot. One panel at the Centennial Congress included the previous and current assistant secretaries with oversight of the Forest Service in the Department of Agriculture. All the living chiefs of the Forest Service were present. Also featured was the premier of an award-winning, two-hour high-definition video called *The Greatest Good*, which included short interviews with several Region 6 employees, as well as photographs from a number of forests around the Region.

On September 4-9, 2005, the Old Smokeys (Region 6 retirees) and the Region hosted the national Forest Service reunion in Portland. The reunion event, at the Red Lion Inn on the River, was attended by around twelve hundred retirees and others from almost all Forest Service offices. The event, which also marked the centennial of the agency, was attended by Chief Dale Bosworth and some of the current staff from the Washington Office and the Department of Agriculture. The R-10 ranger boat *M/V Chugach* from Alaska's Tongass NF visited and gave short tours on the Columbia River. The event also featured a restored 1925 Dodge Brothers business car from Region 3, the oldest vehicle that has always been owned by the Forest Service.

Honor Poles from the Mt. Baker-Snoqualmie NF

The Region was honored when the Mt. Baker-Snoqualmie NF was asked to provide the two large cedar poles for carving by the Lummi Indian Nation to honor the citizens who died in the terrorism attacks on September 11, 2001. The Mt. Baker-Snoqualmie NF news release on August 26, 2004, noted that the forest "partnered with the Lummi Indian Nation in 2003 to provide the carving wood for the Honoring Pole, which was placed in Shanksville, Pennsylvania, at the site of the 9-11 crash of United Flight 93. In April 2004, the Mt. Baker-Snoqualmie National Forest provided two large western redcedar poles for the Liberty and Freedom Poles." These latter two poles, along with a third Sovereignty pole mounted horizontally linking the other poles, were carved by Lummi master carver Jewell "Praying Wolf" James and crossed the country on a trailer truck, making several stops along the way. The poles were presented to the Pentagon on September 20, 2004. For several years the poles have been temporarily installed at the Historic Congressional Cemetery; it is expected that they will become part of the Forest Service's 9-11 Memorial Tree Groves Project on Kingman Island in the Anacostia River in Washington, DC. On September 7, 2002, a fourth pole, the Healing Pole, was presented to the families of victims from the World Trade Center attack at the Sterling Forest, about one hour north of New York City.

Service. These partnerships, it was hoped, would enable the nation's federal land managers to deliver one-stop customer service to users of public lands. By pooling resources, the agencies would avoid duplication of effort; the agencies co-located several offices, shared personnel, combined operations, harmonized processes and permits, and standardized public information.

One such effort was in southeastern Oregon between the Fremont NF and the BLM area office. The process worked very well, but there were problems because of departmental differences, such as two different sets of manuals, regulations, policies, and directives. The federal laws are the same, but each department has a few agency-specific laws that govern management. Also, each agency has different ways of handling timber sales, grazing permits, and special use permits.

In spite of such difficulties, the Service First Web site notes that "savings in space costs are substantial and collocation has improved relationships among employees. BLM expects to combine 11 out of 15 of its Oregon/Washington offices within 15 years." In the Regional Office some operations have been successfully combined. Public affairs, fire, fleet management, heritage, American Indian coordination, printing and graphics, and others have been integrated with employees from the two agencies working side by side or with one person handling one job for both agencies.

Tied to Service First recently has been an effort to combine the Forest Service with the Department of the Interior. The Government Accountability Office (GAO) studied this idea in late 2008 and early 2009. The Congress took comments from a variety of interested parties, including Mike Dombeck, former chief of the Forest Service. The GAO report (GAO-09-223) found that in the short term there would be few cost savings, considering that the Service First initiative has been successful in many cases of combining public services for the BLM and USFS. In the long term there could be significant savings, but the program could take a decade or more to implement. Little effort on part of the Congress or the affected agencies has been brought to the effort since the GAO report was published in late February 2009.

Stewardship Contracting

In the 1990s, the Forest Service took steps to end the divisive debates surrounding timber harvest on the national forests by making watershed health and sustainable forest ecosystems agency priorities. Timber harvest continued in the Region, but at a fraction of the peak levels reached in 1987, 5.6 billion board feet per year to just over 400 million board feet per year 20 years later. The primary purpose of timber harvest on the national forests became vegetation management, rather than furnishing wood fiber for the nation. In 1996, the Forest Service began to consider the alternative of stewardship contracting. Stewardship contracts entail multifunction procurements over multiple years, are oriented toward achieving end results, promote collaboration with local community groups and other stakeholders, and exchange goods for services.

The Omnibus Consolidated Appropriations Act of 1999, also referred to as the Stewardship End Result Contracting Projects Act, approved a series of stewardship projects, including:

(1) road and trail maintenance or obliteration to restore or maintain water quality;

(2) soil productivity, habitat for wildlife and fisheries, or other resource values;

(3) setting of prescribed or controlled fires to improve the composition, structure, condition, and health of stands or improve wildlife habitat;

(4) removing vegetation or other activities to promote healthy forest stands, reduce fire hazards, or achieve other land management objectives;

(5) watershed restoration and maintenance;

(6) restoration and maintenance of wildlife and fish habitat; and

(7) control of noxious and exotic weeds and reestablishing native plant species.

The act was amended in 2003 and provisions of the act were extended until 2013. The Region 6 Web site in August 2008 listed over 100 stewardship contracts in the two states. The Colville NF, with the most stewardship projects in the Region, had 14 projects, followed by the Siuslaw NF with 13.

Economic Recovery Grants

The rural economic assistance program of the Northwest Economic Adjustment Initiative provided 18 grants to be distributed among nine counties, for a total of $290,000. This program derived from the Forest Ecosystem Management Assessment Team (FEMAT) report of 1993 and Northwest Forest Plan of 1994. The Northwest Economic Adjustment Initiative pumped $1.2 billion into Northwest communities [for local projects] to offset the decline in timber harvests. More than that, these projects highlight the importance of building local skills and catalyzing community efforts that lead to additional projects and that may leverage millions more for local community development. The Northwest Economic Adjustment Initiative is the socioeconomic companion plan to the Northwest Forest Plan (Yeaworth and Kusel, 2002).

The projects selected to receive economic recovery grants were located in Oregon, Washington, Montana, and Idaho. The federal share of the projects ranged from $75,000 to $100,000; most were $30,000 or less. Projects were selected to give community leaders in small, rural, timber-dependent towns more development (and survival) ideas and provide needed encouragement and training on projects to help with economic recovery and community stabilization. For example, in Sweet Home, Oregon, money was expended for cleaning up the exteriors of downtown businesses, as well as providing portable benches for outdoor events. Most projects were completed during 1993, but some required time extensions. The economic adjustment initiative

Civil Rights

Every administration since the Civil Rights Act of 1964 has sought to ensure that the federal workforce reflected the face of America and served all segments of the American public. The Forest Service's goal was to build a diverse workforce through new recruitment procedures. The agency invested more than $2.3 million per year, in partnership with universities, on national recruitment initiatives. For example, the agency recruited more than a hundred students per year through partnerships with Historically Black Colleges and Universities, Hispanic Serving Institutions (Columbia Basin College in Pasco and Heritage University in Toppenish), Tribal Colleges (Northwest Indian College in Bellingham), and the President's Committee on People with Disabilities to fill natural resource and other professional positions.

The Millennium Bug (Y2K) Program
During the early months of FY 2000, the Forest Service successfully designed a program to prevent disruptions of business processes caused by possible Year 2000 (Y2K) computer failures. Although there was not a crash of computers on January 1, 2000, there had been a huge media blitz over the worst-case scenario, but computer programmers were able to overcome software problems, install temporary fixes, and upgrade older programs. The agency also completed implementation of a new IBM-based computer system and a new office automation suite. Over 32,000 personal computers were installed at over 800 locations throughout the agency.

and economic action program from the State and Private Forestry program are no longer funded by Congress (Saranich personal communication).

Payments to States

Since 1908, states and counties have received 25 percent of Forest Service revenues to help fund schools and roads. In addition, several western Oregon counties received 50 percent of BLM revenues from the revested Oregon and California Railroad and reconveyed Coos Bay Wagon Road grant lands (see Chapter 4). Payments were never stable as they were mostly tied to fluctuating timber sales. With timber sales in decline since 1987, payments to states and counties had dropped dramatically by the mid-1990s. The three counties in Oregon that had benefited most from the so-called timber payments were Lane, Douglas, and Klamath counties; most of their payments derived from timber sales on the Willamette, Umpqua, and Siuslaw NFs as well as from the Eugene and Roseburg BLM Districts. In Washington, the largest recipients were Skamania, Lewis, and Jefferson counties.

In 1999, two bills were introduced in Congress to increase and stabilize payments to states by decoupling payments from federal receipts, but both were blocked by the Republican congressional leadership. Two other bills dealing with these payments were opposed by the Clinton administration because they continued to link payments to timber sales or other Forest Service funds.

The administration responded by stating its willingness to work with Congress based on five core principles: (1) providing a permanent, stable source of funding; (2) allowing flexibility at the local level; (3) promoting noncontroversial projects to build trust and collaboration; (4) promoting strong collaboration; and (5) establishing clear lines of authory. After nearly 11 months of negotiation, a compromise was reached. On October 30, 2000, the president signed the Secure Rural Schools and Community Self-

Determination Act, stabilizing payments to rural counties by providing about $1.1 billion above current payments over five years. The act based payments on the average of the state's three highest payments from FY86 to FY99. It also required counties that receive more than $100,000 to invest 15 to 20 percent in county projects and/or forest restoration, maintenance, or stewardship. Finally, it required the Secretary of Agriculture to create citizen advisory committees representing environmental, commercial, and local interests.

It was assumed that counties would no longer have to depend on controversial timber sales to provide the funding for their local schools and roads. It didn't work out that way. In early 2006, the Bush administration proposed changes in the act. Specifically, and most controversially, the proposal would sell off some 304,370 acres of national forest lands in 35 states. Funds from the sale of these often remote and hard to manage lands would be used to fund the act for the coming years. The sale of these lands would replace the standard formula from the 2000 act, and after the sale of these properties, both the act and the payments would disappear. The state with the largest amount of national forest land for sale was California, while Wisconsin had only 80 acres on the list. As soon as the proposed action was announced on February 28, 2006, opposition quickly mounted from all corners of the country. The land sale proposal died a quiet death before the end of the year.

The original act expired in 2006, after two year-to-year extensions. Continued efforts to renew the act have met with half-hearted success in Congress. What this essentially meant was that money for the counties still comes from sales or from fees paid by timber sale receipts, grazing fees, special use permit fees, and recreation fees. Another two-year extension of the payments to states formula was added to the bill to shore up the nation's financial system in October 2008.

National Monuments

Based partially on earlier efforts to have Mount St. Helens declared a national volcanic monument, a 30-member citizens group near Bend, Oregon, began a similar effort in 1988 to have a new national monument established around the Newberry Crater on the Deschutes NF. Two years later an act was passed by Congress and signed by President George H. W. Bush that established the Newberry Crater National Volcanic Monument. The monument encompasses some 55,500 acres with an additional 10,300 acres established for special management to protect the resources of the volcanic monument.

On June 9, 2000, President Clinton invoked the Antiquities Act of 1906 to establish four new national monuments—Hanford Reach in central

Washington; the Cascade-Siskiyou in southwest Oregon; Canyons of the Ancients in Colorado, and the Ironwood Forest in Arizona. The Cascade-Siskiyou National Monument, some 52,947 acres in southwest Oregon just north of the California border, is next to the Rogue River-Siskiyou NF. The area includes the distinctive Pilot Rock. The area contains spotted owls and peregrine falcons, as well as many species of butterflies, snails, and fish found nowhere else. Management is by the BLM but the area serves as a reminder that both agencies have lands that serve as critical habitat for many species and that each contributes to biological diversity in the Pacific Northwest.

More than a dozen plus areas were reportedly under study as additional possible national monuments by President Clinton before he left office in early 2001. One was around the Steens Mountain area in southeastern Oregon. After considerable public response, state concurrence, and congressional agreement, a 425,000-acre area became the Steens Mountain BLM Cooperative Management and Protection Area in legislation passed by Congress and signed into law on October 24, 2000. No new national monuments within Region 6 have been proposed.

Planning and Controversy

SPOTTED OWL CONTROVERSY AND THE TIMBER WARS

On June 26, 1990, the U.S. Fish & Wildlife Service formally listed the northern spotted owl as "threatened" under the Endangered Species Act throughout Oregon, Washington, and northwest California. This listing impacted private and tribal lands in addition to federal lands since "taking" of a threatened species anywhere is prohibited, and this includes the loss of any essential habitat. The Fish & Wildlife Service delayed implementation of the decision until forest plans could be developed.

On March 7, 1991, U.S. District Court Judge William Dwyer ruled that the Forest Service management plan for spotted owls was inadequate under the law. By that time logging in national forests had been cut by 50 percent in Region 6. In April 1994, the Clinton administration presented Judge Dwyer with the Northwest Forest Plan, which preserved 70 percent of federally owned old-growth forests and reduced timber sales to less than a fourth of those in the 1980s, as well as providing for restoration of watersheds that had been damaged by logging operations. Both sides in the dispute objected to the plan. In December 1994, Judge Dwyer held that the Northwest Forest Plan complied with the National Forest Management Act (Yaffee 1994).

Spotted owl. USFS photograph.

FOREST CONFERENCE AND FEMAT IN 1993

President Clinton, not long after he took office in 1993, called for and attended a forest conference in Portland, Oregon. On April 2, 1993, Clinton and Vice President Al Gore, along with five cabinet members, met representatives of the public in Portland, Oregon, to discuss the spotted owl and timber harvest situation in the Pacific Northwest and northern California. Never in the history of the agency had the administration put such emphasis on resolving problems in the national forests and adjacent BLM districts. The president directed a team of research scientists to write a balanced, comprehensive, and long-term policy for the management of more than 24 million acres of public land administered by the Forest Service and BLM. The scientific analysis and recommendations were to be discussed, analyzed, and concluded within a 90-day window. Thus the issue that was two decades in the making was to have three months to conclude!

The Forest Ecosystem Management Assessment Team (FEMAT), led by Forest Service researcher Jack Ward Thomas, was given the task of organizing an interagency team of scientists to analyze the best scientific evidence and report on the best solutions to the federal forest problems—basically old growth and spotted owls. The FEMAT team comprised research and management employees from the Forest Service and BLM, along with the best university researchers and experts; they were housed on the 14th floor of the U.S. Bank building in downtown Portland. The FEMAT report was given the highest priority of any activity in the Forest Service and the BLM, as authority had come directly from the president. Whatever was needed—computers, software, desks, travel, etc.—was ordered and delivered the same day. Another team of Forest Service and BLM professionals, some removed from the FEMAT team, were to develop an accompanying EIS. The FEMAT scientists prepared an extensive report (over 1,000 pages) that, along with the Supplemental Environmental Impact Statement (SEIS), was printed by the end of July.

There were many stories about the processes used to produce the FEMAT report and SEIS. Jack Ward Thomas flew back to Washington, DC, at the end of the presidentially imposed time frame to meet with President Clinton and other top advisors and to testify before Congress. He was literally making corrections to the draft FEMAT report on the airplane. An amusing situation occurred on his way back:

> My plane from Washington took me to Denver. Then, there was a plane to Portland. By the time I made my way to the bull pen that served as the exit for the commuter airline in Portland, I was exhausted. I sprawled on a bench of chairs. The three men sitting across from me were looking me over carefully and whispering back and forth. Finally, one spoke. "Say, Mister, did anyone ever tell you that you look just like that biologist fellow? What's his name? Jack Ward Thomas?"
>
> "Well, yes, several people have told me that over the years."
>
> "Geez, fellow, that must REALLY piss you off?"
>
> "Sometimes." (Thomas 2005: 517)

The president authorized a 30-day extension for finalizing and printing the FEMAT report. Yet the SEIS did not have such a time extension, which created a dilemma—the SEIS was scheduled to be "on the street" *before* the FEMAT report, with its alternatives or options, was finalized. After some back and forth between Thomas and the president, the SEIS team was also given a 30-day extension.

An interesting issue came up during the social and economic analyses of the FEMAT options. County representatives such as planners were called in to score the possible negative or positive effects of the FEMAT options on their respective counties. However, they were not willing to have their names or evaluations associated with the FEMAT report as they feared retaliation by the county commissioners and the timber industry. The same fear of reprisals was expressed during the issues with Section 318 timber sales (see below).

The following spring, the Northwest Forest Plan was approved and implemented. The timber industry and many of the western Washington and Oregon counties were not happy with the selected outcome—often called Option 9—which severely reduced timber harvest in the affected areas. It caused an immediate uproar in the Northwest, but after several months a relative calm seemed to come over the controversy. Several months after publication of the FEMAT report, Thomas was appointed as chief of the Forest Service. One of his primary goals as chief was to implement ecosystem management not only in the Pacific Northwest but also across the all the agency-administered lands.

NORTHWEST FOREST PLAN OF 1994

By 1992, there were more than a dozen lawsuits ongoing and three court injunctions involving the northern spotted owl, the marbled murrelet, and timber harvest in old-growth forest. In October 1993, the Forest Service signed a memorandum of understanding with several other federal agencies to establish a regional framework for cooperative planning, decision making, and implementation of the Northwest Forest Plan. About 30 percent of the area covered by the plan already had special congressional designations (such as wilderness areas and wild and scenic river areas). Under the plan, the remaining 70 percent was classified into late-successional reserves (30 percent), adaptive management areas (6 percent), managed late-successional areas (1 percent), administratively withdrawn areas (6 percent), riparian reserves (11 percent), and matrix (16 percent). Matrix is a mix of general forest uses where timber management is allowed.

Each allocation had accompanying standards and guidelines that work together to form a comprehensive management strategy including protection measures, restoration activities, and commercial timber harvest. Ongoing monitoring helped determine the plan's effectiveness in promoting habitat for northern spotted owls and marbled murrelets, protecting and restoring watersheds, and promoting a sustainable level of goods and services from federal forests.

The Northwest Forest Plan represented the first time that the Forest Service and the BLM, in conjunction with the National Marine Fisheries Service, Fish & Wildlife Service, and Environmental Protection Agency, developed a common management approach for an entire ecological region. Scientific contributions provided federal land managers with crucial information for environmentally sound management decisions for complex and varied ecological systems throughout the region. The comprehensive ecosystem management strategy developed under the plan served as a prototype for similar regional initiatives in other parts of the United States.

The Northwest Forest Plan required the USFS and the BLM to undertake efforts to protect spotted owls, wild salmon, and over 1,000 other species that inhabit the old-growth forests of the Pacific Northwest. The so-called survey and manage provision required that the federal land agencies that were part of the Northwest Forest Plan would survey old-growth areas for rare plants and wildlife before allowing logging or other developments in the federal forests. The surveys were designed to look for various vascular plants, bryophytes, lichens, fungi, mollusks, reptiles, amphibians, and mammalian species known to inhabit old-growth forests. Despite attempts to indentify as many species as possible in areas slated for development, the reality was soon evident that the USFS and the BLM could not spend years in the field to search for species that sometimes only appear during their mating or birthing seasons.

As part of the settlement of a lawsuit by the logging industry the survey and manage portions of the Northwest Forest Plan were eliminated in 2001. During 2004, a revised SEIS was written to remove or modify the survey and manage guidelines. A year later, Regional Forester Linda Goodman signed a declaration that the Forest Service and the BLM were to protect 57 species that were originally to be studied in the 1994 plan.

Adaptive management areas (AMAs) were designed in the 1993 FEMAT report and carried over to the Northwest Forest Plan. The 10 AMAs range from about 92,000 to nearly 500,000 acres of Forest Service, BLM, and National Park Service administered lands. The boundaries of the AMAs also include state and private lands. Most of the AMAs are associated with sub-regions that were impacted negatively by reduced timber harvest from the federal lands. The AMAs were to work closely with the Pacific Northwest Research Station to test different methods of managing the forest areas, usually by loosening National Forest Management Act plan standards and guidelines (Stankey and Shindler 1997). Tim Tolle, in the Regional Office, helped in the development of AMA plans. The 10 AMA areas support and participate in the International Model Forest Network begun by the Canadian Forest Service. There are 10 international model forests operating in Canada. The 10 AMAs from the Northwest Forest Plan are: Applegate (Oregon), Central Cascades (Oregon), Cispus (Washington), Finney (Washington), Goosenest (California), Hayfork (California), Little River (Oregon), North Coast Range (Oregon), Olympic (Washington), and Snoqualmie Pass (Washington).

The AMAs have been a disappointment for researchers working on adaptive management of ecosystems. For example, the Finney AMA had no constituency to formulate a plan until 2008, 14 years after it was approved. The Cispus AMA plan was essentially "hijacked" by local environmental groups, who believed that the design of the Cispus AMA should not loosen the forest standards and guidelines but tighten them. The approved Cispus AMA plan left few opportunities for researchers to test adaptive management. Other AMAs have been more successful, but few if any, have lived up to their full potential.

In mid-August 2008, the Fish & Wildlife Service announced a 23 percent reduction of the forest lands designated as spotted owl habitat under the plan. Most of the areas to be removed from protection either had no owls, like the forest on Ft. Lewis, or were permanently protected in such management designations as wilderness. However, *The Oregonian* on August 12, 2008, reported that "the move reduced by about one-third the critical habitat acreage within coastal Oregon forests where the U.S. Bureau of Land Management is considering increased logging."

INTERIOR COLUMBIA BASIN ECOSYSTEM MANAGEMENT PROJECT

After completing the Northwest Forest Plan, the administration focused on federal lands in the eastern portion of the Pacific Northwest. In July 1993, the Forest Service and the BLM were directed to "develop a scientifically sound and ecosystem-based strategy for management of eastside forests." In response, the two agencies initiated the Interior Columbia Basin Ecosystem Management Project (ICBEMP). The ICBEMP was designed to provide a long-term, comprehensive strategy for managing public lands on a landscape level, addressing forest and grassland health, fish and wildlife habitat, and regional social and economic issues. The project included completion of a framework for ecosystem management based on a scientific assessment of the interior Columbia River Basin.

The assessment examined the status of ecosystems within the basin, including their historic trends, current status and trends, and projected future outcomes and conditions. The ICBEMP science team gathered information from many sources throughout the basin, including more than 20 databases and 180 geographic information system layers for mapping. Forty expert panels/workshops were convened and more than 130 reports were contracted from independent scientists. Scientists focused on the need to reduce the threat of wildland fire, the spread of noxious weeds, and damage to forest health from disease and insects, while remaining sensitive to local communities who rely on natural resources. Beginning in March 1994, the public was invited to periodic project update meetings. Project staff gave more than 70 special presentations. Two far-reaching video teleconference broadcasts were held to elicit public comments and provide information; they reached more than 87 communities in six states, eliciting thousands of public comments.

In 1997, the ICBEMP finally released two draft EIS, one for the east side of the Cascades in Washington and Oregon and the other in the upper Columbia River basin. The ensuing 335-day public comment period included more than 100 meetings and generated more than 83,000 comments. Project staff met with interested citizens, resource advisory councils, provincial advisory committees, special interest groups, and Forest Service and BLM employee groups. The executive steering committee participated in several tribal summits with American Indian tribes. However, there were widespread criticisms of the two drafts. In October 1998, Agriculture Secretary Dan Glickman and Secretary of the Interior Bruce Babbitt directed the ICBEMP team to develop a new approach for the project and to issue a supplemental draft EIS for public comment. The supplemental draft EIS was to focus only on critical, broad-scale issues—landscape health, aquatic and terrestrial habitat, and socioeconomics within the project area.

Released in May 2000, the supplemental draft EIS received about 500 public comments. A final EIS and proposed decision were released in December 2000. In January 2003 the regional executives for the Forest Service, Forest Service research branch, Bureau of Land Management, Fish & Wildlife Service, the National Marine Fisheries Service, and the Environmental Protection Agency signed a memorandum of understanding completing the project. The regional federal managers decided in this memorandum that, instead of a formal, basin-wide decision on management of the basin, they would adopt a strategy of incorporating the science learned from the decade-old assessment into on-going land use planning efforts. The decision makers believed that this would lead to greater flexibility than the FEMAT process on federal, state, and private lands in the basin.

The ICBEMP scored two notable successes. It completed the first comprehensive, landscape-level scientific assessment of its size in the United States, and it was the first major federal resource management planning process to include, from the beginning, a multi-state coalition of counties that served as a model for county involvement in other regions.

NFMA PLANNING RULE OF 2000

By the 1990s, the public was demanding more involvement in policy planning, and new ecological insights were revolutionizing natural resource management through ecosystem-based approaches. A planning rule revision was overdue.

For more than 10 years, the Forest Service worked on revising the National Forest Management Act (NFMA) rules, which had been adopted in 1979 and amended in 1983. Agriculture Secretary Glickman facilitated the process by appointing a new committee of scientists, a group of eminent specialists commissioned to review and evaluate the Forest Service's planning procedures. The committee published its report and recommendations in March 1999. The Forest Service released the proposed rule changes in October. The agency held 23 national public workshops to discuss them and received more than 10,000 public comments. The agency analyzed the comments and used them in preparing the final NFMA rule, released in November 2000.

The final NFMA rule, as it was referred to by the Forest Service, affirmed ecological, social, and economic sustainability as the overall goal for managing the national forest system. Maintaining and restoring ecological sustainability was confirmed the highest priority. In the spirit of Gifford Pinchot, first chief of the Forest Service, the rule also facilitated greater public collaboration in all phases of the planning process. Under the new rule, national forest management was to be based on cooperatively developed landscape-level goals. The post-decisional appeal process was replaced with a pre-decisional objection process to increase public input into the decision making.

The final NFMA rule also placed greater emphasis on the use of science in planning, partly by promoting the use of regional ecosystem assessments in planning and decision making. The rule emphasized three elements: sustainability, collaboration, and science. Each forest or grassland supervisor must have access to an advisory committee with knowledge of local conditions and issues, although an advisory committee is not required for each national forest or grassland—advisory committees used by other agencies may be utilized through proper agreements. The final planning rule calls for establishing science advisory boards, independent scientific review to determine if land management plans meet sustainability goals, and, when appropriate, science consistency evaluations to determine whether the planning process is consistent with the best available science.

Another issue, first raised as early as 1995 by Chief Jack Ward Thomas and later Chief Mike Dombeck, was what became called the "process predicament" or "analysis paralysis." Chief Dale Bosworth in 2001 called together a team to study the problem of excessive "red tape." By the following summer, the team presented its report; the executive summary addressed three problem areas:

> 1. Excessive analysis—confusion, delays, coast, and risk management associated with required consultations and studies;
> 2. Ineffective public involvement—procedural requirements that create disincentives to collaboration in national forest management; and
> 3. Management inefficiencies—poor planning and decision-making, a deteriorating skills base, and inflexible funding rules, problems that are compounded by the sheer volume of the required paperwork and the associated proliferation of opportunities to misinterpret or misapply required procedures.

It was thought that by addressing these issues, the agency could save as much as $1 million annually. Although it was widely touted at the time, the effort largely failed because of agency inertia and the inability to get Congress to respond. Yet, in the aftermath of the huge fires in the summer and fall of 2002, the Forest Service was able to respond to the situation with the healthy forests initiative (HFI). This initiative was intended to reduce the risks that wildfires pose to people, structures, and the environment. The following year, Congress responded by passing the Healthy Forests Restoration Act of 2003, which was intended to better enable the Secretary of Agriculture and the Secretary of the Interior to conduct hazardous fuels reduction projects.

Timber Management

Timber cut volume (see Appendix E) was reduced to around 3.9 billion board feet in 1990, decreased again in 1991 to under 3.2 billion board feet; in 1992 the volume cut was further reduced to 2.1 billion board feet. These reductions were necessary to meet obligations under the new forest plans in the Region and increasingly because of the spotted owl restrictions. By 1995, the timber cut was below 1 billion board feet (877 million board feet). That was the lowest it had been since 1942, during World War II. In 2000, the Regional timber cut level was down again, to 409 million board feet and in 2002 it fell again, when 276 million board feet were harvested. The latest figure, for fiscal year 2007, shows a slight rise in timber cut to 419 million board feet, which is still under the projected cut levels of 1.1 billion board feet in the Northwest Forest Plan.

Reforestation work remained mostly funded through K-V funds deriving from timber sales. In 1991, a record high 120,000 acres were reforested, reflecting the high volume of timber harvesting from 1986 to 1989.

During the late 1980s and mid-1990s, at the peak of the summer operating season K-V funds were transferred for emergency firefighting purposes. Many K-V reforesting projects had to be closed down or not completed.

> During fiscal years 1988 through 1995, the Forest Service transferred almost $1.5 billion of K-V funds to help pay for about $3 billion in firefighting costs. In fiscal years 1988 and 1989, the Forest Service transferred a total of $723 million from the K-V fund for emergency firefighting, which was fully restored through supplemental appropriations by fiscal year 1990. In fiscal years 1990 and 1992, the Forest Service again used K-V funds—$295 million in total—for firefighting purposes. In fiscal year 1994, the Forest Service transferred

> $350 million of K-V funds for firefighting, which was restored to the fund at the beginning of fiscal year 1995 after the agency had requested and received supplemental appropriations for that purpose. In fiscal year 1995, the Forest Service transferred $125 million from the K-V fund to the firefighting fund (General Accounting Office 1996: 27).

The K-V and other funds to support forest projects relied on something of an honor system: the Forest Service would transfer funds to fight summer fires, then Congress would restore the funds with a supplemental appropriations bill. By the early 1990s, the Office of Management and Budget (OMB) in yearly audits of appropriations and expenditures was not happy with how the Forest Service was reporting the cost accounting for each year. Much of the trouble related to fire-related spending and money transfers near the end of the fiscal year. How the firefighting funds were allocated each year also caused problems. Funding was based on a 10-year average; fire costs were rising rapidly, though they were quite low in some years. Several times the OMB accountants wanted to stop all firefighting once the authorized fire funds were exhausted, but congressional supporters were able to put a stop to such fiscal nonsense. It took several years for the agency to come up with an acceptable method to get a "clean audit." Then the funding restoration stopped because the OMB and the Department of Agriculture refused to submit the funding request to Congress:

> In the past, when such transfers were made, the Department of Agriculture requested and received supplemental appropriations to restore the transferred moneys generally within 2 years of the original transfer. However, until recently, the Department of Agriculture

Ten-year-old Douglas-fir trees planted with K-V funds growing in the Pacific Northwest, c. 1995. USFS photograph.

> has not submitted a request for a supplemental appropriation to the Congress. While the Forest Service acknowledges that failure to restore this amount will be potentially disruptive to the K-V program, forest and district offices continue to operate and plan for future reforestation projects as if the transfers had not occurred (General Accounting Office 1996: 26).

1990 APPROPRIATIONS ACT—SECTION 318

The Forest Service solicited competitive bids for the sale and harvest of timber in the Region under Section 318 of the Appropriations Act of 1990. Section 318, also known as the Northwest Timber Compromise, mandated that the "Forest Service shall offer ... an aggregate timber sale level of seven billion seven hundred million board feet of net merchantable timber from [all] the national forests of Oregon and Washington for fiscal years 1989 and 1990." Section 318 directed the Forest Service to offer, during fiscal years 1989 and 1990, 5.8 billion board feet of timber from the so-called spotted owl forests. This provision responded to a timber shortage in the Pacific Northwest resulting from clashes between environmentalists and the timber industry; the timber industry thought it was necessary to provide a stable supply of timber to the saw mills. Many of the sales came from older sales that were cancelled for a variety of reasons. Most of these sales were brought into compliance before they were offered to timber sale bidders.

Section 318 required advisory boards of five people or more, representing the timber industry, environmental groups, and the community. This was the first time in agency history that a broad spectrum of the public had the authority to advise directly the forest supervisor on timber sales. For many forests that had Section 318 sales the six-member advisory board approved most timber sales. For example on the Umpqua NF, the local school principal, who served on the advisory board for the Umpqua NF, usually sided with the industry, since schools were heavily funded by timber sale receipts and the timber companies were major employers in the county. The board also included a businessman from Cottage Grove and a tribal representative. The advisory members "disagreed on about 40% of the sales and where environmental groups disagreed with the rest of the board, they sued through other environmental groups." Former Forest Supervisor Robert Devlin noted that the advisory group had one instance of internal controversy:

> One of the advisory board members went on record at a board meeting that a proposed sale in Medicine Creek on the Diamond Lake District

> was a very important and special place for the tribes. According to him, this area had deep religious meaning for the tribe. Sue Shaffer, also a board member asked this person, which tribe it was important area to as she was tribal chair of the Cow Creek [Band of Umpqua Indians] and as far as she knew, the area wasn't important to her tribe. The board approved the sale to go forward. Found out years later that the board member who was making statements about the importance of the area was leading "eco-system" tours at a fee in that area (Devlin personal communication).

The environmental community in the Pacific Northwest was unhappy about Section 318. It immediately attacked the provision as unconstitutional as they believed that these sales did not have to comply with environmental laws and regulations in place since the original sales were offered. Environmentalists protested planned timber sales in the Region, winning a number of cases. They also decided to take a different approach—to take their case out of the Pacific Northwest and nationalize the issue. "Feature stories appeared in *The New Yorker* and *National Geographic*, network news ran stories of activists sitting in trees in protest, and the issue reached the pinnacle of media exposure when the spotted owl made the cover of *Time* magazine of June 25, 1990. This successful campaign in the arena of public opinion ... [as] groups ... launched a more targeted political campaign to delegitimize the strategy of using appropriations riders to exempt Northwest forests from the application of environmental statutes." (Hoberg 2001: 69) As a result of this pressure from the environmental community, many of the Section 318 sales were never harvested.

THE 1995 RESCISSION ACT OR SALVAGE RIDER

In 1994, wildfires burned 1.4 million acres on the national forests in the West (4 million acres on all ownerships) and the Forest Service spent over $1 billion for firefighting. The fires left large areas of dead and dying trees, raising the risk of insect infestation and future large fires. In general, the dead trees had to be harvested within two or three years before they lost economic value to decay and insect damage. On July 27, 1995, President Clinton signed into law the 1995 Rescission Act. One section of the act directed the release of timber sales, which has been since referred to as the "salvage rider," subject to provisions in the Appropriations Act of 1990. The program was designed to reduce the risk to forest health by salvaging timber from burned areas and creating jobs in the affected communities.

The Rescission Act gave salvage operations primacy over other laws, specifically federal laws for environmental protection and prohibited appeals by the public. The president directed the affected agencies to implement the Act's salvage provision through December 31, 1996, in accordance with the Northwest Forest Plan of 1994, other existing policies and plans, and existing environmental laws. Implementation of the Rescission Act was immediately surrounded by controversy. Environmental organizations labeled it "logging without laws." As implementation proceeded, the Forest Service found itself in court battling with both the forest products industry, who claimed the agency was not following the law as written, and the environmental organizations, who claimed the law was being implemented beyond the intent set by Congress.

The timber target level, under the Rescission Act of 1995, was 4.5 billion board feet of salvage timber, plus or minus 25 percent, with most of the volume coming from non-Northwest Forest Plan forests. The Forest Service achieved that goal, but generally looked at the act as a law that was forced on the agency. The salvage rider greatly harmed relations between the Forest Service, the forest products industry, and the environmental community. Future attempts at providing better forest health in the national forests, regardless of their scientific basis, would have to withstand the challenge that they were "just another salvage rider."

Salvage sales in roadless areas were quickly becoming a contentious issue. Secretary Glickman directed the Forest Service to allow salvage sales in roadless areas only where the risk of fire was high in the vicinity of homes and communities or where trees were susceptible to insect attack within three years. Harvest of green trees during timber salvage was raising further concerns. The Forest Service directed managers to subordinate the harvest of green trees to the salvage of dead and dying trees, limiting green-tree harvest to areas where it was necessary for safety and stand improvement.

The Rescission Act also contained a provision about the earlier Section 318 sales pertaining to old-growth sales in Region 6 that were delayed by lawsuits and by new listings of threatened and endangered species, especially the spotted owl listing in 1990. Because of the listing of the northern spotted owl as a threatened species, a number of Forest Service 318 sales were "modified, eliminated or held in abeyance." Many Section 318 sales did not go forward as a result of concerns about significant impacts to species listed under the Endangered Species Act. Historian Pete Steen noted in his oral history with Forest Service Chief Jack Ward Thomas:

> The salvage rider said that the in-limbo [318 timber] sales "would be released to the original buyers under the original conditions." The Forest Service was able to trade out of some of the sales—"we did everything we could to spare these areas"—but some of them were cut. The enviros yelled, "See what the Forest Service is doing under the guise of salvage—they are cutting old growth forests!" Salvage and the 318 sales were two very different things, and they damn well knew it. But, boy, they used the circumstance most adroitly as a propaganda tool. The Forest Service was the big loser (Steen 2004: 121).

FOREST FRAGMENTATION

Forest researchers/scientists and wildlife professionals in the agency, especially at the H. J. Andrews Experimental Forest, and in universities began to question the ability of the national forests in the Region to provide long-term habitat. The original heavily forested areas had been carved over the decades into many smaller units by clearcuts and roads. Many view forest fragmentation as one of the greatest threats to biodiversity in forests.

The problem of habitat destruction is compounded by the inability of small forest fragments to support viable populations, especially of large vertebrates; the local extinction of species that cannot be supported by the small fragment; and edge effects that alter the conditions of the outer areas of the fragment, greatly reducing the amount of forest interior habitat. Some animals, like deer, thrive along the edge of these fragments, but other species do not. The effect of fragmentation on the flora and fauna of a forest patch depends on the size of the patch, and its degree of isolation. Forest patches that are smaller or more isolated will lose species faster than those that are larger and not isolated.

Some of the effects can be mitigated—for example if a clearcut harvest area is reforested, the increasing number of trees will lessen the isolation of adjoining forest fragments—but it may require years of regrowth to approximate the original forest conditions. In many areas in the Region increasing emphasis is being placed on so-called wildlife corridors. This is where forest "strings" are not harvested, thus providing a travel corridor to other fragments and members of the same species. Such corridors also provide cover while animals travel.

An additional factor related to the fragmentation issue is that the national forests and the three national parks in the Region are virtually the only remaining old-growth habitat areas. Most of the forest areas under private and state control in the two states have been heavily cut over and are generally considered to be "commodity forests." During the 1980s and 1990s, many of the corporate forest lands were changed to a short rotation system (sometimes referred to as the economic rotation) and trees were harvested after only about 40 years. The national forests in the Region generally have at least an 80- to 120-year harvest schedule. The remaining old stands of big trees, as well as those currently growing, on the national forests will play an increasingly important role in biodiversity in the coming decades.

SPECIAL FOREST PRODUCTS

The small Pacific yew, an evergreen tree, is often found in association with Douglas-fir trees on the west-side national forests and BLM lands. In the early 1960s, yew bark was collected from Pacific yew trees near La Wis Wis Campground on the Gifford Pinchot NF. In the late 1980s, the Forest Service began working with the National Institutes of Health and the Centers for Disease Control to collect yew tree bark and protect areas with yew trees. Bristol-Meyers Squibb (BMS), in extensive testing of all types of possible cancer cures, found that taxol, which could be processed from the bark of the Pacific yew tree, was effective at fighting some forms of cancer including breast cancer. The interest in taxol created jobs in the collection, harvest, and processing of yew bark. In 1991 and 1992, the Forest Service, with funding from BMS, established yew inventory teams to find the locations of the largest concentrations of the yew trees. This funding helped, in some cases, to lessen the impact of downsizing caused by the Northwest Forest Plan. However, the increased funding would not last.

Hauser Chemical Research Inc. of Boulder, Colorado, had an agreement with BMS to collect bark on national forest and BLM lands in western Oregon to process it into taxol. Hauser developed a four-level network for the collection and initial processing of bark. The network employed bark harvesters, collectors, processors, and logger-processors. Hauser built a special yew bark processing plant just north of Cottage Grove, Oregon. BMS worked with the Region and the BLM to develop an EIS to cover the gathering of the yew bark. Wendall Jones (personal communication) noted that "there was a move to centralize … this work in the Washington Office … But R6 argued that we could handle the field details much better right here." The draft EIS was completed in 1992, but before the final EIS was published the following year, BMS closed its yew bark/taxol operations because the company found that it could extract taxol from ornamental yew plants more cost effectively than by working with the Forest Service and BLM. The two agencies finished and published the final EIS in 1993, in part because there might be other companies willing to work with the agencies to produce taxol in the future.

Yew bark peeling at BLM site near Creswell, Oregon, 1982. Author's collection.

In 2002, a bronze plaque was erected by the National Cancer Institute and the American Society of Pharmacognosy on a large rock at La Wis Wis Campground that said in part: "Near this location on August 21, 1962, Arthur Barclay and a team of botanists from the U.S. Department of Agriculture collected bark of the Pacific yew … Drs. Monroe Wall and Mansukh Wani … isolated taxol from that sample. Since 1990 taxol has been the drug of choice for treatment of ovarian cancer and is widely used in the treatment of breast cancer."

Range Management

Management of the Forest Service grazing ranges came under heavy fire in the 1990s and early 2000s. This was partly because of the relatively lower fees that the Forest Service had been charging per grazing animal, and partly because of the environmental costs associated with the grazing program on public lands. The 2005 GAO report on grazing fees documented that the Forest Service fails to obtain market prices for grazing and thus loses millions of dollars annually in the program. Effectively, this loss of revenue gives an advantage to those stock owners who have the privilege of having one or more grazing permits.

In Region 6, 11,408,000 acres of land are approved for grazing 550,000 animal units months (AUMs) in 2004. Just under 400,000 AUMs were actually billed or used. Grazing in the Region was highest on the Wallowa-Whitman, Malheur, Okanogan-Wenatchee, Fremont-Winema, Ochoco, Umatilla NFs, while the Siuslaw, Olympic, and Willamette NFs had no grazing.

There have been rising concerns about environmental damage to forest ecosystems that can result from grazing on the public lands. Overgrazing has always been an issue in the management of the grazing animals. Additional issues include:

- Damage to streams by animals walking through the water, affecting water quality and fish populations;
- Reduction of streamside vegetation thus increasing stream temperature, which can be deadly for some fish species, and causing land erosion;
- Replacement of native species such as elk and deer on grasslands by grazing animals;
- Increase in non-point pollution when animals defecate in streams;
- Decrease in the resiliency of native grass and shrub species, especially those that rely on a fire regime for reproduction and health; and
- Encouragement of the spread of invasive species (U.S. Government Accountability Office 2005).

The Rescission Act of 1995 requires that the USFS identify all grazing allotments for which NEPA analysis is needed. Based on this requirement, the Region began in the early 2000s to require that annual grazing permits have to go through the NEPA compliance process before they can be reissued. For the first time, the national forests in the Region have been spending large amounts of time and effort to make these analyses on the range permits. The end date established in the national schedule to complete the NEPA analysis on most allotments without disrupting permitted livestock grazing activities is 2010.

Sheep grazing in the Cascades, Pacific Northwest, 1980. USFS photograph.

The Consolidated Appropriations Resolution of 2003 directed the Secretary of Agriculture to renew grazing permits for those permittees whose permits expired prior to or during fiscal year 2003. The Forest Service was behind schedule on its mandated NEPA compliance process and was also dealing with pending lawsuits. The Interior Appropriations Act of 2004 further directed the Secretary to renew grazing permits that expired or were transferred or waived between 2004 and 2008 (Norbury 2005).

Recreation and Wilderness Management

After World War II, the number of recreational visits to the national forest lands soared; from just 18 million visitor-days in 1946, it climbed to almost one billion in 1999—nearly 50 times more. By the 1990s, recreation dwarfed all other uses of the nation's national forests and grasslands, contributing billions of dollars to the gross national product. Region 6 was no exception: in 1990, the Region experienced over 43 million recreation visitor days (RVDs), a huge increase from the previous year, when an estimated 36 million RVDs were experienced. The two states were virtually equal in recreation use—Oregon with 21 million and Washington with 22 million. The most popular sport in the Region was skiing, followed closely by camping. Multnomah Falls (Mt. Hood NF), Mount St. Helens, and Windy Ridge (both on the Gifford Pinchot NF) received the most recreational visitors at the Forest Service interpretive sites in 1990. However, money to operate the various visitor centers around Mount St. Helens has been difficult for the forest and the Region to obtain. The funding situation became so pronounced that *The Oregonian* editorial board on July 27, 2008, recommended that Mount St. Helens should become a national park.

Hikers at Bald Mountain on the Pacific Crest National Scenic Trail, Mt. Baker-Snoqualmie National Forest, Washington, 1980. USFS photograph.

Controversies continued over certain types of recreational use of the national forests in the Region. Wilderness has for decades raised the ire of the timber industry and they have tried, unsuccessfully, to tie wilderness to single use, not multiple use as mandated by law. The expansion of ski areas into adjacent national forest lands has also been quite controversial. Expansions and new developments at Mt. Ashland, Mt. Hood, Willamette Pass, Early Winters, Klamath Falls, and others have all been in the headlines in recent years. Some of the problems discussed included roadless area intrusions, threatened and endangered species, and land trades.

In 1997, the Forest Service acknowledged the central role of recreation by crafting a "recreation agenda" to protect and maintain the essential wildland character of the national forests and grasslands. The agenda included a new commitment to reaching youth and under-served populations, thereby building future constituencies and extending the benefits of outdoor recreation to all Americans. The recreation agenda was designed to guide Forest Service recreation programs into the twenty-first century, helping the Forest Service live within the limits of the land while increasing visitor satisfaction. Partnerships were key; projects would be prioritized based on feedback from partners and local communities, in accordance with sound science. It was expected that by focusing on core competency—offering outstanding natural settings for dispersed recreation—the recreation agenda would improve customer service, expand conservation education and interpretation, and build community relationships and partnerships.

In 2002, Secretary of the Interior Gale Norton announced that 26 trails were being recognized as national recreation trails. Three were in the Pacific Northwest: the Historic Columbia River Highway State Trail in Oregon, the John Wayne Pioneer Trail in Washington, and the North Cascades National Park/Ross Lake National Recreation Area Segment of the Pacific Northwest Trail in Washington. The Columbia River Highway State Trail is located on the Mt. Hood NF.

Another trail, the Pacific Northwest Trail, has recently been identified as a possible national recreation trail. The proposed trail extends approximately 1,000 miles from the Continental Divide in Glacier National Park, Montana, to the Pacific Ocean at Olympic National Park, Washington. The route selected goes through the Flathead and Kootenai NFs; the Kaniksu NF; and in Region 6 passes through the Colville NF, the Okanogan and Wenatchee NFs, Pasayten Wilderness, Ross Lake NRA, North Cascades NP, Mount Baker-Snoqualmie NF, along the Skagit River, through Deception Pass, across Whidbey Island, onto the Olympic NF, and finally through Olympic NP to the Pacific Ocean.

Disabled skier on Mt. Hood, Mt. Hood National Forest, Oregon, c. 1980. USFS photograph.

Riders in the Pasayten Wilderness, Okanogan National Forest, Washington, c. 1980. USFS photograph.

RECREATION FEES AND RESERVATIONS

All receipts on federal lands were traditionally returned to the U.S. Treasury and could not be invested in the projects, facilities, and services that generated them except for K-V funds. By the 1990s, many underfunded federal facilities were deteriorating. The Forest Service joined other federal agencies in working with Congress to address the problem. In fiscal year 1996, Congress authorized a demonstration program: fees collected from recreational users at a number of sites would be retained by federal agencies for investment at those sites. The fees were to be used to maintain thousands of miles of trails, retrofit hundreds of facilities and sites for accessibility, refurbish hundreds of campsites, upgrade signs and information for visitors, expand office hours for improved visitor services, and more. The Forest Service reinvested at least 90 percent of revenues at the project or site where it was collected. The remainder was distributed to fee projects within the region, at the discretion of the regional forester.

The Forest Service began by testing many kinds of fees to see what the public might best accept. Based on public comments, changes were made to projects. Free days or areas were initiated to accommodate low-income visitors; volunteers were rewarded with free passes; and annual pass systems were designed to reduce per-visit costs. The Forest Service worked to ensure that the program did no damage to relationships with concessionaires and permittees, and that the program was fairly administered to all user groups. However, several groups, including one from Bend, Oregon, formed to protest the imposition of any fees for use of the national forests in the Region. They were unsuccessful in their efforts.

In the 1980s, the Forest Service opened a campsite reservation service because of the growing recreational use of the national forests and grasslands. A national reservation service for campsites across the agency was deemed necessary for the most sought-after camping sites. The national recreation reservation service (NRRS) emerged from a 1995 memorandum of understanding among major federal providers of outdoor recreation opportunities that was designed to provide seamless, one-stop shopping for federal recreation opportunities. Launched in 1999, the NRRS included nationwide recreation facilities managed by the Forest Service, National Park Service, Bureau of Land Management, Bureau of Reclamation, and the U.S. Army Corps of Engineers. Through the NRRS customers benefitted from enhanced recreation information and reservation services through multiple sales channels for more than 2,500 recreation sites and 60,000 facilities. ReserveAmerica of Ballston Spa, New York, was given the contract to develop, implement, and operate the NRRS service. Yet in Region 6 and elsewhere across the country there are many campgrounds that do not fall into the campground reservation system. Many of these are relatively primitive campgrounds, sometimes without water, although most have pit toilets. Rentals of historic cabins and lookout towers/houses in the Region are included in the NRRS program (see below).

The heritage program of the Forest Service includes the unique passport in time (PIT) program. PIT projects in the Forest Service began in the late 1980s when Gordon Peters from the University of Minnesota-Duluth came up with the idea of using students and volunteers on field archaeology projects on the nearby Superior NF in Minnesota.

> In 1989, … we called the program "Passport in Time" In 1990, … the Uinta NF in Utah gave PIT its first national flavor. After that successful

> year, Mike Beckes, then archaeologist for the eastern region of the FS, convinced FS leaders in Washington, D.C., that PIT played a valuable role for heritage and the agency. It officially became a national program in 1991. From its humble beginnings in the Great Lakes states, it has now grown to include 117 national forests in 36 states (Forest Service PIT Web site).

In the Pacific Northwest Region, PIT projects apparently started in 1995 on the Winema NF with a project called "Life at the Hub." Since that time, scores of PIT projects have been conducted on most of the forests in the Region, mostly focused on historic cabins, sites, records, and archaeological surveys and "digs." In 2008, there were seven different projects, two on the Malheur NF, two on the Siuslaw NF, one on the Deschutes NF, one on the Fremont-Winema NF, and one on the Umpqua NF.

One way in which to help with the restoration of historic cabins and lookout buildings has increasingly been to rent the facilities to recreationists. As many of the sites are in remote locations, including mountain tops, it was initially thought that few people would venture into the "wilds" of the Region to enjoy

Fall Mountain lookout rental, Malheur National Forest, Oregon, 2008. USFS photograph.

staying a few nights in a cabin or lookout. In spite of these misgivings, the rental program has been a success. Many old buildings have been restored by money collected from rentals. Currently there are nine rentals in Washington national forests, while there are 56 sites available in Oregon national forests. Reservations for these facilities are made through the National Recreation Reservation Service Web site.

The Region has drawn up implementing regulations for summer homes on the national forests as a result of the Cabin Users Fee Fairness Act of 2000. In part, the regulations concern the appraisal and reappraisals of these cabins, of which there are around 2,800 in Region 6. A Regional information packet in 2008 has a brief description of problems facing the appraisals: "Nine cabins on Lake Wenatchee [Okanogan-Wenatchee NF] are the 'poster children' nationally for 'what could happen to you.' These cabins are scheduled to have fees of $15,000-$18,000/yr. One cabin owner tried selling his cabin over the internet and by auction but had no bids."

The Regional packet also noted that summer cabin owners on the Deschutes NF had been taking water from Crescent Lake without a state water permit. "The local water master says there is no allocation available to them, and the local water district will not sell them any water ... drilling wells would solve the problem but cabin owners do not want to incur the expense." The Deschutes NF cannot legally issue new permits if the cabins are not in compliance with state laws.

A new program began in the mid-2000s called heritage expeditions: short day trips and longer multi-day trips on the national forests designed to highlight recreation and heritage aspects of the national forest. In 2007 only the Willamette NF and the Lolo NF in Montana offered these trips, which are organized at the forest and ranger district levels. On the Willamette NF, there were 10 one-day hikes led by recreation/heritage employees from the Sweet Home RD. A four-day trip was to focus on Judge Waldo and his favorite places in the High Cascades of Oregon which he had visited during the 1890s (see Chapter 2).

ROADLESS AREAS

The 1995 salvage rider controversy highlighted a lingering problem: the lack of special protections for roadless areas on national forest lands. Following passage of the 1964 Wilderness Act, the Forest Service began inventorying roadless areas for possible designation by Congress as wilderness. Many of these areas were chosen for the national wilderness preservation system, including the Opal Creek Wilderness on the Willamette NF, which was established in

1996. (Seidmen 1993) By 1998, remaining inventoried roadless areas covered some 58.5 million acres, or about 31 percent of the national forest system in the U.S.

Many inventoried roadless areas do not meet the strict criteria for wilderness designation, even though they share many wilderness characteristics. In late 1998, recognizing the unique value of roadless areas, the Forest Service began to formalize a vision for their long-term protection. On October 13, 1999, President Clinton formally announced his support when he directed the Forest Service to undertake an open and public process to "provide appropriate long-term protection for most or all of these currently inventoried 'roadless' areas, and to determine whether such protection is warranted for any smaller 'roadless' areas not yet inventoried."

The Forest Service released to the public the Draft Roadless Area Conservation Environmental Impact Statement on May 9, 2000, with the final EIS released on November 13, 2000. One change between the draft and final versions was a ban on new road construction and the inclusion of the Tongass NF in the analysis. The final roadless rule was published in January 2001 and generally limited, with some exceptions, timber harvest, road construction, and road reconstruction within these roadless areas of the national forest system. These nationally applied prohibitions replaced the management prescriptions that had been developed in the land management plans. In Region 6, over 4 million acres of roadless areas were identified, 1,965,000 acres in Oregon national forests and 2,015,000 acres in Washington. The Wenatchee NF had the most identified roadless areas in the Region (579,000 acres), while the Winema NF had the least (32,000 acres).

The roadless rule became the subject of several lawsuits in federal district courts. As part of the legal challenge to the rule by the State of Wyoming, the U.S. District Court for the District of Wyoming issued a permanent injunction and set aside the roadless rule on July 14, 2003, finding that the roadless rule had been promulgated in a manner that was illegal, both procedurally and substantively. The court ruled against the government on five of six claims under NEPA, and also found that the roadless rule violated the Wilderness Act of 1964 because the timber harvest and road construction prohibitions constitute establishment of de facto wilderness (only Congress can designate wilderness areas). Essentially, this meant that the 2001 rule was still in effect.

In July 2004, Agriculture Secretary Ann M. Veneman proposed an even different rule that would establish a process for governors to work with the Forest Service to try to develop statewide support for conserving roadless areas in their states. The Forest Service received approximately 1.8 million comments. On May 5, 2005, Agriculture Secretary Mike Johanns announced the state petitions rule would replace the 2001 roadless rule. Under this 2004 rule, if the secretary, following recommendations from a national advisory committee, accepts a state petition, then the Forest Service would be directed to initiate a state-specific rule making. A few months later, Johanns announced the selection of members to the roadless area conservation national advisory committee.

The governors of Oregon and Washington declined to take part in this 2004 rule process. Essentially their positions were that the 2001 rule decisions should stand. More legal challenges ensued and the 2001 roadless rule was reinstated. Then in mid-August 2008 the U.S. District Court for the District of Wyoming issued a permanent injunction against the reinstated 2001 roadless rule. This effectively did away with both the roadless rules, leaving decisions about roadless areas and wilderness to Congress to resolve.

Wilderness advocates in the Pacific Northwest began pushing Congress in the early 2000s to expand several existing wilderness areas and create new ones. The congressional proposal centered on the Mt. Hood area. Tied to any wilderness additions, there was the need to exchange land in the Cooper Spur Ski Area for other land in the Government Camp area. Haggling in Congress, along with the recalcitrance of a few members over the exchange, especially the value of the land in question, killed the wilderness proposal for several years. Finally, in March of 2009, Congress passed the Omnibus Public Land Management Act. This act designated many wildernesses and special areas across the U.S., including several in Oregon, including 119,590 acres of new or expanded wildernesses, a 34,550-acre Mt. Hood National Recreation Area, as well as several other special management areas on the Mt. Hood NF, several land exchanges also on Mt. Hood, 81 miles of wild and scenic rivers additions, and two national trails (see Appendices F, G, and H).

Wildlife and Fish Management

The Forest Service from its inception in 1905 tended to think of wildlife management as providing for big game, upland and waterfowl bird species, and various salmon, trout, and warm-water species of fish. Traditionally, the agency has managed national forests for wildlife and fish habitat, leaving population management, including reintroductions, up to state and other federal agencies. Occasionally, there were efforts to protect or enhance other species, but this was seldom the main job of the district resource technicians.

That would have to wait until the agency was able to hire specialists in fish and wildlife. It was only in the 1980s and especially the 1990s that these ideas began to change.

At Diamond Lake on the Umpqua NF, tui chub were found again in 1992 (see Chapter 6). Within a few years, an estimated 90 to 100 million tui chub had taken over one of the most productive lakes in the Cascades, almost totally eliminating the rainbow trout population. In September 2006, the Oregon Department of Fish and Wildlife successfully treated Diamond Lake again with rotenone to rid the lake of the invasive chub. Since the poisoning and restocking, not only has the water quality improved but the stocked rainbow trout have flourished.

The Endangered Species Act of 1973 helped to speed the change with the listing of a few key species—such as the bald eagle and peregrine falcon—as well as more controversial species such as the much-debated northern spotted owls and wolves, wolverines, bull trout, and several species of salmon in the Pacific Northwest. Hunted to extinction in the Pacific Northwest by the late 1930s, wolves have been slowly making a return to their traditional areas. In the Umatilla NF in northeastern Oregon state biologists distinctly heard the howls of two adults and two pups in late July 2008. In the same month, wolves were photographed and two were captured in western Okanogan County on the Okanogan and Wenatchee NFs in Washington. The female was found to be a lactating female that obviously was nursing pups. The wolves are back in the Washington Cascades.

Bull trout, for years lumped with dolly varden, are found in many parts of the Region, but are especially prominent in rivers flowing through the Columbia Basin. Bull trout, according to the U.S. Fish & Wildlife Service's species profile Web site (at ecos.fws.gov/speciesProfile) have distinctive stream habitat requirements: "They need cold water to survive....They also require stable stream channels, clean spawning and rearing gravel, complex and diverse cover, and unblocked migratory corridors….Bull trout exhibit two forms: resident and migratory. Resident bull trout spend their entire lives in the same stream/creek. Migratory bull trout move to larger bodies of water to rear young and then migrate back to smaller waters to reproduce. An anadromous form of bull trout also exists in the Coastal-Puget Sound population, which spawns in rivers and streams but rears young in the ocean."

On October 30, 1992, the Fish & Wildlife Service received a petition from three environmental organizations to list the species as endangered throughout its range. After several years of court review, the court issued an order and opinion remanding the Fish & Wildlife Service to further consider placing the bull trout on the ESA list. But it was not until June 10, 1998, that bull trout were listed as a threatened species under the Endangered Species Act. This listing was for those distinct populations found in the Columbia Basin in Washington, Oregon, Montana, and Idaho, as well as in the Klamath River in south-central Oregon. All populations of bull trout in the Region have declined because of stream degradation through land erosion caused by roads and other developments, blockage of rivers and creeks, poor water quality and pollution, fishery management by the states, and the introduction of nonnative species of fish. On November 1, 1999, other populations were included in the listing —Jarbridge River in Idaho and Nevada, St. Mary-Belly River in Montana, and coastal Puget Sound in western Washington. All bull trout in the lower 48 states are now listed as threatened.

However, there have been a few success stories, including one from the Willamette NF, where the "McKenzie River Ranger District replaced a culvert blocking Bull Trout access to a spawning stream ... Also, Bull Trout fry were successfully relocated from the McKenzie River to the Middle Fork Willamette River and Hills Creek, on the Willamette NF east of Oakridge." (Rolf Anderson personal communication)

In January 1999, the Fish & Wildlife Service convened a team to develop a recovery plan for the bull trout. The team consisted of Fish & Wildlife Service personnel and representatives of state fish and wildlife agencies (from Washington, Oregon, Idaho, and Montana) and American Indian tribes (Nez Perce and Upper Columbia River United Tribes includingd the Confederated Tribes of the Colville Reservation, Coeur d'Alene Tribe, Kalispel Tribe, Kootenai Tribe of Idaho, and Spokane Tribe). By late 2002, the draft recovery plans were completed for the varied populations of bull trout throughout the Pacific Northwest.

As part of a 2003 federal court settlement, the Fish & Wildlife Service was ordered to develop a plan designating critical habitat in the Pacific Northwest for bull trout and this was completed in 2005. Bull trout recovery areas include all the national forests in Washington and all in Oregon except for the Siuslaw, Umpqua, and Rogue River-Siskiyou. The Fish & Wildlife Service in April 2004 initiated a review on the status of bull trout and completed it in April 2008. The review made two recommendations—retain threatened status for the species and evaluate if genetically distinct populations of bull trout exist, that would merit separate protection under the act.

SALMON AND STEELHEAD LISTING

The National Marine Fisheries Service (NMFS) is responsible for conducting Endangered Species Act status reviews and making listing determinations for anadromous fishes on the West Coast, including Pacific salmon and steelhead. Pacific salmon and steelhead are fishes with a life-cycle that spans a variety of fresh and saltwater habitats. Salmon are born in creeks and rivers, then migrate downstream to ocean waters, where they grow and mature. The salmon return to freshwater, to the very stream where they were born, to reproduce then die. The eggs laid by the salmon will begin the cycle again.

Most salmon stocks throughout the Northwest are at a fraction of their historic levels. Historically, overfishing by trawlers, seining, and fish wheels in the Columbia had been major causes of decline. More recently, the major causes of salmon depletion have included unfavorable ocean conditions over the past two decades, loss of freshwater and estuary habitats, farm and city water pollution, and devastating floods. In addition many other factors have contributed to salmon declines, including Mount St. Helens ash that clogged the Toutle, lower Cowlitz, and lower Columbia Rivers; off-shore fishing pressures; sport fishing on the rivers and streams; American Indian rights granted under treaty and by the Boldt decision in 1974; and hatchery-raised salmon competing for ocean food and diluting the genetic diversity of the species.

In October of 1993, environmentalists filed a petition with the NMFS for protection of Oregon coastal coho under the ESA. Almost two years later the NMFS proposed placing the fish on the endangered species list, but on April 25, 1997, the agency decided not to list the species, citing the Oregon Salmon Plan (see below) as the primary reason. This decision was followed by court action to compel a listing to gain ESA protection for the salmon. On June 1, 1998, a federal district court in Portland ruled that NMFS's failure to list Oregon coastal coho was arbitrary and capricious and violated the law, and as a result, the agency listed the fish in August. However, both NMFS and the state of Oregon filed an appeal of the district court decision with the U.S. Court of Appeals for the 9th Circuit. As a result, the NMFS in March of 1999 listed seven additional salmon populations as endangered or threatened under the act.

In June 2005 the lower Columbia coho was added as a threatened species and in May 2007 the Puget Sound steelhead was also listed as threatened. By 2008, the following spring, Chinook salmon species in several rivers had been named on the ESA list: Upper Columbia River Chinook were listed as endangered, while the Chinook salmon in Puget Sound, lower Columbia River, and upper Willamette River were listed as threatened. Chum salmon were listed as threatened in the Columbia River and Hood Canal summer run. Coho salmon were listed as threatened in southern Oregon and northern California rivers, lower Columbia River, and the Oregon coast rivers. Coho were also listed as a species of special concern in the Puget Sound and Strait of Georgia.

In 2007, the State of Oregon released the Oregon Coast Coho Conservation Plan. The planning process had started in 2004 with a collaborative agreement with the NMFS to address the conservation and rebuilding of the coho runs in rivers and streams that flow into the Pacific Ocean from the Columbia River south to the California border. Many of these rivers have their headwaters on the Siuslaw and Rogue River-Siskiyou NFs, as well as the Tillamook State Forest. Cooperation between the state, federal agencies, and private landowners has been key to program implementation. Technical information was provided by several federal agencies, including the Forest Service. The plan remains in operation today. (See www.oregon-plan.org/OPSW/cohoproject for a description of the plan.)

Basically, the salmon listings affect the national forests in the Region that have streams flowing into the Pacific. Special procedures have been put into effect by the Northwest Forest Plan, as well as ICBEMP, to further limit stream degradation, protect stream banks, restore woody debris and fallen trees in streams, and remove obstacles to fish migration. In some cases, wooden boxes with fertilized salmon eggs have been placed in creeks to enhance the survivability of young salmon stocks. On the eastside forests and BLM districts, special administrative reviews or "screens" and an "aquatic conservation strategy" were put in place to better manage the salmon, but in 2007, the aquatic conservation strategy EIS was blocked by the courts, thus reverting management to operate under the earlier screens.

Salmon eggs in water, Pacific Northwest, 1980. USFS photograph.

Steelhead are a distinct variety of large rainbow trout that are born in the freshwater streams of the Pacific Northwest, migrate downstream to the Pacific Ocean, then return after several years to the freshwater to spawn. Unlike salmon, steelhead return to the ocean after spawning to complete another life-cycle. The NMFS listed steelhead as endangered in 1997 and they were thereby afforded some protections under the ESA, but legal issues prevented further efforts at listing or recovery. In September 2001 a district court ordered the NMFS to update all of its salmon and steelhead listings.

NMFS announced its final listing decisions for 10 populations of West Coast steelhead on December 23, 2005, part of the agency's effort to comply with the order. On January 5, 2006, NMFS changed the listing status of upper Columbia River steelhead to threatened, and applied the protective regulations for threatened West Coast salmon and steelhead to this population. By 2008, the following populations of steelhead had been placed on the ESA list: upper, middle, and lower Columbia River, upper Willamette River, and Puget Sound populations listed as threatened, and Oregon coast rivers population listed as species of concern.

Robert McClure wrote about the steelhead listing in the Cedar River municipal watershed in the Mt. Baker-Snoqualmie NF:

> One of the Puget Sound area's most battered runs of steelhead spawns [is] in the Cedar River, a source of Seattle's drinking water. Although some think the additional protections ... could spell trouble for that drinking-water supply, city officials say they could help the steelhead without reducing Seattleites' water supplies. The National Marine Fisheries Service's proposal to protect the embattled fish came in response to a petition by Sam Wright, a retired state fishery manager. Steelhead stocks in this region "are all on downward trends, and in most of the rivers, the numbers are getting so small you have to question whether the population is even viable," Wright said. "In the Lake Washington system, they're functionally extinct." Fifty or fewer adult steelhead have returned to the Lake Washington/Cedar River ecosystem in the past four years (*Seattle Post-Intelligencer*, March 30, 2006).

WILDLIFE AND FISH PROGRAMS

Several new wildlife and fish programs were developed in the Washington Office including the "join us" program—an initiative to strengthen public-private partnerships in wildlife and fisheries management; "taking wing"—a waterfowl and wetlands initiative to enhance habitat and support the North American waterfowl plan; and the very successful "animal inn" program that was designed to communicate the importance of managing dead standing and fallen trees for wildlife habitat. Other wildlife initiatives included "partners in flight" (involving neotropical migrant birds), "get wild!" (cooperation to manage and improve healthy ecosystems and high-quality wildlife habitat for terrestrial and semi-aquatic wildlife), "answer the call" (providing food, cover, and water supplies for quail), and many others.

For fish, a program called "fish watch" (an education program designed to elevate the importance of fish and clean water) was started by presidential executive order in 1995. The highly successful "rise to the future!" was another program designed to enhance the production of fish on the national forests through partnerships. Related special programs included "bring back the natives" (conservation of native fish species) and "every species counts" (manage, protect, and conserve plant communities and aquatic species). Other fishing efforts included "pathway to fishing," "adopt a watershed," and "future fisherman's foundation," all led by key fish and wildlife organizations and federal agencies.

Engineering

Authorization for forest road funding declined nationally from $600 million in 1980 to less than $200 million in 2000. The road system was designed primarily for timber hauling, but many roads were no longer needed for that purpose. As a result, the Forest Service could no longer afford the upkeep of its vast road system. Deteriorating forest roads were causing landslides, soil erosion, and stream siltation, destroying habitat for sensitive species and reducing safe public access. The agency began the process of revising its road management rule and policy. The intent was to ensure that new and existing roads were essential for resource management and use; that construction, reconstruction, and maintenance of roads minimized adverse environmental impacts; and that unneeded roads were decommissioned and natural processes restored.

In Region 6, during 1991, there were more than 70,100 miles of roads that could accommodate passenger cars and high-clearance vehicles. In that year, 358 miles of new roads and 786 miles of reconstructed roads were built, as well as two new bridges, and six bridges were reconstructed. But during the same year, the Region began closing some unneeded roads; 19,400 miles of roads were closed and 1,270 miles of existing roads were obliterated. An article by Bill Monroe noted:

> State wildlife agencies in Oregon, Washington and Idaho have voiced concern about the Forest Service's aggressive road-building programs for more than a decade, often voicing frustration about their inability to control access to important winter range and the protection once afforded by the then wilderness that existed before roads became cobwebbing the forests to get out the harvest ... Phil Hirl, assistant director of engineering for Region 6, said more than 15,000 of the 68,000 miles of Forest Service roads in Oregon have been permanently barricaded and an additional 900 miles have been eliminated—either torn up or planted with new trees (*The Oregonian*, December 19, 1991).

In February 1999, the Forest Service announced an interim rule that temporarily suspended road construction and reconstruction in certain unroaded areas on national forests and grasslands. The interim rule gave the agency 18 months to draft a new road policy and develop new analytical tools, and on March 2, 2000, the Forest Service outlined a proposed road management policy. The new policy was designed to help the agency prioritize its road maintenance and reconstruction work so the national forest road system would be more affordable to manage in the future. All roads were to meet new standards designed to comply with resource objectives and sustainability at likely funding levels. Unneeded roads were to be identified for decommissioning, starting with those that posed the greatest risk to public safety or environmental health.

In recent years, off-road vehicle use has become a huge management concern in the Region. The increasing number of off-road all-terrain-vehicles (ATVs) on existing roads has not, for the most part, caused any significant resource damage. However in some areas a few people with large four-wheel drive pickups and ATVs have deeply rutted roads as well as non-roaded forest areas in critical-soil areas, destroyed plants, and harassed animals, while some using snowmobiles have illegally entered wilderness or non-motorized areas. Even the coastal areas on the Oregon Dunes National Recreation Area have not been immune from sand or dune buggies causing resource damage and disturbing birds (such as the endangered snowy plover). As a result, many of the national forests in the Region have designed off-road vehicle plans that generally restrict these popular vehicle types to designated corridors or special management areas.

SCENIC BYWAYS

Congress established the scenic byways program under the Intermodal Surface Transportation Efficiency Act (ISTEA) of 1991, and strengthened it further in 1998 and 2005. National Scenic Byways designations recognize those roads across the country that exhibit one or more of the six categories—scenic, natural, historic, recreational, archaeological, or cultural—that contribute to a unique travel experience. As of 2006, 126 roads in 44 states had been designated as either National Scenic Byways or All-American Roads. The ISTEA (referred to as the "Ice Tea") act contained two items that helped the scenic byway program; the act established a federal national scenic byways system, while encouraging each state to do the same, and required the states to spend 10 percent of their federal highway allocations for roadside enhancements, including bike paths.

This was not a requirement for Forest Service road and highway program but, according to Phil Hirl (personal communication), it "sounded like a great idea to me, so I went to Jim Hall of FHWA [Federal Highway Administration] project office in Vancouver and suggested we do the same 10% with Forest Highway Funds … Then we had to convince the State and County people. This was $1.5 million that could improve highways." The various scenic byways that benefitted from these funds were the historic Columbia River

In **Oregon** 10 scenic byways have been established since 1991:
— Cascade Lakes near the eastern crest of the Cascade Range on the Deschutes NF;
— Hells Canyon along the upper rim of the Snake River on the Wallowa-Whitman NF;
— Historic Columbia River Highway above the Columbia River in the national scenic area on the Mt. Hood NF;
— McKenzie Pass-Santiam Pass on the Willamette NF and Deschutes NF;
— Outback on the edge of the Deschutes NF and Fremont-Winema NF;
— Over the River and Through the Woods on the Willamette NF;
— Pacific Coast on the Siuslaw NF and Rogue River-Siskiyou NF;
— Rogue-Umpqua on the Rogue River-Siskiyou NF and Umpqua NF;
— Volcanic Legacy on the Deschutes and Fremont-Winema NFs; and
— West Cascades on the Willamette NF and Mt. Hood NF.

In **Washington**, there are six scenic byways:
— Chinook on the Mt. Baker-Snoqualmie NF;
— Coulee Corridor (not near any forest);
— International Selkirk on the Colville NF and Panhandle NF;
— Mountains to Sound on the Mt. Baker-Snoqualmie NF;
— Stevens Pass on the Okanogan-Wenatchee and Mt. Baker-Snoqualmie NFs;
— and the Strait of Juan de Fuca Highway near the northern edge of the Olympic NF.

Highway, McKenzie-Santiam Loop, and Rogue-Umpqua scenic byway. Phil Hirl continued: "I was on the Oregon Scenic Byway committee when I worked [before retirement]. We sat [wrote] criteria, got nominations for routes from Forest and local agencies. They made plans to visit nominated routes. Then I retired … I was asked to come back and represent the FS though hired by ODOT [Oregon Department of Transportation]. We drove all routes and rated … Almost all of the NF Scenic Byways are now state Scenic Byways."

Watershed Management

In 1991, 211 cities and communities in the Region were totally or partially dependent on water from the national forest watershed. It was estimated that 2.6 million people in the Region were served by this national forest water. That same year, 2,853 acres were treated to improve watershed conditions.

In 1999, the Forest Service launched a series of collaborative large-scale watershed restoration projects. Around the country, 15 large watersheds in 23 states that provide water for millions of people and habitat for numerous sensitive and threatened species were chosen to become national prototypes. In FY2000, the agency invested 24 million in 12 of these watershed restoration projects across the country. Federal, state, tribal, and private partners put up about $22 million in matching funds. The projects ranged from the three-million-acre Blue Mountains Demonstration Area in northeastern Oregon to the multi-state "Chesapeake Bay watershed partnership" in the mid-Atlantic region.

The Blue Mountains Demonstration Area incorporated portions of eight ranger districts on the Malheur, Umatilla, and Wallowa-Whitman NFs, and BLM's Baker Resource Area, as well as state, private, and tribal lands. It is a mixture of many ownerships covering around 2.6 million acres within the watershed of the Middle Fork of the John Day River, Desolation Creek, and Grande Ronde rivers. On June 30, 1999, Governor John Kitzhaber, Forest Service Chief Mike Dombeck, and Under Secretary Jim Lyons announced the establishment of the "Blue Mountains Demonstration Area (BMDA). The Blue Mountains were selected as a Demonstration Area because its watersheds are considered high priorities for restoration, its communities are considered to be at risk economically and socially, and because of the demonstrated support for restoration that exists. Specifically, several ecological assessments, including those conducted by the Governor and the Interior Columbia Basin Ecosystem Management Project (ICBEMP), identified the Blue Mountains as an area of poor aquatic and forest health where wildlife, fish, water quality, recreation and forest resources are at risk from wildfires, forest insect and diseases, noxious weeds and roads." (BMDA Annual Report 2002:1) From 1999 until 2002, more than $6.5 million in contracts were awarded. Most were to contractors living in the demonstration area.

AMERICAN HERITAGE RIVERS

In his 1997 State of the Union address, President Clinton announced the American heritage rivers initiative (AHR) to provide special recognition to outstanding stretches of America's rivers. This initiative was an innovative response to communities seeking federal assistance in revitalizing their economies, protecting natural resources, and preserving the history and culture of their rivers. The Environmental Protection Agency (EPA) administers the program. The president chose to designate 14 heritage rivers, one of which was the Willamette River in western Oregon.

The heritage river designation for the Willamette River included a nearly 200 mile reach from Eugene north to Portland. This stretch of the river flows through the three largest cities in the state: Eugene/Springfield, Salem, and Portland. While this portion of the Willamette River does not cross any national forest, activities including timber harvesting and road construction on national forest lands in the basin affect water quality. The upper Willamette, Coast Fork Willamette, McKenzie, Santiam, and Clackamas rivers flow into this heritage river segment. The tributary rivers and creek drain national forest and BLM lands in the Cascades and the Coast Range.

Increases in industrial use, sewage disposal, and population during the early and mid-twentieth century resulted in serious pollution of the river. Major efforts at the state level have begun to reduce municipal (including sewage and street runoffs), farm, and industrial sources of pollution in the last 50 years. For the most part, the efforts have been successful in transforming the river from a sewage- and industrial waste-laden river to a relatively clean showcase. The heritage rivers Web site (www.epa.gov/rivers/98rivers/willamette.html) describes activities by the governor of the state to help the river:

> The Willamette River Basin Task Force, appointed by Governor John Kitzhaber in 1996, developed both short-term and long-term action plans for the river. As a result of this work, Governor Kitzhaber created the Willamette Restoration Initiative (WRI), a public/private partnership to promote, integrate and coordinate efforts to protect and restore the health of the Willamette watershed. Composed of key stakeholders throughout the basin, the WRI coordinates the AHR program. Designation as a heritage river will help carry out an integrated vision for the Willamette.

Fire Management

Wildland fire management changed in the 1990s as several issues caused concern: there were fewer employees, as well as logging and road crews, who could be called upon to fight fires; the impact of Colorado's fatal 1994 South Canyon Fire; the rising number of large fires (100,000 acres burned or more); and the growing number of homes built in fire-prone rural areas, the so-called wildland-urban interface. Wendall Jones, retired director of timber management in the Region, reported that there were also many changes in the firefighting business during the last two decades. "The Northwest Fire Coordination center at the Portland Airport was fully developed in this period. There were changes in the smokejumper business—types of planes used, better equipment. Communication on fires was vastly improved with current technologies. [The Region] Really got full use of women in fire positions in this era." As many of the fire crews derived from forest and ranger district employees, the downturn in logging and road building in Region 6 has resulted in shortages of firefighters. Jones continued, "[The] downturn in [the] timber business resulted in fewer organized crews … who all had BD [brush disposal] crews, KV [K-V tree planting] crews, and foresters and forestry techs working on timber sales made up a big part of sector boss, division boss, and other key positions on major fires. Result—turned to organized contract crews made up to a great extent by agricultural workers." (Jones personal communication)

The South Canyon Fire near Glenwood Springs in Colorado will long be remembered, next to such firefighting calamities as the 1910 Big Blowup and the 1949 Mann Gulch Fire, as a pivotal event in the annals of wildland firefighting. The fire on BLM-managed land at the base of Storm King Mountain burned some 2,200 acres in a huge blow-up on July 6th, 1994. Forty firefighters had to deploy their fire shelters, but 14 died. Nine of the fatalities were from the Prineville hot shot crew on the Ochoco NF stationed at Redmond, Oregon. The tragedy riveted the attention of firefighters and the general public nationwide.

An interagency team was formed to investigate the fatalities and what had gone wrong. The subsequent report set in motion a series of reviews and an interagency effort to effect fundamental change, culminating in the 1995 federal wildland fire management policy and program review. Fire scientists assessed what occurred and suggested guidelines that may help firefighters avert such a tragedy in the future (Butler et al. 1998).

Beginning in the 1930s, increasing firefighting effectiveness had excluded all fire from the forests, even surface fires. Small trees and brush, no longer kept down by fire, had built up in lower-elevation western forests. Dense coniferous thickets commonly added 200 to 2,000 small trees per acre in old-growth stands and 2,000 to 10,000 per acre where the forest canopy had been removed through timber harvest. These dense fuels could make the fires so severe that they destroyed entire forest stands. The trend was clear: the fires that had been postponed for 70 years through a policy of systematic fire exclusion would no longer wait. Large fires were returning to the West.

The heightened fire risk was exacerbated by the growing wildland-urban interface. A rising population density in many suburban and rural areas placed people and communities in forest ecosystems naturally prone to fire, increasing the threat to life and property. Areas surrounding the metropolitan areas in the Pacific Northwest, as well as fast-growing areas like the Bend-Redmond-Sisters area of central Oregon, face increasing threats from fire next to housing developments. Wildland-urban interface fires, such as the 1991 Oakland Hills Fire in the San Francisco Bay Area, illustrated the destructive power of fire in an urban environment. After decades of efforts directed at extinguishing every fire that burned on public lands, research has shown that our best intentions to stop all fires have disrupted the fire regimes that existed for thousands of years. Moreover, as more and more communities develop and grow in areas that are adjacent to fire-prone lands in the wildland-urban interface, wildland fires pose increasing threats to people and their property.

Following a landmark wildland fire season in 2000, on September 8 of that year, Secretary of Agriculture Glickman and Secretary of the Interior Babbitt delivered a national plan to the president outlining steps that could be taken to better manage fire for the health of communities and the environment. The report, *Managing the Impact of Wildfires on Communities and the Environment*

President George W. Bush with Leslie Weldon at B&B Fire briefing in Redmond, Deschutes National Forest, Oregon, August 23, 2003. USFS photograph.

(referred to as the National Fire Plan 2000) called for increasing national firefighting capabilities; rehabilitating and restoring lands and communities affected by fire; using techniques such as prescribed fire, a low-level fire set and managed by land managers, to reduce hazardous fuel levels; planning for fire preparedness; and increasing cooperative programs in support of local communities. In addition, the plan was to secure funding to replenish and enhance the fire-suppression funding accounts, which had been depleted by the year's extraordinary fire costs, and to repay FY 2000 emergency transfers from other appropriations accounts.

Congress provided the Forest Service with funding to hire 3,500 firefighters in 2001 to increase firefighting capability. That same year saw a major fire on the Okanogan and Wenatchee NFs. The Thirtymile Fire burned in the Chewuch River drainage north of Winthrop, Washington. Four young firefighters lost their lives while fighting the fire. Following this tragedy, the Forest Service implemented a series of policies to improve firefighter safety. On April 24, 2003, the revised firefighting safety policies were incorporated into the Forest Service directives system.

One of the most difficult issues with implementing the fire plan has been the need to reduce fuel buildup in the national forests and BLM lands. Many times the greatest need is immediately adjacent to communities and homes next to the federal forest areas. Reduction of the fuels, either through prescribed burning or by mechanical means, tends to aggravate people who have specifically built homes and summer cabins to be next to the trees. In addition, the use of prescribed fires results in smoke and the possibility of escaped fires.

Another especially dangerous fire season in 2002 resulted from several years of severe drought in the Rocky Mountains, with fires starting in June and July. Destructive fires in Arizona, Colorado, and New Mexico set the stage for one of the most disastrous fire season in decades. Then on July 13, a fire started in southwest Oregon that soon became Oregon's largest fire in recorded history. It was also the most expensive fire to fight. The Biscuit Fire blackened 499,965 acres and encompassed most of the Kalmiopsis Wilderness and a good portion of the earlier Silver Fire of 1987. Yet the whole area was not intensively burned, as the fire burned in a mosaic pattern—approximately 20 percent of the area was only burned lightly, with less than 25 percent of the vegetation killed. Another 50 percent of the area burned very hot, with more than 75 percent of the vegetation killed. Many acres of critical habitat for wildlife and plant species burned, and the late seral and old-growth stands that remain are very precious.

In the 2002 fire season, the Forest Service spent around $500 million on the four largest fires of the season. In all jurisdictions, the 2002 fires claimed the lives of 23 firefighters, destroyed more than 800 homes, scorched about 7 million acres, and cost $1.5 billion. Much of that money went to the fire aviation budget for aerial firefighting with the use of air tankers and helicopters.

There were 13 aviation accidents fighting wildfires, resulting in six fatalities. The Forest Service grounded some of its large air tankers after several air accidents where the wings on the tankers were ripped off during fire operations. But this decision faced opposition from the western states, which were faced with another bad fire season without aerial protection. Several of the larger tankers were approved for use in the 2003 fire season, but fire crews had to rely more on smaller air tankers, helicopters, and ground crews. In May 2004, the Forest Service and BLM terminated contracts for 13 heavy air tankers as a result of the National Transportation Safety Board's concerns about the airworthiness of the aircraft. In July 2004, the agencies determined the airworthiness of eight P3 aircraft and returned these planes to service.

HEALTHY FORESTS INITIATIVE

President George W. Bush visited the Biscuit Fire site in 2002 and announced a program to remove brush and ladder fuels from the national forests, as well as to restore and salvage log burned-over areas. The Healthy Forests Initiative (HFI) that was launched in August was designed to reduce the risks that severe wildfires pose to people, communities, and the environment. The HFI has been used to expedite timber sales on federal lands, which has led to much opposition from the environmental community. Supporters, including many western legislators, communities, and timber industry interests, believed that the expedited process would reduce the fire risk through careful logging and brush removal operations. But opponents, including many environmentalists and wildlife biologists, view the HFI as simply an attempt to use the fire issue to increase logging, much as the Timber Salvage Act of 1985 did (Vaughn and Courtner 2005).

A year later, the HFI program was codified into law. President Bush, at the December 3, 2003, signing ceremony for the Healthy Forests Restoration Act of 2003, said,

> Today, about 190 million acres of forest and woodlands around the country are vulnerable to destruction. Overgrown brush and trees can serve as kindling, turning small fires into large, raging blazes that burn with such intensity that the trees literally explode ... The bill expedites

> the environmental review process so we can move forward more quickly on projects that restore forests to good health. We don't want our intentions bogged down by regulations … The new law directs courts to consider the long-term risks that could result if thinning projects are delayed … It places reasonable time limits on litigation after the public has had an opportunity to comment and a decision has been made.

Mineral Management

In the twentieth century, hazardous wastes or sources of hazardous substances accumulated at thousands of sites on the national forests and grasslands. In the 1990s, through its environmental compliance and protection program, the Forest Service launched cleanup and restoration projects pursuant to the Comprehensive, Environmental Response, Compensation, and Liability Act (CERCLA) of 1980 and the Resource Conservation and Recovery Act (RCRA) of 1976. The agency cleaned up most sites contaminated by its own past actions and established cleanup policies. Most cleanup sites were abandoned or inactive mines. A 1998 inventory identified about 39,000 abandoned or inactive mines on national forest lands. Some 5,000 mine sites required cleanup, with about 1,700 sites qualifying for action under CERCLA. In addition, the national forests and grasslands harbored numerous landfills, dumps, and illegal drug laboratories, adding to the cleanup burden. Since 1995, Forest Service actions resulted in more than $200 million worth of cleanup work performed or paid for by the responsible parties.

One such cleanup was the Holden Mine on the Okanogan-Wenatchee NF. Howe Sound Company's Holden Mine was one of the largest copper mines in the United States during its productive life from 1938 to 1957. The mine abruptly closed in 1957 when profits declined. During the life of the mine, about 8.5 million tons of mill tailings had been placed on the national forest, covering approximately 90 acres, and another 1.5 million tons of tailings had been put back into the mine. The total area of disturbance is approximately 120 acres, and contamination extends downstream to the mouth of Railroad Creek where it enters Lake Chelan.

After Howe Sound closed the mining operation, the mining interests, including the Holden townsite, were deeded to the Lutheran Bible Institute. Holden Village was incorporated in 1962 for the purposes of managing the Holden site as a nonprofit and interdenominational retreat center. Holden

Village is the largest retreat center in the nation operating under a special use permit from the Forest Service.

The Forest Service undertook a major project from 1989 to 1991 to stabilize the movement of contaminants found in the mine tailings, reduce erosion into Railroad Creek, end wind-blown dust, and to reestablish vegetation over the tailing piles. During 1996 and 1997, federal and state agencies, lead by the Forest Service, conducted discussions with Intalco (successor to Howe Sound Company) about the Holden Mine site. On April 11, 1998, the agencies reached agreement with Intalco, directing the company to perform a detailed cleanup study of the site under the direction of the Forest Service. Flooding in north-central Washington in October 2003 caused damage at the site that warranted immediate action to avoid further damage to the environment. The Forest Service approved the operation that began on November 14, 2003, and was successfully completed within 10 days.

Law Enforcement

In early 1991, because of the effort to get taxol to the market, bark from Pacific yew trees became increasingly valuable. As a result, trees were stripped illegally of their bark. For example, on May 14, 1991, 56 trees were found on the Willamette NF that had been stripped of their bark. The total wet bark was estimated at more than 700 pounds. Once Hauser Northwest and parent company Bristol-Meyers Squibb pulled out of agreements with the BLM and the Forest Service in Region 6, the problem went away.

Timber theft was a continuing problem in the Region during the 1990s. Congress established the Timber Theft Investigations Branch (TTIB) in 1991 as an independent unit after concerns had been voiced about the methods and oversight of timber theft investigations. The TTIB was given a three-year charter, which was extended. But the TTIB still experienced difficulties doing their assigned work because of what they saw as interference by Forest Service managers. A Regional Office news release dated April 6, 1995, announced that the TTIB would be moved from being an independent unit to be part of the Forest Service law enforcement organization. In early 1995, the USDA Office of Inspector General report found the charges of management interference unwarranted and unproven.

Wendall Jones reported that two timber sale default investigators—Jerry Hofer from the Regional Office timber management and Chris Anderson from the Olympic NF—were given special honors by the Department of

USFS law enforcement K-9 Unit with Andy Cariell and Laurence Olson at the Multnomah Falls Lodge, Mt. Hood National Forest, Oregon, 2006. Author's collection.

Justice "for being a major factor in putting together the case that resulted in guilty decisions." (Jones personal communication)

Violence touched the Region in 1996. It began on October 28, when arsonists torched a Forest Service pickup truck in Detroit on the Willamette NF. They also attempted to burn down the Detroit Ranger Station, but the plastic jug filled with gasoline failed to ignite. Spray-painted on the building was a tag previously unseen: "ELF," which stood for the Earth Liberation Front. Two days later arsonists struck the Oakridge Ranger Station. The building was burned from the four corners into the middle. It appeared that a professional arsonist was at work. The Willamette NF and local communities were shocked, to say the least. It took years to find the culprits, but by 2008 they were all serving time in prison.

As an act of civil disobedience, environmental activist Tre Arrow (his given name was Michael J. Scarpitti) in July 2000 scaled the front of the Regional Office where he spent 11 days perched on a nine-inch window ledge to protest the logging of the Eagle Creek timber sale on the Mt. Hood NF. Later that year, while protesting another timber sale in the Oregon Coast Range, he fell while he was sitting in a tree. Subsequently he was implicated in several arson attacks on Forest Service, Oregon Department of Forestry, and private property. He was indicted on four felony counts in 2002, then fled to Canada, where he unsuccessfully sought political asylum; he pleaded guilty to two arson charges in August 2008 and was sentenced to 78 months in prison. He moved to a halfway house in Portland in June 2009.

Hadley Hawkins on the Bend-Fort Rock RD, Deschutes NF, was named the sole winner of the Forest Service's nationwide Law Enforcement Officer of the Year in 2008. Kate Ramsayer described an interview with Hawkins about

Protester Tre Arrow sitting on Regional Office ledge, Portland, Oregon, 2000. USFS photograph.

an incident when, despite his training, he knew when he needed backup: "Hawkins recalled a [rave] party near the Deschutes River, where he stopped counting at 60 cars on the road leading up to the scene. At that point, he said, he backed up, called in fellow officers to help patrol outside of the area and stopped partygoers for drunken driving or other violations (*The Bulletin*, August 9, 2008).

On September 20, 2008, the murder of a Forest Service law enforcement officer occurred at the Dungeness Forks Campground on the Olympic NF south of Sequim. Kristine Fairbanks, who had worked for the Forest Service as law enforcement officer for more than two decades, was shot by Shawn Matthew Roe. Fairbanks and her K-9 German shepherd companion "Radar" patrolled the public lands on the northern Olympic Peninsula. The killer also was responsible for killing Richard Ziegler, a 59-year-old retiree, the same day, then sheriff's deputies killed Roe in a confrontation at a convenience store later in the day. The memorial service at Port Angeles was attended by more than 3,000 people, including Governor Chris Gregoire, Chief Gail Kimbell, new Regional Forester Mary Wagner, and former Regional Forester Linda Goodman.

Research Work on the National Forests

Research took on a new and demanding role in the 1990s. The Forest Service began to understand that decades of forest research was now coming into the policy arena. Forest managers, the public, and politicians were looking for scientifically credible answers to how to manage old-growth habitat. New terms, such as new forestry and new perspectives, were applied to these complicated issues. But in reality, research had few answers that would satisfy everyone.

Since 1991, the J. Herbert Stone Nursery has become a nationally recognized leader in the production of native plants for public lands. The nursery began a native grass program with 15 species, and today the nursery produces over 12 tons of seed a year representing over 40 grass species and dozens of western forbs, including wetland plant species. Many of these native seeds are used in fire restoration work around the Region and on other public lands.

H. J. Andrews Experimental Forest's 200-year study of down logs, Willamette National Forest, 1990. Author's collection.

Wind River canopy crane for studying the old-growth forest, Gifford Pinchot National Forest, Washington, c. 2000. USFS photograph.

Increasing money was spent on basic research in the Region. A good example was the start-up of the Wind River Canopy Crane on the T. T. Munger Research Natural Area in the Gifford Pinchot NF in 1994. The unique 285-foot crane stands over old-growth Douglas-fir and offers many opportunities for researching the top of the tree canopy. The crane can lift an eight-person gondola 244 feet into the forest heights and extend the horizontal reach to a 125-foot radius from the crane tower.

NEW FORESTRY AND NEW PERSPECTIVES

New concepts for land management/environmental planning came to the forefront of discussion in the 1990s: new forestry and new perspectives. Jerry Franklin, formerly with the Forest Service, and Chris Maser, formerly with the Bureau of Land Management, are often considered gurus of a "new forestry" approach to forest management. The findings of decades of important scientific forest research provided much-needed clues to the long-term health and productivity of the coniferous forests of the Northwest. They led the Forest Service into new forestry as a search for alternative ways to manage the federal forests. Interestingly, this push for change has come from inside the agency, rather than from external pressure from interest groups or Congress.

Although some felt that new forestry was nothing more than "old forestry" as it should have been applied, most believe that it includes many elements that were not in the mainstream of forest management techniques or philosophy. A revolution in management practices was taking place based on research from within the agency. Timber industry representatives who were shown the Ennis timber sale on the Willamette NF in 1990 as an example of new forestry management were puzzled, as they observed a number of dead and live trees left standing with the tops blown off (for cavity nesters), down logs that had been left to deteriorate on the ground (for nutrient replenishment), and that only light fire had been used to clear the site (so as to not damage the soils). This was a different kind of management than their forestry school training and long-time experiences of clearcutting, broadcast burning the site, then replanting with genetically superior trees.

In the summer of 1990, the Forest Service embraced a concept called "new perspectives" as a top-down idea that would be complementary to the new forestry bottom-up idea. New perspectives was an attempt to upgrade the 1960s concept of multiple-use for the 1990s. New perspectives quickly became a new goal for the national forests that was more philosophical and addressed the larger societal questions and values surrounding the management of the national forests.

New forestry practices, Ennis Creek Timber Sale area, Willamette National Forest, Oregon, 1990. Author's collection.

It was hoped that new forestry and new perspectives would be able to effectively address the NFMA planning requirements for biological diversity, but more traditional methods were employed in the forest plans in the 1990s including the FEMAT and ICBEMP projects (see above).

New forestry and new perspectives were replaced in 1992 with the more comprehensive term "ecosystem management." The Forest Service, Bureau of Land Management, Fish & Wildlife Service, National Park Service, and other land management and regulatory federal agencies embraced ecosystem management as a new paradigm for management in the 1990s and into the next century. Ecosystem management has been closely associated with issues related to roadless areas, old-growth forests, and spotted owls.

Epilogue

The Future of the U.S. Forest Service in the Next 100 Years

Over the past 100 years, the U.S. Forest Service in the Pacific Northwest has experienced a series of daunting changes, legal challenges, shifting societal values, as well as a seemingly endless number of new laws. Some of these came from the agency itself, while others were from external sources. The conservation movement of the early twentieth century created the national forests and instituted scientific management of the national forests. The environmental movement of the late twentieth century changed society by becoming more environmentally aware, as well as support laws that have mandated change in the management of the national forests.

Most of management by the Forest Service for resources other than "timber wars" and "wilderness battles" has remained unseen and unheralded by the forest users and the press. This type of everyday management does not make headlines, but it is the basis for this history, which, sometimes steeped in controversy, serves as the backdrop for another hundred years of management of the national forests in the Pacific Northwest.

Changing Priorities During the First 100 Years

At the turn of the twentieth century, the Forest Service's emphasis was on organizing a work force to manage millions of acres of new national forests. Often termed "white-hat" or custodial forestry, the management of the national forests in the Region was most concerned about fire protection and grazing. By the 1930s, management's priority was changing to economic and social relief of workers hit hard by the Great Depression. The use of the CCC to build a magnificent recreational base was a once-in-a-lifetime opportunity for the agency. World War II saw another change of priorities: to protect the forests from fire and produce increasing amounts of wood products.

After the end of the war, timber management and increasing timber access and supplies became the most prominent management issue—the agency moved into the "hard hat" or industrial forestry era. Employment was up. "Get out the cut" became the primary management goal, while employees were imbued with the sense that anything was possible with enough money and people. Monetary awards and grade increases were given for meeting and exceeding management-set timber targets and related management objectives.

By the 1970s there was growing national concern about the environment, especially wilderness protection. After the timber boom of the 1970s and 1980s came the timber bust caused, in part, by the spotted owl dilemma. Employment in the timber-dependent communities and the Forest Service fell in the 1990s because of decreased timber cutting and road building. Law enforcement became necessary because of the growing number of illegal activities on the national forests and for the protection of visitors and employees. By the early 2000s, fighting fires and budget restrictions were the major concerns of Forest Service management.

For the most part, these changing priorities were met effortlessly by employing new personnel to meet specific new national goals (e.g., minorities), laws (e.g., heritage and wildlife specialists), and internal direction (e.g., planners). But with the downturns in federal programs and congressional funding in the 1990s, these same new employees have, in some cases, been the first ones fired.

The Forest Service defines itself in the tradition of Gifford Pinchot—its goal is the conservation of forests. The idea of multiple-use in the 1960s was believed to embody the Pinchot principle of conservative use and scientific management. Ecosystem management in the 1990s has taken the multiple use effort further to manage the forests and all natural systems.

In Pinchot's book *Breaking New Ground* (1947) he noted that the national forests were to be used and enjoyed "for the greatest good for the greatest number in the long run." Each decade, this goal has been redefined. This is as it should be if the national forests are to survive for another century.

National Forests Have Become the Last Holdouts of the Original Forest and Working Ecosystems

The old-growth areas in the Pacific Northwest's national forests are quickly becoming the last vestiges of the original forest that was here before the coming of the settlers. For the most part, private and state forests have, once again, become the primary source for wood supplies. National forests, due to various laws and plans, are in many instances the only places where old forests and the creatures that depend on them can survive. The days of extensive logging of the national forests are over. Yet if the goal remains to both produce wood fiber and protect species, the forests cannot be just left alone and much remains to be done.

The fire issue continues to plague the Forest Service. The agency's efforts to reduce the occurrence and severity of fires have resulted in overgrown forests that are primed for fire on a massive scale. Many Region 6 national forests are at high risk for destructive fires that will severely damage trees, habitats, and water supplies for communities. Extensive use of prescribed fire in restoration work to reduce forest fuels could yield positive effects in the Pacific Northwest. But there are many opponents to using fire to reduce the threats of wildfires.

Changes on a massive scale based on projections of global climate change will have implications for forests, fishes, animals, and people in the region. While the Pacific Northwest Research Station and thousands of other researchers around the world are debating the actions needed to reduce climate change, it is becoming increasingly evident that changes to forest and grassland communities will be severe over the next 100 years. There will be many opportunities for research scientists to study the effects of global climate change on the ecosystems of the Pacific Northwest.

In addition, huge budget deficits at the national level will undoubtedly cause significant funding problems, and opportunities, for all natural resources agencies over the next decade. Cuts in services and employees will be felt by the agency for a decade or more. It is unlikely that funding will ever approach the levels of the 1980s. However, there is at least one opportunity that has appeared in 2009—the stimulus package provided funds, mostly managed through the State & Private Forestry branch, for projects on the national forests and in the private sector. Of course this increase in funding will dry up as Congress will undoubtedly reduce funding for such projects in the future.

Forest Service Changes in the Future

The Forest Service prides itself on being apolitical. The truth is that for a century the agency has been at the mercy of changing national events and concerns. Fortunately, regional foresters have not been chosen to fill political agendas from the White House. Yet at least one regional forester was fired by the president in the 1930s (see Chapter 5). Not every chief forester and regional forester was a career Forest Service employee.

The future of the Forest Service can be summed up in one word—funding. The agency has traditionally had strong support from the timber industry, but since the early 1990s, timber output from the Pacific Northwest Region's national forests has dropped significantly. Sawmills and plywood mills have closed, mill employment has dropped, fewer loggers and truck drivers have been needed.

Schools and timber towns that once were supported by federal dollars related to logging have, in many cases, been badly affected. Entire counties in Washington and Oregon are no longer dependent on wood from the national forests. The economic and political support base for the Forest Service is quickly disappearing. Without support from outsiders, funding for Forest Service programs will disappear as well.

Fighting forest fires and preventing fire from burning ranches, homes, and summer homes has become a defining management concern in the agency at the start of the twenty-first century. The increased amount of building in the wildland-urban interface has had an adverse effect on management, especially because of increased costs to protect structures and people from fire. At the present, firefighting costs have risen to half the entire Forest Service budget; much like medical costs, they are rising faster than Congress can approve money.

As more and more people live in urban and suburban sites in the Northwest, fewer have direct connections with the land and resources. Food comes from supermarkets and wood comes from lumber super stores. Water comes from the faucet. There is little connection to the national forests.

In the future, the Forest Service may well be located in the Interior Department, not in the Department of Agriculture. Although efforts to return the management of the national forests to Interior have persisted since 1905, none have yet been successful (Williams 2008).Without strong backing from the timber industry, recreation users, and conservation and environmental groups, the Forest Service may become another agency in the giant Interior Department or possibly a Department of Natural Resources. The Forest Service will fight this change, but they may not succeed. An alternative, as the author has suggested to Chief Gail Kimbell, would have the Forest Service *lead* the combination effort such that the agency would be positioned to direct a new department or change the existing Department of the Interior. Only time will tell.

Appendix A

Pacific Northwest Region Regional Foresters

Regional Forester	Years Served
Edward T. Allen	1908-1909
Charles S. Chapman	1909-1911
George H. Cecil	1911-1925
Christopher "Chris" M. Granger	1925-1930
Clarence J. "C. J." Buck	1930-1939
Lyle F. Watts[1]	1939-1943
Horace "Hoss" J. Andrews	1943-1951
J. Herbert "Herb"Stone	1951-1967
Charles A. Connaughton	1967-1971
Rexford "Rex" Resler	1971-1972
Theodore "Ted" Schlapfer	1972-1977
Richard "Dick" E. Worthington	1977-1981
Jeff M. Sirmon	1982-1985
Charles T. "Tom" Coston	1985-1986
James F. Torrence	1986-1989
John F. Butruille	1989-1992
John E. Lowe	1992-1996
Robert W. Williams	1996-1999
Harv Forsgren	1999-2002
Linda Goodman	2003-2008
Mary Wagner	2008-

1. Lyle F. Watts left Region 6 to become Chief of the Forest Service, 1943-1952.

Appendix B

Establishment and Discontinuation of Forest Reserves and National Forests in the Pacific Northwest Region

Forest Reserve/ National Forest	**State**	**Year of Action**	**Effect of Action**
Ashland	Oregon	1893	Established Reserve
			Added with other land to establish Crater NF
Baker City	Oregon	1904	Established Reserve
		1906	Combined with Blue Mountains Reserve
Blue Mountains	Oregon	1906	Established Reserve
		1908	Distributed among Whitman, Malheur, Umatilla, and Deschutes NFs
Bull Run	Oregon	1892	Established Reserve
		1908	Combined with lands from Cascade Reserve to establish Oregon NF
Cascade	Oregon	1907	Name changed from Cascade Range Reserve
		1933	Combined with Santiam to establish Willamette NF
Cascade Range	Oregon	1893	Established Reserve
		1907	Name changed to Cascade Reserve
Chelan	Washington	1908	Established Forest from lands from Washington NF
		1955	Name changed to Okanogan NF
Chesnimus	Oregon	1905	Established Reserve
		1907	Combined with Wallowa Reserve to create the Imnaha Reserve
Columbia	Washington	1908	Established Forest from lands from Rainier NF
		1949	Name changed to Gifford Pinchot NF
Colville	Washington	1907	Established Reserve
Coquille	Oregon	1907	Established Reserve
		1908	Land combined with Siskiyou NF
Crater	Oregon	1908	Established Forest
		1932	Name changed to Rogue River NF
Deschutes	Oregon	1908	Established Forest from Blue Mountains NF
Fremont	Oregon	1906	Established Reserve

Forest Reserve/ National Forest	State	Year of Action	Effect of Action
Gifford Pinchot	Washington	1949	Established Forest. Name change from Columbia NF
Goose Lake	Oregon	1906	Established Reserve
		1908	Combined with Fremont NF
Heppner	Oregon	1906	Established Reserve
		1909	Combined with Blue Mountains NF lands to establish Umatilla NF
Imnaha	Oregon	1907	Established Reserve
		1908	Name changed to Wallowa NF
Malheur	Oregon	1908	Established Forest from Blue Mountains NF
Maury Mountain	Oregon	1905	Established Reserve
		1907	Combined with Blue Mountains Reserve
Minam	Oregon	1911	Established Forest
		1920	Combined with Whitman NF
Mt. Baker	Washington	1924	Name changed from Washington NF
Mt. Hood	Oregon	1924	Name changed from Oregon NF
Mt. Rainier	Washington	1897	Name changed from Pacific Reserve
		1907	Name changed to Rainier Reserve
Ochoco	Oregon	1911	Established Forest
Okanogan	Washington	1911	Established Forest
		1920	Combined with Chelan; name discontinued
		1955	Re-established from Chelan NF
Olympic	Washington	1897	Established Reserve
Oregon	Oregon	1908	Established Forest from Cascade and Bull Run NFs
		1924	Name changed to Mt. Hood NF
Pacific	Washington	1893	Established Reserve
		1897	Name changed to Mt. Rainier Reserve
Paulina	Oregon	1911	Established Forest
		1915	Distributed among Crater, Deschutes and Fremont NFs
Rainier	Washington	1907	Name changed from Mt. Rainier Reserve
		1933	Distributed to Columbia, Snoqualmie and Wenatchee NFs
Rogue River	Oregon	1932	Established Forest from Crater NF
Santiam	Oregon	1911	Established Forest
		1933	Combined with Cascade NF to form Willamette NF

Forest Reserve/ National Forest	State	Year of Action	Effect of Action
Siskiyou	Oregon	1906	Established Reserve
Siuslaw	Oregon	1908	Established Forest from Tillamook and Umpqua NFs
Snoqualmie	Washington	1908	Established Forest from lands from Washington NF
Tillamook	Oregon	1907	Established Reserve
		1908	Distributed among Siuslaw and Umpqua NFs
Umatilla	Oregon	1908	Established Forest from Blue Mountains NF
	OR/WA	1927	Land added
Umpqua	Oregon	1907	Established Reserve
Wallowa	Oregon	1905	Established Reserve
		1907	Combined with Chesmimus Reserve to create the Imnaha Reserve. Name discontinued
		1908	Re-established Forest from Imnaha NF
Washington	Washington	1897	Established Reserve
		1924	Name changed to Mt Baker NF
Wenaha	OR/WA	1905	Established Reserve
		1920	Combined with Umatilla NF
Wenatchee	Washington	1908	Established Forest from lands from Washington NF
Whitman	Oregon	1908	Established Forest from Blue Mountains NF
Willamette	Oregon	1933	Established Forest from Santiam and Cascade NFs
Winema	Oregon	1961	Established Forest from Rogue River NF and Klamath Indian Reservation

This document was prepared from Forest Service document FS-612, 1997; *Establishment and Modification of National Forest Boundaries and National Grasslands – A Chronological Record 1891 – 1996.* A more complete listing of all reserve and forest boundary modifications can be found in the same publication. Forests grouped together for administration, but not officially consolidated, are not listed.

Appendix C

National Forest Acres 2006

Forest	Gross Acres	NFS Acres	Other Acres
Colville	1,029,621	954,413	75,208
Deschutes	1,853,289	1,597,192	256,097
Fremont	1,713,819	1,207,039	506,852
Gifford Pinchot	1,409,415	1,319,649	89,766
Malheur	1,541,730	1,465,293	76,437
Mt. Baker	1,329,950	1,301,787	28,163
Mt. Hood	1,117,775	1,069,109	48,666
Ochoco	979,089	851,095	127,994
Okanogan	1,535,028	1,499,013	36,015
Olympic	692,038	627,695	64,343
Quinault Unit	5,460	5,460	0
Rogue River	688,644	628,443	60,201
Siskiyou	1,163,247	1,094,736	68,511
Siuslaw	835,752	634,207	201,545
Snoqualmie	1,570,462	1,256,973	313,489
Umatilla	1,512,867	1,407,002	105,865
Umpqua	1,027,372	983,131	44,241
Wallowa	1,076,553	997,022	79,531
Wenatchee	1,918,890	1,735,298	183,592
Whitman	1,316,974	1,266,902	50,072
Willamette	1,790,967	1,678,014	112,953
Winema	1,096,986	1,045,548	51,348
TOTAL R-6	27,206,000	24,625,021	2,580,979

Note: NFS refers to national forest system lands and correspond to the officially proclaimed national forests, not the various forest names that currently apply. Source: USDA Forest Service. 2007. *Land Areas of the National Forest System as of September 2006*. FS-383. Washington, DC: USDA Forest Service.

Appendix D

Purchase Units (PU), Land Utilization Projects (LUP), National Grasslands (NGL), and Others

Unit	Gross Acres	NFS Acres	Other Acres
Bogachiel PU	951	807	144
Brazier PU	320	320	0
Cougar Bar PU	363	363	0
Drift Creek PU	1,865	1,047	818
Fifteenmile Creek PU	594	555	39
Gold Basin PU	80	80	0
Golden Phoenix PU	274	274	0
Illabot Creek PU	1,151	140	1,011
Leeds Island PU	329	0	329
Mt. Hood PU	980	354	626
Ramsey Creek PU	2,278	2,278	0
Skagit PU	820	579	241
Yachats PU	5,838	748	5,090
Crooked River NGL	173,629	112,357	61,272
Northeast Washington LUP	498	498	0
Western Oregon LUP	856	856	0
McQuinn Strip	62,533	59,068	3,465
Other Unit Total	237,516	172,779	64,737

Source: USDA Forest Service. 2007. *Land Areas of the National Forest System as of September 2006*. FS-383. Washington, DC: USDA Forest Service.

Appendix E

Timber Harvest Cut and Sold 1909-2007
Volumes in thousand board feet

Year	# Timber Sales	Vol. Harvested	Vol. Sold
1909	158	32,428	80,938
1910	238	48,869	103,315
1911	315	47,789	227,021
1912	301	66,869	161,924
1913	406	81,511	462,054
1914	663	134,490	226,471
1915	866	91,792	574,920
1916	923	136,344	212,524
1917	1,117	186,391	982,891
1918	1,210	174,608	174,798
1919	1,078	196,229	128,007
1920	1,339	200,721	254,479
1921	1,384	219,273	179,383
1922	1,408	243,356	182,894
1923	1,507	320,265	1,374,502
1924	1,560	335,451	922,156
1925	1,378	314,782	678,694
1926	1,147	405,277	650,048
1927	1,060	477,722	221,325
1928	1,323	443,678	370,466
1929	1,344	514,313	1,041,513
1930	1,556	510,605	1,721,433
1931	1,491	391,765	687,316
1932	1,439	157,352	73,880
1933	1,564	130,593	42,276
1934	1,746	279,791	88,935
1935	2,121	230,941	71,222
1936	2,102	324,591	149,048
1937	2,126	446,841	722,066
1938	2,009	428,663	184,457
1939	2,258	498,315	1,005,954
1940	2,234	665,006	688,298
1941	1,984	862,900	325,575
1942	1,683	807,611	1,103,104
1943	1,777	985,376	1,393,045
1944	1,907	1,487,855	1,016,972
1945	1,935	1,284,641	958,545
1946	2,037	968,330	687,206
1947	2,910	1,533,628	1,508,017
1948	2,750	1,530,019	1,415,424
1949	2,427	1,484,874	1,156,039

Year	# Timber Sales	Vol. Harvested	Vol. Sold
1950	2,902	1,495,754	1,383,879
1951	2,905	2,071,363	1,795,011
1952	3,112	1,793,505	1,895,640
1953	3,665	2,446,031	2,015,868
1954	3,744	2,415,019	2,301,331
1955	3,638	2,633,124	2,392,174
1956	2,861	2,632,905	2,445,699
1957	3,129	2,775,468	2,996,821
1958	4,001	2,616,955	3,919,020
1959	3,814	3,650,596	4,379,052
1960	3,586	4,194,647	3,825,205
1961	4,004	3,577,414	3,812,811
1962	4,059	4,156,976	4,800,424
1963	5,253	4,539,802	5,509,927
1964	5,215	4,831,813	4,703,393
1965	4,633	5,095,758	4,602,854
1966	4,317	5,426,859	4,636,370
1967	4,421	4,464,972	1,485,046
1968	4,094	5,137,453	4,610,140
1969	3,942	4,752,339	4,200,752
1970	5,338	4,759,401	5,357,159
1971	4,972	3,990,439	4,770,062
1972	5,228	4,869,467	4,767,886
1973	7,877	5,791,353	4,653,033
1974	7,047	4,675,425	4,670,126
1975	6,285	3,991,295	4,901,907
1976	11,315	3,950,180	4,942,782
1977*	1,661	1,465,742	677,577
1977	8,257	4,342,797	4,765,183
1978	10,599	4,340,661	4,987,730
1979	11,295	4,528,703	5,138,244
1980	11,958	3,628,738	5,126,422
1981	12,029	3,381,516	5,482,065
1982	11,936	2,263,760	4,641,557
1983	57,922	3,868,226	4,915,550
1984	80,308	4,538,924	4,962,147
1985	82,998	4,760,265	4,752,483
1986	93,106	4,965,202	5,056,926
1987	69,753	5,597,225	5,272,680
1988	52,833	5,407,778	4,918,857
1989	59,969	5,230,765	2,811,251
1990	57,462	3,879,259	3,997,067
1991	61,723	3,166,492	2,105,761
1992	47,322	2,140,057	740,488
1993	48,419	1,666,200	786,934
1994	44,113	1,126,867	433,850

Year	# Timber Sales	Vol. Harvested	Vol. Sold
1995	41,214	877,078	401,029
1996	34,262	775,741	939,419
1997	34,889	767,897	870,532
1998	28,740	662,396	652,258
1999	30,430	569,399	434,328
2000	22,404	409,156	242,304
2001	25,679	307,010	268,472
2002	26,154	276,715	306,149
2003	25,228	320,871	400,231
2004	23,601	499,974	491,029
2005	20,745	489,642	475,084
2006	22,071	382,106	546,357
2007	24,774	419,381	522,952

* New fiscal year (FY). The FYs before 1977 ended June 30th, but since 1977 the federal FY ends September 30th.

Appendix F

Wilderness Areas in Region 6 in 2009

Wilderness	National Forest	NFS Acres
Badger Creek	Mt. Hood NF	28,140
Black Canyon	Ochoco NF	13,400
Boulder Creek	Umpqua NF	19,100
Bridge Creek	Ochoco NF	5,400
Bull of the Woods	Mt. Hood NF	37,607
Clackamas	Mt. Hood NF	9,470
Cooper	Salmon	13,700
Cummins Creek	Siuslaw NF	9,173
Diamond Peak	Deschutes NF	34,413
	Willamette NF	19,772
Drift Creek	Siuslaw NF	5,798
Eagle Cap	Wallowa NF	204,184
	Whitman NF	146,277
Gearhart Mountain	Fremont NF	22,809
Grassy Knob	Siskiyou NF	17,200
Hells Canyon	Wallowa NF	118,247
	Whitman NF	11,848
Kalmiopsis NF	Siskiyou NF	179,755
Lower White River	Mt. Hood NF with BLM	2,970
Mark O. Hatfield	Mt. Hood NF	64,960
Menagerie	Willamette NF	4,800
Middle Santiam	Willamette NF	7,500
Mill Creek	Ochoco NF	17,400
Monument Rock	Malheur NF	12,620
	Whitman NF	7,030
Mount Hood	Mt. Hood NF	47,160
Mount Jefferson	Deschutes NF	32,734
	Mt. Hood NF	5,021
	Willamette NF	69,253
Mount Thielsen	Deschutes NF	7,107
	Umpqua NF	21,593
	Winema NF	25,567
Mount Washington	Deschutes NF	14,116
	Willamette NF	38,622
Mountain Lakes	Winema NF	23,071
North Fork John Day	Umatilla NF	107,058
	Whitman NF	14,294
North Fork Umatilla	Umatilla NF	20,435
Opal Creek	Willamette NF	20,733
Red Buttes	Rogue River NF	350
	Siskiyou NF	3,400
Roaring River	Mt. Hood NF	36,550

Wilderness	National Forest	NFS Acres
Rock Creek	Siuslaw NF	7,486
Rogue-Umpqua Divide	Rogue River NF	6,850
	Umpqua NF	26,350
Salmon-Huckleberry	Mt. Hood NF	61,180
Sky Lakes	Rogue River NF	75,695
	Winema NF	40,605
Strawberry Mountain	Malheur NF	69,350
Three Sisters	Deschutes NF	94,370
	Willamette NF	192,338
Waldo Lake	Willamette NF	39,200
Wenaha-Tucannon	Umatilla NF	66,375
Wild Rogue	Siskiyou NF	25,658
Total Oregon Wildernesses		2,086,504

Wilderness	National Forest	NFS Acres
Alpine Lakes	Snoqualmie NF	117,862
	Wenatchee NF	244,927
Boulder River	Mt. Baker NF	48,674
Buckhorn	Olympic NF	44,258
Clearwater	Snoqualmie NF	14,192
Colonel Bob	Olympic NF	11,961
Glacier Peak	Mt. Baker NF	283,252
	Wenatchee NF	287,321
Glacier View	Snoqualmie NF	3,123
Goat Rocks	Gifford Pinchot NF	70,817
	Snoqualmie NF	36,201
Henry M. Jackson	Mt. Baker NF	27,436
	Snoqualmie NF	47,504
	Wenatchee NF	25,416
Indian Heaven	Gifford Pinchot NF	20,960
Lake Chelan-Sawtooth	Okanogan NF	95,021
	Wenatchee NF	56,414
Mount Adams	Gifford Pinchot NF	46,353
Mount Baker	Mt. Baker NF	117,528
Mount Skokomish	Olympic NF	13,015
Noisy-Diobsud	Mt. Baker NF	14,133
Norse Peak	Snoqualmie NF	52,180
Pasayten	Mt. Baker NF	108,336
	Okanogan NF	421,141
Salmo-Priest	Colville NF	29,386
	Kaniksu NF	11,949
Tatoosh	Gifford Pinchot NF	15,750
The Brothers	Olympic NF	16,682
Trapper Creek	Gifford Pinchot NF	5,970
Wenaha-Tucannon	Umatilla NF	111,048
William O. Douglas	Gifford Pinchot NF	15,596
	Snoqualmie NF	152,636
Wonder Mountain	Olympic NF	2,349
Total Washington Wildernesses		2,569,391

Source: USDA Forest Service. 2007. *Land Areas of the National Forest System as of September 2006.* FS-383. Washington, DC: USDA Forest Service. Also,the 2009 Omnibus Public Land Management Act of 209

Appendix G

Special Areas in Region 6 in 2009

Area	National Forest	NFS Acres
Cascade Head National Scenic Research Area	Siuslaw NF	6,637
Columbia River Gorge National Scenic Area	Gifford Pinchot NF	27,274
	Mt. Hood NF	43,574
Crystal Springs Special Resources Management Area	Mt. Hood NF	no acres given
Hells Canyon National Recreation Area	Wallowa NF	355,762
	Whitman NF	45,212
Mt. Baker National Recreation Area	Mt. Baker NF	8,473
Mt. Hood Recreation Area	Mt. Hood NF	34,559
Mount St. Helens National Volcanic Monument	Gifford Pinchot NF	112,605
Newberry Crater National Volcanic Monument	Deschutes NF	54,822
Opal Creek Scenic Recreation Area	Willamette NF	12,645
Oregon Dunes National Recreation Area	Siuslaw NF	27,232
Upper Big Bottom and Cultus Cree Protection Areas	Mt. Hood NF	1,860

Source: USDA Forest Service. 2007. *Land Areas of the National Forest System as of September 2006*. FS-383. Washington, DC: USDA Forest Service. Also, the 2009 Omnibus Public Land Management Act of 2009.

Appendix H

Wild and Scenic Rivers in Region 6 in 2009

Wild, Scenic, and Recreation River	National Forest	NFS Acres/Miles
Big Marsh Creek	Deschutes NF	4,800
Chetco River	Siskiyou NF	11,930
Clackamas River	Mt. Hood NF	15,040
Collawash River	Mt. Hood NF	17.8 miles
Crescent Creek	Deschutes NF	3,200
Deschutes River	Deschutes NF	17,408
Eagle Creek	Mt. Hood NF	8.3 miles
Eagle Creek	Whitman NF	7,680
	Wallowa NF	960
East Fork Hood River	Mt. Hood NF	13.5 miles
Elk River	Siskiyou NF	6,080
Elkhorn River	Willamette NF	1,280
Fifteen Mile Creek	Mt. Hood NF	11.1 miles
Fish Creek	Mt. Hood NF	13.5 miles
Grande Ronde	Umatilla NF	6,048
Illinois River	Siskiyou NF	15,105
Imnaha River	Wallowa NF	24,640
Joseph Creek	Wallowa NF	2,752
Little Deschutes River	Deschutes NF	3,840
Lostine River	Wallowa NF	5,120
Malheur River	Malheur NF	4,384
McKenzie River	Willamette NF	4,064
Metolius River	Deschutes NF	9,152
Middle Fork Hood River	Mt. Hood NF	3.7 miles
Minam River	Wallowa NF	6,269
	Whitman NF	6,405
North Fork Crooked River	Ochoco NF	5,704
North Fork John Day River	Umatilla NF	13,792
	Whitman NF	3,520
North Fork Malheur River	Malheur NF	8,160
North Fork Smith River	Siskiyou NF	4,160
North Fork Sprague River	Fremont NF	4,800
North Powder River	Whitman NF	1,920
North Umpqua River	Umpqua NF	8,128
North/Middle Fork Willamette River	Willamette NF	13,536
Roaring River	Mt. Hood NF	4,384
Rogue River	Siskiyou NF	11,845
Salmon River	Mt. Hood NF	3,360
Sandy River	Mt. Hood NF	3,968
Snake River	Wallowa NF	12,000
South Fork Clackamas	Mt. Hood NF	4.2 miles
South Fork Roaring River	Mt. Hood NF	4.6 miles

Wild, Scenic, and Recreation River	National Forest	NFS Acres/Miles
Squaw Creek	Deschutes NF	11,456
Sycan River	Fremont NF	4,270
	Winema NF	14,610
Upper Rogue River	Rogue River NF	12,896
Wenaha River	Umatilla NF	7,054
White River	Mt. Hood NF	7,072
Zig Zag River	Mt. Hood NF	4.3 miles
Total Oregon		318,902 acres
		81 miles
Klickitat River	Gifford Pinchot NF	17
Skagit River	Golden Phoenix PU	74
	Mt. Baker NF	2,306
White Salmon River	Gifford Pinchot NF	95
Total Washington		20,582

Source: USDA Forest Service. 2007. *Land Areas of the National Forest System as of September 2006*. FS-383. Washington, DC: USDA Forest Service. The 2009 Omnibus Public Land Management Act designated wild and scenic river by miles only.

Appendix I
Research Natural Areas in the Pacific Northwest

RNA Name	National Forest	Ecoregion
Oregon		
Cache Mountain	Deschutes NF	East Cascades, Oregon Ecoregion
Metolius	Deschutes NF	East Cascades, Oregon Ecoregion
Moskt Butte	Deschutes NF	East Cascades, Oregon Ecoregion
Pringle Falls	Deschutes NF	East Cascades, Oregon Ecoregion
Torrey-Charlton	Deschutes and Willamette NFs	West Cascades, Oregon Ecoregion
Vance Knoll	Fremont NF	Blue Mountains, Oregon Ecoregion
Augur Creek	Fremont NF	East Cascades, Oregon Ecoregion
Goodlow Mountain	Fremont NF	East Cascades, Oregon Ecoregion
Silver Lake Exclosure	Fremont NF	East Cascades, Oregon Ecoregion
Vee Pasture	Fremont NF	East Cascades, Oregon Ecoregion
Canyon Creek	Malheur NF	Blue Mountains, Oregon Ecoregion
Haystack Butte	Malheur NF	Blue Mountains, Oregon Ecoregion
Gumjuwac-Tolo	Mt. Hood NF	East Cascades, Oregon Ecoregion
Mill Creek	Mt. Hood NF	East Cascades, Oregon Ecoregion
Bagby	Mt. Hood NF	West Cascades, Oregon Ecoregion
Bull Run	Mt. Hood NF	West Cascades, Oregon Ecoregion
Dry Mountain	Ochoco NF	Blue Mountains, Oregon Ecoregion
Ochoco Divide	Ochoco NF	Blue Mountains, Oregon Ecoregion
Ashland	Rogue River-Siskiyou NF	Klamath Mountains Ecoregion
Abbott Creek	Rogue River-Siskiyou NF	West Cascades, Oregon Ecoregion
Coquille River Falls	Rogue River-Siskiyou NF	Coast Range Ecoregion
Port Orford Cedar	Rogue River-Siskiyou NF	Coast Range Ecoregion
Wheeler Creek	Rogue River-Siskiyou NF	Coast Range Ecoregion
Cedar Log Flat	Rogue River-Siskiyou NF	Klamath Mountains Ecoregion
Hoover Gulch	Rogue River-Siskiyou NF	Klamath Mountains Ecoregion
Lemmingsworth Gulch	Rogue River-Siskiyou NF	Klamath Mountains Ecoregion
Cummins Creek	Siuslaw NF	Coast Range Ecoregion
Flynn Creek	Siuslaw NF	Coast Range Ecoregion
Neskowin Crest	Siuslaw NF	Coast Range Ecoregion
Reneke Creek	Siuslaw NF	Coast Range Ecoregion
Sand Lake	Siuslaw NF	Coast Range Ecoregion
Tenmile Creek	Siuslaw NF	Coast Range Ecoregion

RNA Name	National Forest	Ecoregion
Cougar Butte	Umpqua NF	West Cascades, Oregon Ecoregion
Limpy Rock	Umpqua NF	West Cascades, Oregon Ecoregion
Squaw Flat	Umpqua NF	West Cascades, Oregon Ecoregion
Gerald M. Strickler	Wallowa-Whitman NF	Blue Mountains, Oregon Ecoregion
Indian Creek	Wallowa-Whitman NF	Blue Mountains, Oregon Ecoregion
Little Granite Creek	Wallowa-Whitman and Nez Perce NFs	Blue Mountains, Oregon Ecoregion
Vance Knoll	Wallowa-Whitman NF	Blue Mountains, Oregon Ecoregion
Gold Lake Bog	Willamette NF	West Cascades, Oregon Ecoregion
Hagan	Willamette NF	West Cascades, Oregon Ecoregion
McKenzie Pass	Willamette NF	West Cascades, Oregon Ecoregion
Middle Santiam	Willamette NF	West Cascades, Oregon Ecoregion
Olallie Ridge	Willamette NF	West Cascades, Oregon Ecoregion
Rigdon Point	Willamette NF	West Cascades, Oregon Ecoregion
Three Creek	Willamette NF	West Cascades, Oregon Ecoregion
Torrey-Charlton	Willamette and Deschutes NFs	West Cascades, Oregon Ecoregion
Wildcat Mountain	Willamette NF	West Cascades, Oregon Ecoregion
Bluejay	Fremont-Winema NFs	East Cascades, Oregon Ecoregion
Cannon Well	Fremont-Winema NFs	East Cascades, Oregon Ecoregion
Cherry Creek Basin	Fremont-Winema NFs	East Cascades, Oregon Ecoregion
Washington		
Halliday Fen	Colville NF	Canadian Rockies Ecoregion
Maitlen Creek	Colville NF	Canadian Rockies Ecoregion
Round Top Mountain	Colville NF	Canadian Rockies Ecoregion
Salmo	Colville NF	Canadian Rockies Ecoregion
Monte Cristo	Gifford Pinchot NF	East Cascades, Washington Ecoregion
Butter Creek	Gifford Pinchot NF and Mt. Rainier NP	West Cascades, Washington Ecoregion
Cedar Flats	Gifford Pinchot NF	West Cascades, Washington Ecoregion
Goat Marsh	Gifford Pinchot NF	West Cascades, Washington Ecoregion
Sister Rocks	Gifford Pinchot NF	West Cascades, Washington Ecoregion
Steamboat Mountain	Gifford Pinchot NF	West Cascades, Washington Ecoregion
Thornton T. Munger	Gifford Pinchot NF	West Cascades, Washington Ecoregion

RNA Name	National Forest	Ecoregion
Lake Twenty-two	Mt. Baker-Snoqualmie NF	North Cascades Ecoregion
Long Creek	Mt. Baker-Snoqualmie NF	North Cascades Ecoregion
North Fork Nooksack	Mt. Baker-Snoqualmie NF	North Cascades Ecoregion
Perry Creek	Mt. Baker-Snoqualmie NF	North Cascades Ecoregion
Ronald J. Taylor	Mt. Baker-Snoqualmie NF	North Cascades Ecoregion
Chewuch River	Okanogan and Wenatchee NFs	Okanogan Ecoregion
Roger Lake	Okanogan and Wenatchee NFs	Okanogan Ecoregion
Wolf Creek	Okanogan and Wenatchee NFs	Okanogan Ecoregion
Quinault	Olympic NF	Pacific Northwest Coast Ecoregion
Wet Weather Creek	Olympic NF	Pacific Northwest Coast Ecoregion
Meeks Table	Mt. Baker-Snoqualmie NF	East Cascades, Washington Ecoregion
Pataha Bunchgrass	Umatilla NF	Blue Mountains, Washington Ecoregion
Rainbow Creek	Umatilla NF	Blue Mountains, Washington Ecoregion
Thompson Clover	Okanogan and Wenatchee NFs	East Cascades, Washington Ecoregion
Eldorado Creek	Okanogan and Wenatchee NFs	West Cascades, Washington Ecoregion
Fish Lake Bog	Okanogan and Wenatchee NFs	West Cascades, Washington Ecoregion

Appendix J
Bibliographic Notes

The National Archives and Records Service in College Park, Maryland, and the regional archives in Seattle hold the official historical records of the Forest Service. Various universities and colleges such as the University of Oregon, Oregon State University, and the University of Washington hold many archival boxes of historical information from retired Forest Service employees, as well as special interest groups. The Forest History Society (FHS) in Durham, North Carolina, is a rich source of information on the history of the agency in the Pacific Northwest. The official history collection of the Forest Service's national historian was transferred to the FHS in 1998. Since that time, the records and books have been supplemented by scores of boxes collected by the author before he retired.

One of the many portions of the author's collection includes the original bound volumes of the Pacific Northwest Region's newsletters *Six Twenty Six*, *R-6 Administrative Digest*, and *Greensheet*, dating from 1917 to around 1985, and the *Timber Line*s publication of the Thirty-Year Club (now called OldSmokeys). Along with a mix of current news about retirees and their activities, each issue of the latter included short reminiscences by club members about their days in the Forest Service. The *Timber Lines* issue of May 1955 had a special section called "memoirs," the first of many longer accounts of the early days. The *Timber Lines* supplement of December 1962 was a lengthy memoir by Thornton T. Munger, the first important federal forest researcher and administrator in the Pacific Northwest.

Other materials derived from efforts by the author to collect stories and documents about the Forest Service in the Northwest. Many of these materials were transferred in the summer of 2007 to Oregon State University Library, where they are part of the Gerald W. Williams Collection. An inventory of records and photographs is underway. The OSU Library is in the process of digitizing thousands of photographs from this collection that illustrate the history of the Forest Service, the Army's Spruce Production Division (World War I), the Civilian Conservation Corps, logging, and scenic views from the West, especially those from the Pacific Northwest.

Magazines and journals have important articles and stories about the Forest Service. Magazines that are national in scope include the *Journal of Forest History* and *Forest History Today* (Forest History Society), *Environmental History* (American Society for Environmental History), *American Forests* (American Forestry Association, published under various titles), and the *Journal of Forestry* (Society of American Foresters). Regional historical journals include the *Oregon Historical Quarterly* and the *Pacific Northwest Quarterly*. Journals relating particularly to recreational groups are the *Mazama* and the *Sierra Club Bulletin.*

The following references are useful bibliographies of forest history books:

Davis, Richard C. (ed). 1977. *North American Forest History: A Guide to Manuscripts and Archives in the United States and Canada.* Santa Barbara, CA: Forest History Society.

Fahl, Ronald J. (ed). 1977. *North American Forest and Preservation History: A Bibliography.* Santa Barbara, CA: Forest History Society.

Munns, E.N. 1940. *A Selected Bibliography of North American Forestry.* 2 volumes. Washington, DC: U.S. Government Printing Office.

Ogden, Gerald (ed). 1976. *The United States Forest Service: A Historical Bibliography, 1876-1972.* Davis, CA: Agricultural History Society.

Williams, Gerald W. 1995. *Selected References Concerning the USDA Forest Service.* Publication R6-SP-UP-04-95. Portland: Pacific Northwest Region. Fourth revision.

A number of history books and manuscripts, several never published, have influenced the research and writing on the present book. The majority of the following books were printed by the Forest Service, and several were published by local printers. Two of the books originated as masters or doctoral theses—Holstine's thesis on the Colville National Forest and Rakestraw's doctoral thesis that deals with the early history of the Region. The rest touch on the history of individual national forests:

Bach, Melva, with Ward Tonsfeldt (ed.). 1981. *History of the Fremont National Forest.* Bend, OR: Tonsfeldt Consulting.

Baker, Gail C. 1950. *Historical Writeup of the Deschutes National Forest [w/yearly history updates to 1965].* Bend, OR: Deschutes National Forest.

Brown, Carroll E. 1960. *History of the Rogue River National Forest.* Vol. 1 1893-1932 and Vol. 2 1933-1969. Medford, OR: Rogue River National Forest.

Colville National Forest. 1965. *History of the Colville National Forest, Compiled by the Forest Staff in 1941, Updated to 1965.* Colville, WA: Colville National Forest.

Connaughton, Charles. 1973. "Forestry in Mid-Century." Manuscript. Portland, OR: Charles Connaughton.

Connaughton, Charles, and Elwood R. Maunder (interviewer). 1976. *Forty-Three Years in the Field with the U.S. Forest Service.* Durham, NC: Forest History Society.

Cooper, Loran J. 1940. *History of the Siskiyou National Forest.* Grants Pass, OR: Siskiyou National Forest.

Engles, Harold J. 1982. *Interview with Harold J. Engles, Retiree, USFS Career: 1919-1958, Mt. Baker-Snoqualmie National Forest.* Seattle, WA: Mt. Baker-Snoqualmie National Forest.

Hodgson, Allen Harrison, et al. c.1950. *A History of the Ochoco National Forest: Different Times and Perspectives.* Prineville, OR: Ochoco National Forest.

Holstine, Craig E. 1987. *Forgotten Corner: A History of the Colville National Forest, Washington.* Colville, WA: Colville *Statesman-Examiner.*

Holstine, Craig E. (ed.). 1994. *An Historical Overview of the Wenatchee National Forest Washington.* Portland, OR: U.S. Forest Service, Pacific Northwest Region.

Jackson, Royal G., and Jennifer A. Lee. 1980. *Siuslaw National Forest Oral History Program: Lloyd James Lyne.* Corvallis, OR: Heritage Associates, Inc.

Langston, Nancy. 1995. *Forest Dreams, Forest Nightmares: The Paradox of Old Growth in the Inland West.* Seattle: University of Washington Press.

McClure, Rick, and Cheryl Mack. 1999. *"For the Greatest Good" Early History of Gifford Pinchot National Forest.* Vancouver, WA: Gifford Pinchot National Forest.

Mosgrove, Jerry L. 1980. *The Malheur National Forest: An Ethnographic History.* John Day, OR: Malheur National Forest.

Rakestraw, Lawrence. 1955. *A History of Forest Conservation in the Pacific Northwest, 1891-1913.* Ph.D. dissertation. Seattle: University of Washington. Reprinted by Arno Press in 1979.

Rakestraw, Lawrence, and Mary Rakestraw. 1992. *History of the Willamette National Forest.* Eugene, OR: Willamette National Forest.

Rasmussen, Boyd L. 1939. "A History of the Siuslaw National Forest, Oregon, as of December 31, 1939." Corvallis, OR: Siuslaw National Forest.

Rooney, Jack R. 1997. *Frontier Legacy: History of the Olympic National Forest 1897 to 1960.* Seattle, WA: Northwest Interpretative Association. Reprinted in 2007.

Tucker, Gerald J. 1954. *Historical Sketches of Wallowa National Forest.* Baker, OR: Wallowa National Forest.

Other sources of material, many out of print, have been published that cover the history of the Forest Service at the national level. Several are general while others are program specific:

Cameron, Jenks. 1928. *The Development of Governmental Forest Control in the United States.* Baltimore, MD: Johns Hopkins Press.

Clary, David A. 1986. *Timber and the Forest Service.* Lawrence: University Press of Kansas.

Dana, Samuel T. 1956. *Forest and Range Policy.* New York: McGraw-Hill.

Fedkiw, John. 1999. *Managing Multiple Uses in National Forests, 1905-1995: A 90-Year Learning Experience and It Isn't Finished Yet.* FS-628. Washington, DC: U.S. Forest Service.

Foss, Phillip O. 1960. *Politics and Grass: The Administration of Grazing on the Public Domain.* Seattle: University of Washington Press.

Hays, Samuel P. 1959. *Conservation and the Gospel of Efficiency: The Progressive Conservation Movement.* Cambridge, MA: Harvard University Press.

Hirt, Paul W. 1994. *A Conspiracy of Optimism: Management of the National Forests Since World War Two.* LincolnE: University of Nebraska Press.

Ise, John. 1920. *The United States Forest Policy.* New Haven, CT: Yale University Press.

Lewis, James G. 2005. *The Forest Service and the Greatest Good: A Centennial History.* Durham, NC: Forest History Society.

Pinchot, Gifford. 1947. *Breaking New Ground.* New York: Harcourt Brace. (Still in print.)

Morgan, Jr., George T. 1961. *William B. Greeley: A Practical Forester.* St. Paul, MN: Forest History Society.

Pyne, Stephen J. 1982. *Fire in America: A Cultural History of Wildland and Rural Fire.* Seattle: University of Washington Press.

Robbins, William G. 1985. *American Forestry: A History of National, State, and Private Cooperation.* Lincoln: University of Nebraska Press.

Roberts, Paul H. 1963. *Hoof Prints on Forest Ranges: The Early Years of National Forest Range Administration.* San Antonio, TX: Naylor Company.

Robinson, Glen O. 1975. *The Forest Service: A Study in Public Land Management.* Baltimore, MD: Johns Hopkins Press for Resources for the Future.

Rowley, William D. 1985. *U.S. Forest Service Grazing and Rangelands: A History.* College Station: Texas A&M University Press.

Runte, Alfred. 1991. *Public Lands, Public Heritage: The National Forest Idea.* Niwot, CO: Roberts Rinehart Publishers.

Smith, Darrell H. 1930. *The Forest Service: Its History, Activities and Organization.* Washington, DC: Brookings Institution.

Steen, Harold K. 1976. *The U.S. Forest Service: A History.* Seattle: University of Washington Press. Latest edition was in 2005.

Steen, Harold K. 1991. "The Beginning of the National Forest System." FS-488. Washington, DC: U.S. Forest Service.

Steen, Harold K. 2004. *The Chiefs Remember: The Forest Service, 1952-2001.* Durham, NC: Forest History Society.

West, Terry L. 1992. "Centennial Mini-Histories of the Forest Service." FS-518. Washington, DC: U.S. Forest Service.

Williams, Gerald W. 2000. *The USDA Forest Service—The First Century.* FS-650. Washington, DC: U.S. Forest Service. Updated in 2005.

Williams, Gerald W. 2007. *The Forest Service: Fighting for Public Lands.* Westport, CT: Greenwood Press.

This history is written for a general audience but includes many terms that are in common use in the forestry profession. For readers who are not familiar with such terms, the following reference books should be consulted:

Society of American Foresters. 1983. *Terminology or Forest Science, Technology, Practice and Products.* Washington, DC: Society of American Foresters.

Davis, Richard C. (ed). 1983. *Encyclopedia of American Forest and Conservation History.* Two volumes. New York: Macmillan.

McCulloch, Walter F. 1958. *Woods Words.* Portland: Oregon Historical Society.

Bibliography

Adams, Mary Beth; Linda Loughry; Linda Plaugher (compilers). 2004. "Experimental Forests and Ranges of the USDA Forest Service." GTR-NE-321. Newtown Square, PA: USFS Northeastern Research Station.

Arnst, Albert. 1985. "We Climbed the Highest Mountains [Lookout Panorama Photographs]." Pamphlet. Portland, OR: Fernhopper Press.

Baker, Gail C. 1950. *Historical Writings of the Deschutes National Forest.* Bend, OR: Deschutes National Forest.

Barnes, Will C. 1929. "Poem." *The Ranger,* Vol. 1, #3 (April): 44.

Beals, Herbert K., Scott Cline, and Julie M. Kohler. 1981. "On the Mountain's Brink: A Forest Service History of the 1980 Mount St. Helens Volcanic Emergency." Pamphlet. Vancouver, WA: Gifford Pinchot NF.

Beuter, John H., K. Norman Johnson, and H. Lynn Scheurman. 1976. *Timber for Oregon's Tomorrow: An Analysis of Reasonably Possible Occurrences.* Research Bulletin 19. Corvallis: Oregon State University, Forest Research Laboratory. Updated in 1990.

Binus, Joshua. 2002. "McQuinn Strip Land Dispute." Oregon History Project Web site, www.ohs.org/education/oregonhistory

Blake, Grover C. c1960. "Sowing the Seed." Manuscript of 33 pages in the Umpqua NF archives, Roseburg, OR.

Blue Mountains Demonstration Area. 2002. *Blue Mountains Demonstration Area FY2002 Annual Report.* Web site at www.fs.fed.us/bluemountains

Boyd, Robert. 1999. *The Coming of the Spirit of Pestilence.* Seattle: University of Washington Press.

Bradbury, Bill. 2007-08. *Oregon Blue Book 2007-2008.* Salem, OR: Office of the Secretary of State.

Center for the Study of the Pacific Northwest Web site (http://www.washington.edu/uwired/outreach/cspn/Website/index.html) "A History of Treaty-Making and Reservations."

Brant, Irving. 1988. *Adventures in Conservation with Franklin D. Roosevelt.* Flagstaff, AZ: Northland Publishing.

Bright, George A. 1913. "An Extensive Reconnaissance of the Wenaha National Forest in 1913." Reprinted in 2008, David C. Powell (ed.), as publication F14-SO-06-08. Pendleton, OR: Umatilla National Forest.

Brooks, Karl Boyd. 2006. *Public Power, Private Dams: The Hell's Canyon High Dam Controversy.* Seattle: University of Washington Press.

Brown, Carroll E. 1960. *History of the Rogue River National Forest.* Vol. 1 1893-1932. Medford, OR: Rogue River National Forest.

Brown, Carroll E. 1969. *History of the Rogue River National Forest.* Volume 2 1933-1969. Medford, OR: Rogue River National Forest.

Brown, Greg G., and Charles C. Harris. 1991. "The Forest Service: Changing of the Guard?" Pp. 339-44 in *Proceedings of the 1991 Society of American Foresters*

National Convention, San Francisco, California, August 4-7. Bethesda, MD: Society of American Foresters.

Bruckart, John "Ray." 1980. "Memoirs of John "Ray" Bruckart, Sr." Manuscript of 45 pages. Portland, OR: USDA Forest Service, Regional Office.

Buck, C. J. 1918. "Lands Problems." *Journal of Forestry*, Vol. 16, #7 (Nov): 807-13.

Bulletin, The. Newspaper published in Bend, OR.

Butler, Bret W., et al. 1998. *Fire Behavior Associated with the 1994 South Canyon Fire on Storm King Mountain, Colorado*. RMRS-RP-9. Ft. Collins, CO: Rocky Mountain Research Station.

Chesnut, Alma. 1942. "Scott Leavitt—The Forest Ranger Who Came to Congress." *American Forests*, Vol. 38, #3 (March): 148-49, 190-91.

Chriswell, Harold C. 1989. *Memoirs: Harold C. (Chris) Chriswell 1933 to 1971*. Bellingham, WA: Harold Chriswell.

Clark, Cleon L. 1987. *History of the Willamette Valley and Cascade Mountain Wagon Road*. Bend, OR: Deschutes County Historical Society.

Clark, R. C. 1927. *History of the Willamette Valley, Oregon*. Vol. I. Chicago, IL: S. J. Clark Publishing Company.

Clary, David A. 1986. *Timber and the Forest Service*. Lawrence: University Press of Kansas.

Cleator, Fred W. 1939. "Recollections of Bob Marshall: Bob Marshall's Last Wilderness Adventure." Manuscript dated October 2, 1939, in author's personal files.

Cleator, Fred W. 1955. Letter dated January 17, 1955, to Charles Paul Keyser in Portland, in author's personal files.

Congressional Record. 1967. November 14. Washington DC: U.S. Congress.

Connaughton, Charles A. 1966. "The Revolt Against Clearcutting." *Journal of Forestry*, Vol. 68, #5 (May): 264-65.

Connaughton, Charles A. 1973. "Forestry in Mid-Century." Manuscript. Portland, OR: Charles Connaughton.

Corning, Howard McKinley. 1952. "The Lookout's Mechanical Eye." *American Forests*, Vol. 38, #3 (March): 18-19, 51.

Cornutt, John M. 1971. *Cow Creek Valley [OR] Memories: Riddle Pioneers Remembered in John M. Cornutt's Autobiography*. Eugene, OR: Industrial Publishing Co.

Cornwall, George M. 1911. "Utilizing Troops in the National Forests." *American Forests*, Vol. 17, #10 (Oct): 587-89.

Cottrell, Robert C. 2006. *Smokejumpers of the Civilian Public Service in World War II: Conscientious Objectors as Firefighters for the National Forest Service*. Jefferson, NC: McFarland & Company, Inc.

Coville, Frederick V. 1898. *Forest Growth and Sheep Grazing in the Cascade Mountains of Oregon*. Bulletin No. 15. Washington, DC: USDA Division of Forestry.

Coville, Frederick V. 1898. "Sheep Grazing in the Forest Reserves." *The Forester*, Vol. 4, #2 (Feb): 832-35.

Daily Journal. Newspaper published in Salem, OR.

District 6 Committee. 1915. "Report to Mr. Graves of a District VI Committee, Regarding 'Causes for Increasing Costs of Administration and Recommendations for Promoting Economy and Efficiency in the Forest Service'." Report dated December 9, 1915. Portland, OR: U.S. Forest Service, District Office.

Doig, Ivan. 1975-76. "The Murkey Annals of Clearcutting: A 40-Year-Old Dispute." *Pacific Search*, Vol. 10 (Dec-Jan): 12-14.

Egan, Timothy. 1992. "Shelter for the Rural Homeless: Trees and Sky." *New York Times*, May 11, 1992.

Ellis, David Maldwyn. 1948. "The Oregon and California Railroad Land Grant, 1866-1945." *Pacific Northwest Quarterly*, Vol. 39, #4 (Oct): 253-83.

Evans, Gail E. H., and Gerald W. Williams. 1985. "Over Here, Over Here: The Army's Spruce Production Division During the 'War to End All Wars'." Pp. 139-79 in William Woodward and David Hansen (eds.) *Military Influences on Washington History: Proceedings of a Conference*. Tacoma, WA: Military Influences on Washington History Conference.

Everson, William. 1978-9. "Waldport [Conscientious Objectors Camp]: An Interview with William Everson." Interview by Guido Palandri and George Wickes. *Imprint: Oregon*, Vol. 5, #1-2 (Fall-Spring): 1-41. Published in Eugene, OR: University of Oregon Library.

Fedkiw, John. 1999. *Managing Multiple Uses on National Forests, 1905-1995: A 90-Year Learning Experience and it Isn't Finished Yet*. FS-628. Washington, DC: U.S. Forest Service.

Forbes, Robert H. 1965. "Here Comes 'Skyhook Logging!' " *American Forests*, Vol. 71, #1 (Jan): 10-13, 53.

Forest Worker. 1928. "Snag Shooting Tests." *Forest Worker*, Vol.4, #4 (July): 12.

Forest Worker. 1930. "Nelson Ferris MacDuff." *Forest Worker*, Vol.6, #3 (May): 25.

Forest Worker. 1930. "Secretary Hyde Willing to Agree to Construction of Cableway on Mount Hood." *Forest Worker*, Vol.6, #4 (July): 17.

Forest Worker. 1931. "Forest Service Pledged to Conserve Scenic Beauty of Forest Roads." *Forest Worker*, Vol. 7, #3 (May): 9-11.

Forest Worker. 1931. "When Douglas Found the Sugar Pine." *Forest Worker*, Vol. 7, #1 (Jan): 17-18.

Forest Worker. 1932. "Douglas Fire Region Gets Experimental Forest." *Forest Worker*, Vol. 8, #6 (Nov): 10.

Forest Worker. 1933. "Freezing Out the Western Pine Beetle." *Forest Worker*, Vol. 9, #2 (March): 10.

Forester, The. 1898. "War on the Forest Reservations." *The Forester*, Vol. 4, #5 (May): 96-101.

Freidel, Frank. 1985. "Franklin D. Roosevelt in the Northwest: Informal Glimpses." *Pacific Northwest Quarterly*, Vol. 76, #4 (October): 122-31.

Frost, Dale "Jack." 1990. "Recollections of Forest Service Engineering—1945 to 1985." Pp. 561-65 in *The History of Engineering in the Forest Service (A Compilation of History and Memoirs, 1905-1989)*. EM 7100-13. Washington, DC: U.S. Forest Service, Engineering Staff.

Gannett, Henry. 1898. "Report on the Survey and Examination of Forest Reserves (March), 1898." Pp. 3-34 in *Surveys of Forest Reserves*, Senate Document No. 189, 55th Congress, 2nd Session. Washington DC: U.S. Congress.

Gano, Ward. 1990. "Recollection of Forest Service Engineering." Pp. 499-509 in *The History of Engineering in the Forest Service (A Compilation of History and Memoirs, 1905-1989)*. EM 7100-13. Washington, DC: U.S. Forest Service, Engineering Staff.

Geier, Max G. 2007. *Necessary Work: Discovering Old Forests, New Outlooks, and Community on the H. J. Andrews Experimental; Forest, 1948-2000*. PNW-GTR-687. Portland, OR: U.S. Forest Service, Pacific Northwest Research Station.

Gorte, Ross W. 2004. "Below-Cost Timber Sales: An Overview." Washington, DC: Congressional Research Service.

Gray, Gary Craven. 1982. *Radio for the Fire Line: A History of Electronic Communication in the Forest Service 1905-1975*. Washington, DC: U.S. Forest Service.

Grefe, Ray F. 1990. "Engineering in Region 6: USDA Forest Service." Pp. 274-80 in *The History of Engineering in the Forest Service (A Compilation of History and Memoirs, 1905-1989)*. EM 7100-13. Washington, DC: U.S. Forest Service, Engineering Staff.

Guthrie, John D. 1931. "R-6 Forest Ranger Training Camp." *Six Twenty-Six*, Vol. 15, #10 (Oct): 3-4.

Herrett, Wendy Milner, with Jacqueline S. Reiner (interviewer). 2001. "An Interview with Wendy Milner Herrett." Durham, NC: Forest History Society.

Hirt, Paul W. 1994. *A Conspiracy of Optimism: Management of the National Forests Since World War Two*. Lincoln: University of Nebraska Press.

Hoberg, George. 2001. "The Emerging Triumph of Ecosystem Management: The Transformation of Federal Forest Policy." Pp. 55-86 in Charles Davis (ed.), *Western Public Lands and Environmental Politics*. Boulder, CO: Westview Press. 2nd edition.

Holstine, Craig E. (ed.). 1984. *An Historical Overview of the Wenatchee National Forest, Washington*. Portland, OR: U.S. Forest Service, Pacific Northwest Region.

Holstine, Craig E. 1987. *Forgotten Corner: A History of the Colville National Forest, Washington*. Colville, WA: Coville *Statesman-Examiner*.

Hyman, Harold M. 1963. *Soldiers and Spruce: Origins of the Loyal Legion of Loggers and Lumbermen*. Los Angeles, CA: Institute of Industrial Relations.

Jensen, Derrick, and George Graffan with John Osborn. 1995. *Railroads and Clearcuts: Legacy of Congress's 1864 Northern Pacific Railroad Land Grant*. Spokane, WA: Inland Empire Public Lands Council.

Jones, Ira E. n.d. "Recollections" on the National Association of Forest Service Retirees Web site (http://www.fsx.org/).

Justice, Joyce. 1991. "World War II Civilian Public Service: Conscientious Objector Camps in Oregon." *Prologue: Quarterly of the National Archives*, Vol. 23, #3 (Fall): 266-73.

Kaufman, Herbert. 1960. *The Forest Ranger: A Study in Administrative Behavior*. Baltimore, MD: Johns Hopkins Press for the Resources for the Future.

Keele, Denise M., Robert W. Malsheimer, Donald W. Floyd, and Jerome E. Perez. 2006. "Forest Service Land Management Litigation 1989-2002." *Journal of Forestry*, Vol. 104, #6 (June): 196-202.

Kerrick, Michael A. 1985. "Ecotage from Our Perspective: An Explanation of the Willamette National Forest's Policy on Environmental Sabotage Known as 'Ecotage'." Report dated September 1985. Eugene, OR: U.S. Forest Service, Willamette National Forest.

Kresek, Ray. 1988. *Fire Lookouts of the Northwest*. Spokane, WA: Historic Lookout Project. 3rd edition.

Lackey, Robert T. 1997. "Restoration of Pacific Salmon: The Role of Science and Scientists." Pp. 35-40 in *Proceedings of the Sixth Biennial Watershed Management Conference*. Water Resources Center Report No. 92. Davis: University of California, Davis.

Lackey, Robert T. 2003. "Pacific Northwest Salmon: Forecasting Their Status in 2100." *Reviews in Fisheries Science,* Volume 11, #1: 35-88.

Langille, Harold D. 1956. "Mostly Division 'R' Days: Reminiscences of the Stormy, Pioneering Days of the Forest Reserves." *Oregon Historical Quarterly*, Vol. 57, #4 (Dec): 301-13.

Langston, Nancy. 1995. *Forest Dreams, Forest Nightmares: The Paradox of Old Growth in the Inland West*. Seattle: University of Washington Press.

LeMaster, Dennis C. 1984. *Decade of Change: The Remaking of Forest Service Statutory During the 1970s*. Westport, CT: Greenwood Press.

Lien, Carsten. 1991. *Olympic Battleground: The Power Politics of Timber Preservation*. San Francisco, CA: Sierra Club Books. 2nd edition in 2000 by The Mountaineers Book, Seattle, WA.

Loehr, Rodney C. (ed.). 1952. *Forests for the Future: The Story of Sustained Yield as Told in the Diaries and Letters of David T. Mason, 1907-1950*. St. Paul, MN: The Forest Products History Foundation.

McArthur, Lewis A., and Lewis L. McArthur. 2007. *Oregon Geographic Names*. Portland, OR: Oregon Historical Society Press. 7th edition.

McCloskey, Michael. 2005. *In the Thick of It: My Life in the Sierra Club*. Washington, DC: Island Press.

McLean, Herbert E. 1990. "A Very Hot Potato." *American Forests*, Vol. 96, #3&4 (March/April): 30-31, 65-67.

Mason, David T., and Donald Bruce. 1931. *Sustained Yield Forest Management as a Solution to American Forest Conservation Problems*. Portland, OR: Mason, Stevens, and Bruce.

Mayer, Heike. 2005. "Planting High-Technology Seeds: Tektronix's Role in the Creation of Portland's Silicon Forest." *Oregon Historical Quarterly*, Vol. 106, #4 (Winter): 568-93.

Meany, Edmond S. 1923. *Origin of Washington Geographic Names*. Seattle: University of Washington Press.

Merritt, Melvin L. 2005. Reminiscences. Pp. 5-6 in Rolf Anderson's (ed.) *"We Had an Objective in Mind": The U.S. Forest Service in the Pacific Northwest 1905 to 2005*. Portland, OR: The Pacific Northwest Forest Service Association.

Messing, John. 1966. "Public Lands, Politics, and Progressives: The Oregon Land Fraud Trials, 1903-1910." *Pacific Historical Review*, Vol. 35, #1 (Feb): 35-66.

Miller, Char. 2000. "What Happened in the Rainier Grand's Lobby? A Question of Sources." *The Journal of American History*, Vol. 86, #4 (March 2000): 1709-14.

Miller, Char. 2001. *Gifford Pinchot and the Making of Modern Environmentalism*. Washington, DC: Island Press.

Mills, Richard C. n.d. "History of the Founding and Organization of the Waldport [Oregon Conscientious Objectors] Camp: C.P.S. #56." Unpublished manuscript. Eugene, OR: University of Oregon Library, Special Collections.

Minto, John. 1908. "From Youth to Age as an American: Chapter II Learning to Live on the Land." *Oregon Historical Quarterly*, Vol. 9, #2 (June): 127-72.

Moody, William D. 2007. *Spittin' in the Wind: North Cascades Smokejumper Base, The Birthplace of Smokejumping 1939-2007*. Book 1—History and Tales. Missoula, MT: National Smokejumpers Association.

Morning Oregonian. Newspaper published in Portland, OR.

Morse, William B. 1962. "Recipe for Plantation." *American Forests*, Vol. 68, #7 (July): 24-25, 49.

Morse, William B. 1965. "Ann Webb [Heisler]—Girl Forester." *American Forests*, Vol. 71, #4 (April): 28-29, 59-60.

Norbury, Fred. 2005. "Statement of Fred Norbury, Associate Deputy Chief ... Before the Subcommittee on Public Lands and Forests ... September 28, 2005, Concerning the Forest Service Livestock Grazing Program." www.fs.fed.us/congress/109/senate/oversight/norbury/092805.html.

Northwest Forest Service News. Newsletter published by the USFS Pacific Northwest Region.

Oregon Motorist. Newlcetter published by the Oregon State Motor Association. Salem, OR.

The Oregonian. Newspaper published in Portland, OR.

O'Toole, Randal. 1988. *Reforming the Forest Service*. Washington, DC: Island Press.

Pacific Northwest Report. Annual USFS report by the Pacific Northwest Region. Portland, OR: Pacific Northwest Region.

Parker, Louise. 1965. "The Big Flood." *American Forests*, Vol. 71, #3 (March): 14-17, 61-62.

Pease, Lute. 1907. "The Way of the Land Transgressor. I. The West and the President's Land Policies." *Pacific Monthly*, Vol. 18, #2 (Aug): 144-63.

Peterson, R. Max. 1988. "Land Management Planning on the National Forests: In Retrospect and Prospect." In *Forest Lands of the Northwest: Selecting Among Alternative Uses—The 1988 Starker Lectures*. Corvallis: Oregon State University, College of Forestry: 27-34.

Pinchot, Gifford. 1947. *Breaking New Ground*. New York: Harcourt, Brace and Company.

Pioneer of 1847. 1911. "Indian vs. Pinchot Conservation—Pioneer of [18]'47 Upholds Aborigines' Plan of Burning Underbrush—Oregon City, OR." Letter to the editor dated January 24th. *The Oregonian*, January 26, page 10, column 6.

Pomeroy, Kenneth B. 1959. "Forester's Notebook." *American Forests*, Vol. 65, #3 (March): 37.Puter, Stephen A. Douglas, and Horace Stevens (collaborator). 1908. *Looters of the Public Domain.* Portland, OR: The Portland Printing House.

Pyne, Stephen J. 1982. *Fire in America: A Cultural History of Wildland and Rural Fire.* Princeton, NJ: Princeton University Press.

Pyne, Stephen J. 2001. *Year of the Fires: The Story of the Great Fires of 1910.* New York: Viking.

Rakestraw, Lawrence. 1955. "A History of Forest Conservation in the Pacific Northwest, 1891-1913." Ph.D. dissertation at the University of Washington. Reprinted in 1979 by the Arno Press, New York.

Rakestraw, Lawrence. 1981. *A History of the United States Forest Service in Alaska.* Juneau: USFS Alaska Region. Reprinted in 1994.

Rakestraw, Lawrence, and Mary Rakestraw. 1991. *History of the Willamette National Forest.* Eugene, OR: Willamette National Forest.

Ranger, The. Pacific Northwest Regional newsletter for USFS retirees and the public. Printed in Portland, OR.

Remington, C. E. 1990. "Me & the Forest Service—1935-1961." Pp. 460-76 in *The History of Engineering in the Forest Service (A Compilation of History and Memoirs, 1905-1989).* EM 7100-13. Washington, DC: U.S. Forest Service, Engineering Staff.

Resettler. Newsletter of the U.S. Resettlement Administration published in Portland, OR.

Richardson, Elmo R. 1968. "Olympic National Park: 20 Years of Controversy." *Journal of Forest History*, Vol. 12, #1 (April): 6-15.

Richardson, Elmo R. 1980. *BLM's Billion Dollar Checkerboard: Managing the O&C Lands.* Santa Cruz, CA: Forest History Society, Inc.

Richardson, Elmo R. 1983. *David T. Mason, Forestry Advocate: His Role in the Application of Sustained Yield Management to Private and Public Forest Lands.* Santa Cruz, CA: Forest History Society, Inc.

Robertson, Daniel. 1984. "The Awakening of Conservation: The Oregon Land Fraud Trials." *The Umpqua Trapper*, Vol. 20, #2 (Summer Hunt): 27-39.

Roland, Barbara. 1995. "Barbara Roland." Pp. 53-62 in Beverly A. Brown, *In Timber Country: Working People's Stories of Environmental Conflict and Urban Flight.* Philadelphia, PA: Temple University Press.

Rooney, J. R. 2007. *Frontier Legacy: History of the Olympic National Forest 1897 to 1960.* Seattle, WA: Northwest Interpretive Association.

Ross, Homer. 1986. Reminiscences. Pp. 1-3 in *Sampler of the Early Years.* Volume II. Washington, DC: Forest Service Wives.

Roth, Dennis M. 1988a. *The Wilderness Movement and the National Forests, 1980-1984.* FS-410. Washington, DC: U.S. Forest Service.

Roth, Dennis M. 1988b. *The Wilderness Movement and the National Forests*. College Station, TX: Intaglio Press.

Rowland, Mary M.; Larry D. Bryant; Bruce K. Johnson; James H. Noyes; Michael J. Wisdom; and Jack Ward Thomas. 1997. The Starkey Project: History, Facilities, and Data Collection Methods for Ungulate Research. PNW-GTR-396. Portland, OR: USFS Pacific Northwest Research Station.

Rowley, William D. 1985. *U.S. Forest Service Grazing and Rangelands: A History*. College Station: Texas A&M University Press.

Ruby, Robert H., and John A. Brown. 1981. *Indians of the Pacific Northwest: A History*. Norman: University of Oklahoma Press.

Sadler, Russell. 1984. "Forester's Cuts go Against Political Grain." *The Oregonian*, February 5, 1984.

Say, Harold Bradley, and Nort Baser. 1952. "The Plan by the Century." *American Forests*, Vol. 58, #2 (Feb): 27-34.

Schlapfer, Ted, with Linda Dodds (interviewer). 1984. Oral history with Ted Schlapfer. February 1, 1984. Author's personalfiles.

Schneider, Keith. 1989. *New York Times*, July 2, 1989, section 1, page 14. (www.nytimes.com/1989/07/02)

Schwering, Manena Sparks. n.d. "Sparks, Samuel C." Manuscript in the History Notes. McKenzie Bridge, OR: McKenzie River Ranger Station.

Seattle Post-Intelligencer. Newspaper published in Seattle, WA.

Seattle Times. Newspaper published in Seattle, WA.

Seideman, David. 1993. Showdown at Opal Creek: The Battle for America's Last Wilderness. New York, NY: Carroll & Graf Publishers, Inc.

Sherwood, Dan. 1990. "Pickin' Seed." *American Forests*, Vol. 96, #11/12 (Nov/Dec): 33-35.

Silva, Nikki, and Davia Nelson. 2006. "An Unexpected Kitchen." Pp. 124-132 in Holly Hughes (ed.), *Best Food Writing 2006*. New York: Marlowe & Company.

Simons, David R. 1959. "These Are the Shining Mountains." *Sierra Club Bulletin*, Vol. 44, #7 (October): 1-13.

Sirmon, Jeff M. 1990. "Chronology and Comments." Pp. 675-87 in *The History of Engineering in the Forest Service (A Compilation of History and Memoirs, 1905-1989)*. EM 7100-13. Washington, DC: U.S. Forest Service, Engineering Staff.

Six Twenty-Six. Newsletter published by the USFS Pacific Northwest Region. Portland, OR.

Skovlin, Jon M. 1991. "Fifty Years of Research Progress: A Historical Document on the Starkey Experimental Forest and Range." PNW-GTR-266. Dated May 1991. Portland, OR: USFS Pacific Northwest Research Station.

Stankey, George H., and Bruce Shindler. 1997. "Adaptive Management Areas: Achieving the Promise, Avoiding the Peril." PNW-GTR-394. Portland, OR: Pacific Northwest Research Station.

Statesman Journal. Newspaper published in Salem, OR.

Steel, William Gladstone. 1932. "Cascade Range Forest Reserve: A History." Manuscript. Crater Lake, OR: USDI National Park Service, Crater Lake National Park Archives.

Steen, Harold K. 1991. *The Beginning of the National Forest System*. FS-488. Washington, D.C.: USDA Forest Service History Unit.

Steen, Harold K. 2004. *The Chiefs Remember: The Forest Service, 1952-2001*. Durham, NC: Forest History Society.

Tax Liberator. 1915. Newspaper published in Roseburg, OR.

Thayer, Alfred L. 1908. "The Fraudulent Homesteader." *Conservation*, Vol. 14, #11 (Nov): 579-84.

Thomas, Jack Ward. 2005. "Reminiscences." Pp. 515-17 in Rolf Anderson (ed.), *We Had an Objective in Mind: The U.S. Forest Service in the Pacific Northwest 1905 to 2005*. Portland: The Pacific Northwest Forest Service Association.

Tokareczyk, Bob. 2005. Story of Mount St. Helens. Pp. 457-59 in Rolf Anderson's (ed.) *We Had an Objective in Mind: The U.S. Forest Service in the Pacific Northwest 1905 to 2005*. Portland, OR: The Pacific Northwest Forest Service Association.Tucker, Gerald J. 1964. "13 Days in a Blizzard." *Outdoor Life*, Vol. 133, #1 (Jan.): 20-23, 104-7.

Twight, Ben W. 1983. *Organizational Values and Political Power: The Forest Service Versus the Olympic National Park*. University Park: Pennsylvania State University Press.

U.S. Forest Service. 1929. "Forest Figures, North Pacific District (Oregon and Washington)." Portland, OR: USFS Offices of Engineering and Public Relations. See a PDF of this publication at: http://ir.library.oregonstate.edu/jspui/bitstream/1957/10920/1/For_Fig_Nor_Pac_Dis.pdf

U.S. Forest Service, PIT program Web site. http://www.passportintime.com/

U.S. General Accounting Office. 1996. *Forest Service's Reforestation Funding: Financial Sources, Uses and Condition of the Knutson-Vandenberg Fund*. Report GAO/RCED-96-15. Washington, DC: U.S. General Accounting Office.

U.S. Global Change Research Program, National Assessment Synthesis Team. 2000. *Climate Change Impacts on the United States, The Potential Consequences of Climate Variability and Change—Overview: Pacific Northwest*. Updated in 2003. Available at usgcrp.gov.

U.S. Government Accountability Office. 2005. *Livestock Grazing: Federal Expenditures and Receipts Vary, Depending on the Agency and the Purpose of the Fee Charged*. Report GAO-05-869. Washington, DC: U.S. Government Accountability Office.

U.S. Government Accountability Office. 2009. *Federal Land Management: Observations on a Possible Move of the Forest Service into the Department of the Interior*. Report GAO-09-223. Washington, DC: U.S. Government Accountability Office.

Unruh, Bud. 1990. "USDA Forest Service Engineering History." Pp. 555-60 in *The History of Engineering in the Forest Service (A Compilation of History and Memoirs, 1905-1989)*. EM 7100-13. Washington, DC: U.S. Forest Service, Engineering Staff.

Usher, Jack. 1984. Quoted in Sadler 1984.

Van Dresser, Cleveland. 1952. "Abuses Under the Mining Laws—Plunder in Our National Forests" *American Forests*, Vol. 58, #1 (January): 6-9.

Vaughn, Jacqueline, and Hanna J. Courtner. 2005. *George W. Bush's Healthy Forests: Reframing the Environmental Debate.* Boulder: University Press of Colorado.

Waggener Jr., Lawrence A. 1990. "Personal Reminiscence." Pp. 487-98 in *The History of Engineering in the Forest Service (A Compilation of History and Memoirs, 1905-1989).* EM 7100-13. Washington, DC: U.S. Forest Service, Engineering Staff.

Washington State. 2008. "Federally Recognized American Indian Reservations and Trust Lands." 2007 Data Book at www.ofin.wa.gov.databook/population/py06.asp.

Webber, Bert. 1974. "The Forest Service and the Japanese Attacks of World War II." *Timberlines*, Vol. 21 (June): 117-22.

Webber, Bert. 1992. *Silent Siege III: Japanese Attacks on North America in World War II.* Medford, OR: Webb Research Group.

Wiener, Alfred A. 1987. "The Big Buyout." *Journal of Forestry*, Vol. 85, #3 (March): 25-19.

Wiesendanger, Albert. 2005. Reminiscences. Pp. 95-96 in Rolf Anderson (ed.), *"We Had an Objective in Mind": The U.S. Forest Service in the Pacific Northwest 1905 to 2005.* Portland, OR: The Pacific Northwest Forest Service Association.

Williams, Gerald W. 1983. "The Civilian Conservation Corps' (CCC) Contribution to the Forests in the Pacific Northwest Region." Pp. 48-57 in Harold K. Steen (ed.), *History of Sustained-Yield Forestry: A Symposium.* Santa Cruz, CA: Forest History Society.

Williams, Gerald W. 1992. *Judge John Breckenridge Waldo: Diaries and Letters from the High Cascades of Oregon, 1877-1907.* Roseburg and Eugene, OR: Umpqua and Willamette National Forests.

Williams, Gerald W. 1993. "The French Pete Wilderness Controversy, 1937-1978." Unpublished paper.

Williams, Gerald W. 1997. "Early Years of National Forest Management: Implications of the Organic Act of 1897." Paper presented at the biennial meeting of the American Society for Environmental History held in Baltimore, MD, on March 6-9, 1997.

Williams, Gerald W. 2000a. "Aboriginal Use of Fire: Were There Any 'Natural' Plant Communities?" Pp. 179-214 in Charles E. Kay and Randy T. Simmons (eds.), *Wilderness and Political Ecology: Aboriginal Land Management–Myths and Reality.* Logan: University of Utah Press.

Williams, Gerald W. 2000b. "Early Fire Use in Oregon." *Fire Management Today*, Vol. 60, #3 (Summer): 13-20.

Williams, Gerald W. 2000c. "Introduction to Aboriginal Fire Use in North America." *Fire Management Today*, Vol. 60, #3 (Summer): 8-12.

Williams, Gerald W. 2000d. "National Monuments and the Forest Service." Washington, DC: USDA Forest Service. Unpublished MS.

Williams, Gerald W. 2008. "Forest Service Organization and Reorganization: Past to Present." Unpublished MS.

Williams, Gerald W., and Stephen R. Mark (compilers). 1995. *Establishing and Defending the Cascade Range Forest Reserve: As Found in the Letters, Newspapers, Magazines, and Official Reports 1885-1912.* Portland, OR: USDA Forest Service, Regional Office and USDI National Park Service, Crater Lake National Park.

Willingham, William F. 1983. *Army Engineers and the Development of Oregon: A History of the Portland District U.S. Army Corps of Engineers.* Portland, OR: U.S. Army Corps of Engineers.

Wolf, Robert E., with Dick Manning and Dan Hall (interviewers).1990. "Robert Wolf Oral History Project, Nixon 10 Standard Regions and Related Issues." Oral History 227-50 dated November 28, 1990. Missoula: University of Montana, Mansfield Library.

Woodruff, George W. 1906. "Agricultural Settlements in Forest Reserves: Important Law [Forest Homestead Act of 1906] Enacted by Congress Affecting Settlers in [Forest] Reserves." *Forestry and Irrigation,* Vol. 12, #6 (June): 267-71.

Worthington, Richard Endicott, with Linda S. Dodds (interviewer). 1984. Transcript of oral history with Richard Worthington. April 18, 1984. Author's personal files.

Yaffee, Steven Lewis. 1994. *The Wisdom of the Spotted Owl: Policy Lessons for a New Century.* Washington, DC: Island Press.

Yeaworth, David, and Jonathan Kusel. 2002. News release dated December 11, 2002. Web site at www.sierrainstitute.us/neai/PressRelease.pdf

Youngblood, Andrew. 1995. "Research Publications of the Pringle Falls Experimental Forest, Central Oregon Cascade Range 1930 to 1993." PNW-GTR-347. Portland, OR: USFS Pacific Northwest Research Station.

Note: Personal communications can be found in the Gerald W. Williams Collection at the Oregon State University Valley Library

Index